A. Fiechter H. Okada R. D. Tanner (Eds.)

Bioproducts and Bioprocesses

Second Conference to Promote Japan/U.S.
Joint Projects and Cooperation in Biotechnology,
Lake Biwa, Japan, September 27-30, 1986

With 182 Figures and 52 Tables

Springer-Verlag Berlin Heidelberg NewYork
London Paris Tokyo

Coordinating Editor:

Professor Dr. Armin Fiechter

Institut für Biotechnologie, ETH, Hönggerberg,
CH-8093 Zürich, Switzerland

Local Editors:

Professor Dr. Hirosuke Okada

Department of Fermentation Technology,
Faculty of Engineering, Osaka University,
2-1 Yamada-oka, Suita-shi, Osaka 565, Japan

Professor Dr. Robert D. Tanner

Department of Chemical Engineering,
Vanderbilt University,
Nashville, Tennessee 37235, USA

ISBN-13:978-3-642-74229-3 e-ISBN-13:978-3-642-74227-9
DOI: 10.1007/978-3-642-74227-9

Library of Congress Cataloging-in-Publication Data.
Conference to Promote Japan/U.S. Joint Projects and Cooperation in Biotechnology (2nd : 1986 : Moriyama-shi, Japan)
Bioproducts and bioprocesses : Second Conference to Promote Japan/U.S. Joint Projects and Cooperation in Biotechnology, Lake Biwa, Japan, September 27–30, 1986 / A. Fiechter, H. Okada, R. D. Tanner, editors. p. cm.
ISBN-13:978-3-642-74229-3 (U.S.)
1. Biotechnology—Methodology—Congresses. 2. Biological products—Congresses. 3. Biochemical engineering—Congresses. I. Fiechter, A. II. Okada, H. (Hirosuke) III. Tanner, R. D. (Robert D.) IV. Title.
TP248.24.C86 1986 660'.6—dc 19 89-4150

Softcover reprint of the hardcover 1st edition 1989

2151/3020-543210 — Printed on acid-free paper

List of Contributors

Numbers in parentheses refer to the sections within this volume

Aiba, S. (3.6, 4.1)
Andrews, B. A. (4.3)
Asama, H. (5.5)
Asenjo, J. A. (4.3)
Bales, J. C. (2.4)
Beppu, T. (3.5)
Brown, K. (5.3)
Bungay, H. R. (5.8)
Dykstra, K. (4.6)
Effler, W. T. Jr. (4.4)
Endo, I. (5.5)
Evans, P. (4.6)
Follettie, M. T. (3.1)
Fraleigh, S. P. (5.8)
Funahashi, H. (5.7)
Furusaki, S. (2.3)
Goldstein, J. M. (1.3)
Himes, V. B. (1.4)
Ho, N. W. Y. (3.3)
Horvath, J. (5.3)
Hu, W. S. (1.4)
Humphrey, A. E. (5.3)
Hyatt, G. S. (4.6)
Kajiwara, S. (2.5)
Karube, I. (5.1)
Kimura, A. (3.2)
Kobayashi, T. (4.7)
Ladisch, M. R. (4.2)
Lee, S. B. (5.6)
Lin, J. K. (4.2)
Maeda, H. (2.5)
Malaney, G. W. (4.4)
Minowada, J. (1.2)
Nagai, S. (4.8)
Nagamune, T. (5.5)
Nakamura, K. (4.5)
Nigam, S. (4.6)
Nishio, N. (4.8)
Nuchnoi, P. (4.8)
Okada, H. (2.6)
Phillips, J. A. (5.2)
Rudge, S. R. (4.2)
Ruettimann, K. W. (4.2)
Ryu, D. (5.6)
Semerjian, H. (5.3)
Shimizu, S. (5.4)
Sinskey, A. J. (3.1)
Suga, K. (4.1)
Suzuki, T. (5.4)
Taguchi, H. (5.7)
Tamiya, E. (5.1)
Tanaka, A. (1.1)
Tanner, R. D. (4.4)
Tosa, T. (2.8)
Udaka, S. (3.4)
Urabe, I. (2.6)
Vieth, W. R. (2.2)
Wang, D. I. C. (1.3)
Wang, H. Y. (4.6)
Wong, C.-H. (2.7)
Woodward, J. (2.4)
Yamada, H. (2.1)
Yamane, T. (5.4)
Yano, T. (4.8)
Yomo, T. (2.6)
Yoshida, T. (5.7)

Preface

The Japanese biotechnology community has developed a unique system of scientific bilateral communications at the international level. It consists of closed seminars where a restricted number of speakers from Japan and a selected country are invited. Thereby the actual state of biotechnology is visualized very effectively and valuable information exchanged during the discussions and personal contacts. As a result strong relationships between several European and American countries have been built up during recent years and numerous colleagues from abroad have improved their knowledge of Japanese science, economy, and culture.

The present volume reports on an American-Japanese meeting held in Moriyama, which is situated on the lovely Lake Biwa in Japan. The well prepared and efficiently organized event covered the major aspects of bioprocess engineering including animal and plant cell culture, biocatalysts, downstream processing, sensor and bioprocess control, as well as genetic engineering. Altogether a broad spectrum, indicating the strengths and weaknesses of the current efforts being made for the improvement of process technology and the search for new products.

Two sessions were of particular interest to the observer. Animal and plant cell culture has made much progress lately and the papers presented indicated that the process methods have been substantially improved. They are essentially derived from microbial techniques and have reached an impressive level of efficiency. Submerged culture techniques are in the foreground and replacement of serum by medium supplementation is badly needed. In addition, on-line measurement and process control are lagging behind the concepts generally applied in process research and industry.

In view of such needs, the development of biosensors in Japan is given the highest priority. A number of applications were presented indicating the increasing involvement of Japanese industry in the development of on-line measuring systems based on biosensors.

The manuscripts presented in this volume have been drafted carefully by the authors since the meeting took place. Thanks to the critical editorial work of Prof. Okada from Osaka University and Prof. Tanner from Vanderbilt University a most interesting picture of the present day status in bioprocess engineering has become available. It is eminently suitable for promoting cooperation in biotechnology between Japan and the U.S.

A. Fiechter

Contents

1. Plant and Animal Cell Culture

1.1. Cultured Plant Cells as the Catalysts for Bioreactor

Atsuo Tanaka

Laboratory of Industrial Biochemistry, Department of Industrial Chemistry,
Faculty of Engineering, Kyoto University, Kyoto, Japan

Contents

Plant tissue cultures are at present attracting worldwide attention, because plant cells are able to synthesize specific compounds, especially various secondary metabolites useful as medicines, food additives, colouring reagents and so on, which cannot be produced by microorganisms. However, plant cells have several drawbacks such as instability of cell structures and metabolic activities. Immobilization might be a clue solving these problems. This article describes several representative works on cultured plant cells including those on immobilized cells carried out in Japan.

1. Specific Features of Plant Tissue Cultures

Plant tissue cultures have several advantages compared with the corresponding whole plants and organs (Table 1). That is, cultivation of plants is usually limited to specific areas and seasons, while cultured cells are available wherever and whenever. Therefore, desired products will be supplied constantly regardless of climate. Cultured cells can grow far more rapidly than the original plants, this means that increased productivity is expectable. Productivity may also be enhanced by optimizing cultivation conditions, and quality of products will be stabilized or enhanced by selecting excellent cell lines and optimized cultivation conditions. These facts indicate that cultured plant cells will be potent catalysts for bioreactors to produce plant-specific secondary metabolites [1,2]. However, there are several disadvantages inherent in the plant tissue cultures. For example, desired compounds are often

Table 1. Advantages of plant cells as biocatalysts

Plants limited to specific areas and seasons are available whenever
Products can be supplied constantly regardless of weather
Improvement and stabilization of product quality can be expected
Productivity can be increased by controlling cultivation conditions
Cultured cells grow more rapidly than the original plants

Bioproducts and Bioprocesses
Editors: Fiechter, Okada, Tanner

Table 2. Comparison between plant cells and microbial cells as biocatalysts

	Plant cells	Microbial cells
Cell construction	Multicellular	Unicellular
Mechanical strength	Low	Relatively high
Growth	Slow	Fast
Synthesis of metabolite	Relatively difficult	Relatively easy
Productivity	Low	Moderate — high
Product value	Moderate — high	Low — moderate
Control of cultivation	Difficult	Relatively easy
Stability of cell line	Unstable	Relatively stable

localized strictly in specific organs of whole plants, such as roots, flowers and leaves. Differentiation of cells into respective organs is required for the synthesis of such secondary metabolites. Therefore, it is necessary to obtain cell lines having derepressed regulation systems and also to release the cells from such regulations by selecting proper cultivation conditions.

When compared with microbial cells, plant cells have many disadvantages, this means difficulties of cultivation of plant cells (Table 2). Growth of plant cells is usually slow and stabilization of cell lines and control of cultivation are often difficult because plant cells are usually heterogenous mixtures. Contents of desired compounds are low and fluctuate culture by culture. Furthermore, plant cells are more liable to be damaged by agitation than microbial cells because of their large size. In spite of these disadvantages, especially low and unstable productivity derived from slow growth, low content of the metabolites and heterogeneity of cells, plant cells are very useful to produce plant-specific metabolites of high value.

Plant cells are applied not only for de novo synthesis of useful metabolites, but also for salvage synthesis from proper precursors and for bioconversion of various compounds by hydroxylation, glucosylation and so on.

2. Improvement of Plant Cell Cultures

Many efforts have been devoted to stabilize cell lines and to improve productivity of useful compounds. For the improvement of cell lines, several techniques have been applied. These include cell-aggregate selection or cloning, single-cell selection, protoplast culture selection, protoplast fusion and genetic engineering. Especially, the cell-aggregate selection method has been used widely to obtain cell lines showing stable growth and high productivity.

Control of metabolic regulation is also important for the effective production of desired compounds. Hormones often affect growth and differentiation of plant cells and, subsequently, productivity of metabolites. Medium components such as nitrogen source, inorganic salts and metal ions, especially the kind of nitrogen source, that is, ammonium nitrogen or nitrate nitrogen, often give critical effects on growth and metabolism of plant cells. In some cases, synthesis of secondary metabolites is enhanced by stress, such as hyperosmotic conditions and addition of specific reagents. Immobilization might be a kind of stress. Addition of adequate precursors or inducers is sometimes useful for the production of specific secondary metabolites, although some of these additives may affect the viability of cells. Supply of oxygen and, in some cases, illumination of light are also important for the cell growth and the activity of metabolism.

Optimized supply of oxygen with mild agitation is critical for the cultivation of plant cells, because plant cells are more fragile than microbial cells. Although conventional bioreactors for microbial cells are often used for the cultivation of plant cells with a slow agitation in a laboratory scale, special reactors should be designed for a large scale cultivation of plant cells. Method of agitation by holding enough supply of oxygen is one of the problems.

3. Application of Plant Cell Cultures

Several examples of applications of plant cells carried out in Japan are summarized in Table 3. Most of these cells are able to synthesize the respective metabolites de novo, and some of them show the abilities to produce desired compounds from precursors via salvage synthesis or bioconversion, in addition to de novo synthesis.

Cultured cells of *Lithospermum erythrorhizon* have been applied to the industrial production of shikonin and its derivatives (Fig. 1)[3–5], which are utilized as colouring agent of cosmetics. This is a typical example, at present, of the industrial application of plant cell cultures for the production of useful metabolites. Shikonins have been used in Japan as high grade dyes and crude drugs since ancient period. These purple-red pigments with a naphthazalin skeleton are present as fatty acid esters in the root of the plants. This means that acetyl-, isobutyl-, isovaleryl- and 3-hydroxy-isovalerylshikonins are the main components. Japanese scientists developed the commercial production system of shikonins through extensive studies on the selec-

Table 3. Examples of application of plant tissue cultures

Product	Plant cell	References
Shikonins	*Lithospermum erythrorhizon*[a]	3–7
Anthocyanins	*Vitis*	8
	Euphorbia millii	9
Ubiquinone 10	*Nicotiana tabacum*	10–12
Berberine	*Coptis japonica*	13, 14
Reserpine	*Rauwolfia serpentina*	15
Scopolamine	*Hyoscyamus niger*	16–18
Ginseng saponins	*Penax ginseng*	19
Codeine	*Papaver somniferum*[b]	20
Glucosides	*Datura innoxia*	21, 22
Biotin	*Lavandula vera*	25, 26
Blue pigment	*Lavandula vera*[b]	27–29

[a] Industrialized
[b] Immobilized

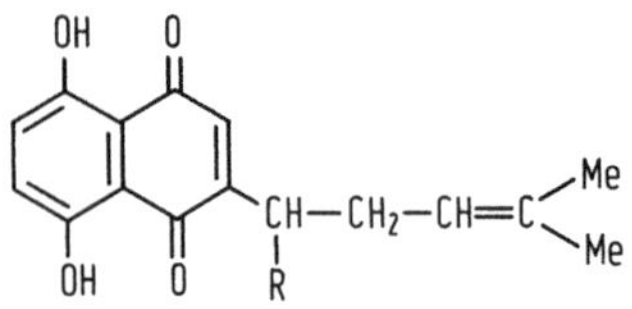

Fig. 1. Structures of shikonin and its derivatives.

Shikonin	R = OH
Deoxyshikonin	R = H
Acetylshikonin	R = $OCOCH_3$
Isobutylshikonin	R = $OCOCH(CH_3)_2$
3,3-Dimethylacrylshikonin	R = $OCOCH{=}C(CH_3)_2$
Isovalerylshikonin	R = $OCOCH_2CH(CH_3)_2$
3-Hydroxyisovalerylshikonin	R = $OCOCH_2C(OH)(CH_3)_2$

tion of cell lines by cell-aggregate cloning method and improvement of the media qualitatively and quantitatively (nitrogen/carbon sources, inorganic salts and metal ions). As the results, they selected two media, each for cell growth and for shikonin production. The growth medium contains high concentrations of both ammonium and nitrate nitrogens, while the production medium contains only a low concentration of nitrate nitrogen [6]. By using these two media in a two-step cultivation, inoculum size, aeration rate and type of reactor containment were optimized for process development. After these optimization experiments, final selection of cell lines was carried out by protoplast culture selection method. The cells selected by this method maintained a stable and high productivity of shikonins for at least 8 months. More than 2 g l^{-1} of shikonins were produced after 14 d of cultivation [7].

Production of anthocyanins by *Vitis* cells was investigated in suspension cultures. High sucrose and low phosphate concentrations and illumination of light facilitated the pigment formation, while the pigment synthesis was repressed by high concentrations of nitrate and 2,4-D. Regulation of aeration was also an important factor [8]. An anthocyanin was also produced by *Euphorbia millii* cells selected by means of cell-aggregate cloning. The cells showed a stable and high productivity of the pigment [9].

Ubiquinone 10 or Coenzyme Q_{10} is an important drug. Although cultured cells of *Nicotiana tabacum* were investigated for their ability to produce this coenzyme in suspension cultures, its content in the cells was relatively low [10, 11]. Several strains with a high productivity were selected by a cell cloning technique from the tobacco cell suspension cultures mentioned above. The selected cells produced 15 mg l^{-1} or about 1.9 mg g^{-1} dry cell of Coenzyme Q_{10} after optimization of cultivation conditions [12].

Cultured cells of *Coptis japonica* produced berberine whose synthesis was repressed by light and stimulated by high aeration [13]. Successively selected cell lines showed a higher productivity of berberine than the non-selected cells. The highest value obtained was 13.2% on a dry weight basis or 1.39 g l^{-1} after 3 weeks of cultivation [14]. Reserpine was produced by *Rauwolfia serpentina* at a content of 0.03–0.06% [15]. The selected cells synthesized more reserpine than ajmaline. Tropane alkaloids such as hyoscyamine and scopolamine were synthesized by cultured cells of *Hyoscyamus niger* [16]. Adventitious roots induced from suspension cultures of *H. niger* produced more scopolamin (0.12–0.30%) than hyoscyamine (0.04–0.08%). Furthermore, cultured roots of *Duboisia leichhardtii* showed much higher contents of alkaloids, that is, 0.53% of hyoscyamine and 1.16% of scopolamine on a dry weight basis [18].

Ginseng saponins, especially ginsenosides Rb_1 and Rg_1, are useful as medicines. Two high saponin-producing cell lines were selected from a callus of *Penax ginseng*.

Shaking cultures showed about 1.8 times higher growth than static cultures, although saponin contents were almost the same. The content reached about 0.7% on a dry weight basis [19].

Glucosylation is one of the interesting reactions catalyzed by plant tissue cultures. For example, umbelliferone (a coumarin), liquiritigenin (a flavonoid), alizarin (an anthraquinone) and/or salicylic acid (a simple phenol) were glucosylated by various cells, such as *Lithospermum erythrorhizon*, *Perilla frutescens*, *Gardenia jasminoides*, *Mallotus japonicus*, *Catharanthus roseus* and/or *Datura innoxia* [21]. In addition to *D. innoxia*, *C. roseus*, *G. jasminoides* and *L. erythrorhizon*, *Nicotiana tabacum*, *Duboisia myoporoides* and *Bupleurum falcatum* also glucosylated salicyl alcohol to form salicin or isosalicin [22].

Cultured cells of *Nicotiana tabacum* were found to catalyze regio- and stereospecific hydroxylation of foreign substrates such as monoterpenoids [23]. Furthermore, these cells catalyzed the interconversion between cycloalkanones and their corresponding alcohols [24]. These oxidation-reduction reactions are very important to obtain specific enantiomers of different structures.

Biotin was synthesized in many species of cultured plant cells, such as *Lavandula vera*, *Nicotiana tabacum* and *Glycine max*. Cultured green *L. vera* cells grown under light illumination contained a high amount of biotin. Selection of potent cell lines was carried out by irradiation of gamma ray, the content reaching about $0.5\ \mu g\ g^{-1}$ fresh weight [25,26]. When these cells were cultivated in a medium supplemented with S-containing compounds, such as L-cysteine, DL-homocysteine, L-cystine or DL-homocystine, a blue pigment was accumulated [27]. Production of this pigment by the immobilized cells [28,29] will be described below.

4. Application of Immobilized Plant Cells

In spite of the interesting characteristics of plant cell cultures mentioned above, several drawbacks are often claimed, that is, low productivity, instability of cell structure as well as metabolic activities, low growth rates and uncontrollable fluctuation in productivity with the change of cultivation scale. Concerning enzymes and microbial cells, immobilization is expected to give favourable effects on the application of cultured plant cells. Immobilization, especially entrapment in an appropriate gel, may be able to protect plant cells from external impacts. Also, it provides optimal growth conditions, increases the productivity of useful compounds and enables repeated or continuous use of the cells. Actually, it has been demonstrated that immobilization is considerably effective in eliminating above-mentioned disadvantages inherent in plant cells [30–32]. Immobilization of plant cells was first reported by Brodelius et al. [33] in 1979. Thereafter, immobilized plant cells have been applied for de novo synthesis, salvage synthesis and bioconversion of various compounds. However, the application of immobilized plant cells is limited because the products should be excreted from the cells for the effective production of these compounds.

In Japan, little information is available at present on immobilized plant cells. Furuya et al. [20] investigated the conversion of codeinone to codeine, an important medicine, by the calcium alginate-entrapped cells of *Papaver somniferum*. Different from the intact plant, cultured cells could not produce codeine de novo, but showed the ability to reduce codeinone to codeine. The entrapped cells exhibited the original conversion activity at least for 24 days when the cells were used repeatedly in a column reactor at 20 °C. Lower temperature seemed to be better for the stability of the activity. Aeration was also an important factor, and increased oxygen supply facilitated the bioconversion, although this conversion was a reduction reaction. However, an optimized aeration might exist for the growth and metabolic activity of the plant cells.

Cultured cells of *Lavandula vera*, which had been screened as a potent biotin-producer by Yamada and his coworkers [25,26], were found to synthesize and excrete a blue pigment, whose formation was induced by the addition of L-cysteine to the medium [27]. Callus of *L. vera* was treated mechanically or enzymatically to obtain a fine suspension of cells suitable for immobilization. However, the viability of the cells thus treated was not sufficient when immobilized and applied for the pigment production. Finely suspended cells were prepared by cultivating the cells successively in the presence of a proper concentration (50 mM) of calcium chloride and this cell suspension could be entrapped efficiently in various gels.

For example, *L. vera* cells entrapped in calcium alginate showed a good growth inside the gels. Growth of the cells was also confirmed by the increases in oxygen uptake of the entrapped cells, chlorophyll content, and cell number after solubilization of gels. When L-cysteine was added to the medium as an inducer of pigment synthesis, the blue pigment was produced and accumulated in the medium holding calcium alginate-entrapped cells [28].

Although we have not characterized the structure of the pigment yet, this pigment is quite different from delphinidin, an agricon of *Lavandula* pigment obtained from

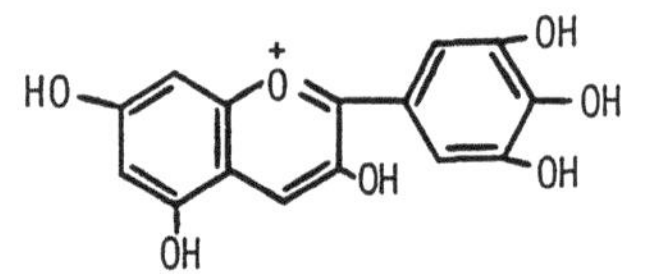

Delphinidin (agricon of Lavandula pigment)

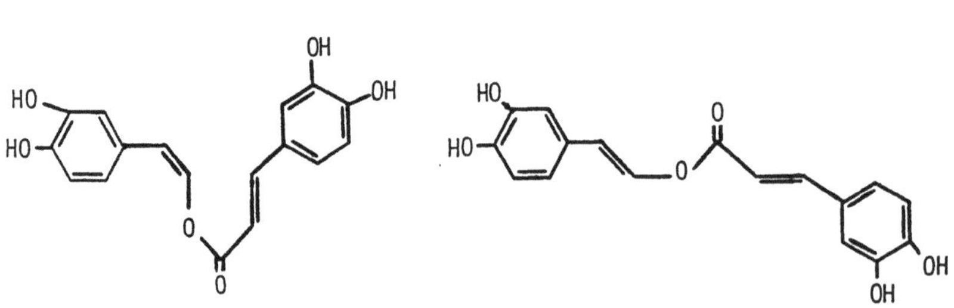

(Z,E)- or (E,E)-2-(3,4-dihydroxyphenyl)ethenyl ester of 3-(3,4-dihydroxyphenyl)-2-propenoic acid

Fig. 2. Structure of pigment produced by *Lavandula* callus culture the flowers. Banthorpe and his coworkers [34] isolated a yellow pigment from *Lavandula* callus culture, which turned blue in the presence of ferrous ion. Our pigment may have a similar structure to that of Banthorpe et al., that is, an ester of 2-(3,4-dihydroxyphenyl)ethenyl alcohol and 3-(3,4-dihydroxyphenyl)-2-propenoic acid (Fig. 2).

In addition to calcium alginate, we entrapped *L. vera* cells with different gel materials (Table 4). Among the natural polysaccharides tested, calcium alginate was found to be the best based on the cell growth, pigment production, mechanical strength and cell holding ability. Although agar gel permitted abundant growth and pigment production, mechanical strength of the gel was not sufficient. Photosensitive resin prepolymer was also suitable gel material as mentioned below.

Although L-cysteine is essential for the induction of the pigment synthesis, this compound inhibits the cell growth severely. After various trials we have succeeded to use the entrapped cells repeatedly for the production of the pigment. For this the entrapped cells were alternatively incubated for growth in the absence of L-cysteine and for production in the presence of L-cysteine. Although the pigment productivity fluctuated fairly, the entrapped cells were alive and active even after 7 months of incubation (Fig. 3). The free cells lost the activity after only one cycle of the incubation.

Table 4. Properties of gel-entrapped *Lavandula vera* preparations

Gel material	Shape	Mechanical strength	Cell growth	Cell leakage	Pigment production
None (free cells)	——	——	+++	——	++
Agar	Block	+	++	+++	+++
Alginate	Bead	++	++	+	++
κ-Carrageenan	Bead	++	+	+	+
Urethane prepolymer	Block	+++	++	+	+
Photo-crosslinkable resin prepolymer	Film	++	—	—	—
Photosensitive resin prepolymer	Film	++	++	+	++

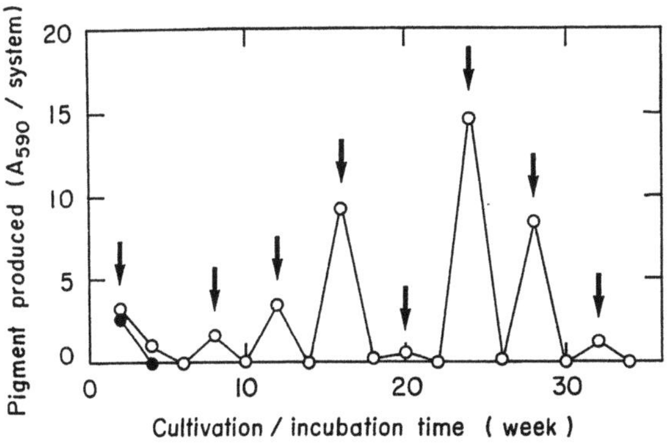

Fig. 3. Repeated use of calcium alginate-entrapped *Lavandula vera* cells for pigment production. *Arrows* indicate production phase in the presence of L-cysteine. ○, Entrapped cells; ●, free cells

Application of synthetic resins as gel materials is very interesting and important because physico-chemical properties of gels, such as network structure, hydrophilicity-hydrophobicity balance and ionic property which will affect the productivity of desired compounds by immobilized biocatalysts, can be selected optionally. *L. vera* cells were also entrapped in different synthetic resin gels and the properties of the entrapped cells were compared with those of the calcium alginate-entrapped cells [29]. The cells entrapped with photosensitive resin prepolymer (PVA-SbQ) (Fig. 4), a derivative of polyvinyl alcohol which had been developed by Ichimura [35], showed an excellent activity of the pigment production (Table 4). Mechanical strength of this resin was also sufficient. On the other hand, photo-crosslinkable resin prepolymer and urethane prepolymer [36] were not good gel materials so far as tested. Although PVA-SbQ-entrapped cells exhibited a high productivity of the pigment, most of the pigment was unfortunately not excreted to the medium, in contrast to the case of the calcium alginate-entrapped cells (Fig. 5). The resin gels tightly held the pigment inside the gel matrices. Difference in the pigment excretion may be due to the different hydrophilicity-hydrophobicity balance between alginate gel and PVA-SbQ gel, since alginate gel has a very hydrophilic nature. Therefore, we examined the PVA-SbQ prepolymers of different degree of saponification of polyvinyl alcohol part. The

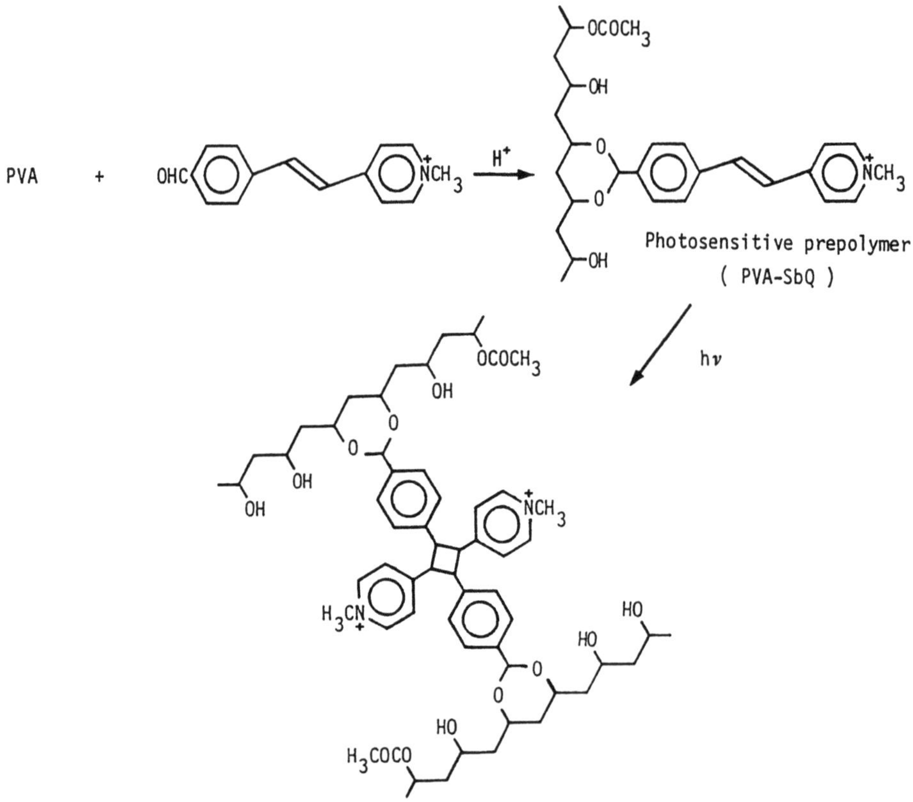

Fig. 4. Structure of photosensitive resin prepolymer, PVA-SbQ

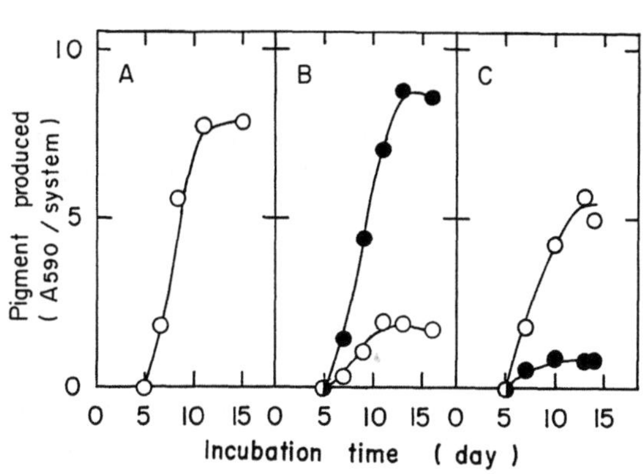

Fig. 5. Time-course of pigment production by *Lavandula vera* cells. *A*, Free cells; *B*, PVA-SbQ-entrapped cells; *C*, calcium alginate-entrapped cells. ○, Pigment in medium; ●, pigment in gels

Table 5. Effect of gel hydrophilicity on pigment production by PVA-SbQ-entrapped *Lavandula vera* cells

Saponification degree of polyvinylalcohol (%)	Total pigment produced (A_{590}/system)	Pigment in medium (%)
100	5.0–7.7	50–61
88	5.3–8.6	14–19
80	3.8–5.3	3– 6

results shown in Table 5 indicated that the pigment excretion was able to be improved by using a prepolymer of high saponification degree, that is, with a gel having a high hydrophilicity.

The cells entrapped with a suitable PVA-SbQ prepolymer could be applied for the pigment production repeatedly. We are still continuing the effort to enhance the pigment productivity by improving medium constituents and cultivation conditions, which will make it possible to use the immobilized plant cells in a continuous production system. The concentration of L-cysteine required for the pigment production was severely dependent on the cell concentration. Limitation of cell growth by omission of ammonium nitrogen was found to be effective for the repeated production of the pigment at a fixed concentration of L-cysteine.

5. Conclusion

As described above, plant cells, in addition to enzymes, microbial cells and animal cells, are becoming more and more important biocatalysts, and, in some cases where the products are excreted from the cells, immobilization will be an excellent technique to improve the properties of plant cells. Development of novel gel materials suitable for plant cells and proper reactors adequate for cultivation of immobilized plant cells are indispensable for the extensive application of immobilized plant cells. Simple methods to maintain cell lines, especially suspended cells, should be developed for easy treatment of plant cells.

6. Acknowledgements

The author would like to express his sincere thanks to Prof. Yasuyuki Yamada and Dr. Fumihiko Sato, Kyoto University, for their generous discussion on preparing this article.

7. References

1. Yamada, Y., Hashimoto, T.: Applications of Genetic Engineering to Crop Improvement (Collins, G. B., Petolino, J. G., eds.), p. 561, Dordrecht/Boston/Lancaster: Martinus Nijhoff/ Dr W. Junk Publishers 1984
2. Collinge, M.: Trends Biotechnol. *4*, 299 (1986)
3. Tabata, M. et al.: Phytochem. *13*, 927 (1974)
4. Fujita, Y. et al.: Plant Cell Rep. *1*, 59 (1981)
5. Fujita, Y. et al.: ibid, *1*, 61 (1981)
6. Fujita, Y. et al.: Proc. 5th Intl. Cong. Plant Tissue and Cell Culture, p. 399 (1982)
7. Fujita, Y., Takahashi, S., Yamada, Y.: Agric. Biol. Chem. *49*, 1755 (1985)
8. Yamakawa, T. et al.: ibid. *47*, 2185 (1983)
9. Yamamoto, Y., Mizoguchi, R., Yamada, Y.: Theor. Appl. Genet. *61*, 113 (1982)
10. Ikeda, T., Matsumoto, T., Noguchi, M.: Agric. Biol. Chem. *41*, 1197 (1977)
11. Ikeda, T., Matsumoto, T., Noguchi, M.: Phytochem. *17*, 1879 (1978)
12. Matsumoto, T. et al.: Agric. Biol. Chem. *45*, 1627 (1981)
13. Yamada, Y., Sato, F.: Phytochem. *20*, 545 (1981)
14. Sato, F., Yamada, Y.: ibid. *23*, 281 (1984)
15. Yamamoto, O., Yamada, Y.: Plant Cell Rep. *5*, 50 (1986)
16. Yamada, Y., Hashimoto, T.: ibid. *1*, 101 (1982)
17. Hashimoto, T., Yamada, Y.: Planta Medica *47*, 195 (1983)
18. Endo, T., Yamada, Y.: Phytochem. *24*, 1233 (1985)
19. Furuya, T. et al.: Planta Medica *48*, 83 (1983)
20. Furuya, T., Yoshikawa, T., Taira, M.: Phytochem. *23*, 999 (1984)
21. Umetani, Y., Tanaka, S., Tabata, M.: Proc. 5th Intl. Cong. Plant Tissue and Cell Culture, p. 383 (1982)
22. Mizukami, H. et al.: Phytochem. *22*, 679 (1983)
23. Suga, T. et al.: Proc. 5th Intl. Cong. Plant Tissue and Cell Culture, p. 381 (1982)
24. Suga, T., Hamada, H., Hirata, T.: Plant Cell Rep. *2*, 66 (1983)
25. Watanabe, K., Yano, S., Yamada, Y.: Phytochem. *21*, 513 (1982)
26. Watanabe, K., Ymada, Y.: Plant and Cell Physiol. *23*, 1453 (1982)
27. Watanabe, K. et al.: Agric. Biol. Chem. *49*, 533 (1985)
28. Nakajima, H. et al.: J. Biotechnol. *2*, 107 (1985)
29. Nakajima, H. et al.: Appl. Microbiol. Biotechnol. *24*, 266 (1986)
30. Brodelius, P., Mosbach, K.: Adv. Appl. Microbiol. *28*, 1 (1982)
31. Prenosil, J. E., Pedersen, H.: Enzyme Microb. Technol. *5*, 323 (1983)
32. Rosevear, A., Lambe, C. A.: Adv. Biochem. Eng./Biotechnol. *31*, 37 (1985)
33. Brodelius, P. et al.: FEBS Lett. *103*, 93 (1979)
34. Banthorpe, D. V., Bilyard, H. J., Watson, D. G.: Phytochem. *24*, 2677 (1985)
35. Ichimura, K.: J. Polym. Sci. Polym. Chem. Edn. *22*, 2817 (1984)
36. Fukui, S., Tanaka, A.: Adv. Biochem. Eng./Biotechnol. *29*, 1 (1984)

1.2 Characteristics and Utility of Human Hematopoietic Cell Lines

Jun Minowada

Fujisaki Cell Center, Hayashibara Biochemical Laboratories, Inc.,
675-1, Fujisaki, Okayama 702, Japan
and Department of Pathology, Loyola University of Chicago, Stritch School of Medicine, Maywood, IL 60153, U.S.A.

Contents

Characteristics and utility of over 100 human leukemialymphoma cell lines are summarized. A brief account of cell line development during the past 25 years depicts advances and developments in both leukemia-lymphoma management and human immunobiology research. The study of these cell-lines led to great improvements in the immunologic diagnosis of leukemias and lymphomas, which has allowed more specific treatment of these malignancies. In the ongoing biotechnology revolution the leukemia-lymphoma cell lines have found new utility, and their potential uses stimulate and necessitate further basic and applied research.

1. Brief Review of Human Leukemia-Lymphoma Cell Cultures

Human leukemias have been known heterogeneous diseases as a malignant disorder of the hematopoietic cells in respect to morphological appearance, clinical presentation and response to the treatment. Based on the resemblance in morphological appearance with those of the normal blood cells the leukemias have been classified into two major types, namely lymphocytic and granulocytic (or myelocytic) leukemias. Based on the clinical manifestation and course of the disease progress, each of the leukemia types is further classified into either acute or chronic leukemia. Upon the advancement of chemotherapy as a curative treatment for acute lymphoblastic leukemia (ALL), the accurate diagnosis of leukemia became an urgent need in the management of patients with leukemia. In 1976, a unified classification system of the acute leukemia was proposed by a group of clinical experts from France, America and Britain (FAB classification system of the acute leukemias for clinical usage) [1]. The FAB classification system is one of the most widely utilized classification system. Nevertheless, it is increasingly recognized that further improvement is needed to correlate the FAB classification with the clinical responses in leukemias.

During the past two decades, significant progress has been made to this end by the immunological characterization of leukemias and of normal hematopoietic cells. The establishment of human leukemia-lymphoma cells into permanent cell cultures has played a major role for this advancement in immunology [2].

1.1 Burkitt Lymphoma

In 1963, Pulfertaft [3] established, for the first time, a series of permanent cell lines from Nigerian patients with Burkitt lymphoma, a malignant lymphoma common among children in certain areas of Africa. Subsequently Epstein et al. [4] and Minowada et al. [5] established similar human lymphoid cell lines from the patients with Burkitt's lymphoma in Uganda and Kenya, respectively. By studying these early cell lines, Epstein et al. [6] discovered that these cultured Burkitt lymphoma cells were infected with herpes type DNA virus which was later designated as Epstein-Barr virus (EB virus). Tanigaki et al. [7] found that the Burkitt lymphoma cell lines were capable of producing immunoglobulins.

Because of increasing evidence accumulated over the years in experimental animal tumor system in which oncogenic virus involvement was established as a causative agent, the fact that the presence of EB virus in all of these Burkitt lymphoma cell lines implicated that human lymphoid tumors might be caused by the virus. Clinical and geographical findings associated with African Burkitt lymphoma supported such a notion. Nevertheless, Burkitt lymphoma cell lines established from cases in the North American region were not associated with EB virus infection [8]. The absence of the EB virus involvement in non-endemic Burkitt lymphoma was further proved by EBNA test [9] as well as molecular hybridation techniques [10] which did not detect the EB virus genome.

1.2 Leukemias

In 1964, Iwakata and Grace [11] reported the successful establishment of a human leukemia cell line (RPMI 6410) from the bone marrow aspirate of a patient with acute myeloblastic leukemia (AML). Most significantly, RPMI 6410 cell line was positive for herpes-type virus infection which was later identified as the EB virus. The finding, which was independently made from the study on African Burkitt lymphoma cell lines where the presence of EB virus was commonly found, was taken as a further evidence for the notion that human leukemia and lymphoma may well be virus-induced malignancies. It was, however, soon found that despite being an AML, the RPMI 1640 cells in culture produced human immunoglobulin. Thus, the identity of RPMI 6410 was established as an EBV positive B lymphoblastoid cell line. Moore and his associates [12] established a large number of similar leukemia-derived B lymphoblastoid cell lines from patients with leukemias. Moreover, they demonstrated that EBV-positive B lymphoblastoid cell lines can be established from the peripheral blood of healthy blood donors [13]. In an isolated case of a patient with multiple myeloma, however, the resultant cell line represented a true myeloma cell line, RPMI 8226 [14].

Foley et al. [15] in 1964 established a leukemic cell line, CCRF-CEM, from a patient with ALL. CCRF-CEM was the first T-cell leukemia cell line, but they failed to establish its identity as a T-cell line at that time. Subsequently, Adams et al. [16] established another cell line, CCRF-HSB-2, but they too were unable to identify the nature of CCRF-HSB-2 as a T-cell line at that time.

1.3 Non-Maligant Hematopoietic Cells

From the first observation by Moore et al. [12,13] that B-lymphoblastoid cell lines with EBV infection can be established from the peripheral blood of non-leukemic donors, numerous cell lines were established from a variety of origins which were such B lymphoblastoid cell lines. In fact, most of the culture attempts with leukemic specimens resulted in B-lymphoblastoid cell lines infected with EBV. RPMI 6410 from an AML specimen turned out to be a non-malignant B cell line with EBV infection rather than a myeloblastic leukemia cell line. Gerber et al. [17] indicated that EBV was an oncogenic virus with which even normal blood cells can be transformed in vitro into an established cell line with sustained growth potential. The issue, however, is still equivocal in strict sence even at the present time. Nevertheless, EBV is an effective means to immortalize human normal B cells into permanently growing cell lines.

1.4 Epstein-Barr Virus and Immunoglobulin

The findings indicating the presence of EBV and of immunoglobulin among most of those lymphoid cell lines during the 1960s led to an extensive study of EBV as a putative etiological agent for human hematopoietic malignancy. Much fundamental knowledge about immunoglobulin biosynthesis resulted from this research on B lymphoid cell lines. In fact, without those cell lines, it would have been impossible or at least extremely difficult to have achieved such advances. Henle et al. [18] have reported that EBV may well be a causative agent for infectious mononucleosis, though this author has some reservation for this conclusion. There is no doubt, however, that utility of these cell lines provided and stimulated feasibility and interest in the research of many disciplines. At the present time, possible relationship between hematopoietic malignancies and EBV is less apparent than one once thought to be.

1.5 T-Cell Leukemias

There are two common characteristics found in most of the early lymphoid cell lines, namely the presence of EBV and the ability to synthesize immunoglobulin. While the presence of immunoglobulin on the lymphocyte membrane was known as a marker of B lymphocytes, a positive marker for T-lymphocytes was not available in the 1960s. Although CCRF-CEM and CCRF-HSB-2 were established earlier, their identity as T-cells was not known at that time. In 1972, Minowada et al. [19] reported, for the first time, E-rosette forming leukemia cell lines, MOLT 1–4, and that instead of being as a leukemia marker, E rosette formation of lymphocytes is a normal T-lymphocyte membrane marker. Jondal et al. [20] reported independently using fresh normal peripheral blood lymphocytes to show that E rosette formation is a normal T cell marker. These observations made possible an extensive investigation of both leukemia membrane phenotypes and cellular immunology in man. Subsequently, a number of T-cell leukemias and lymphomas have been identified, and most significantly Borella et al. [21], for the first time, demonstrated clearly that T-cell ALL in childhood is a minor subtype with poorer prognostic outcome relative to that of non-T ALLs. Many permanent T-cell leukemia-lymphoma cell lines have since been established and have provided evidence for the heterogeneity of this type of hematopoietic malignancy. Some significant new findings from such investigations are described in the following sections.

2. Establishment and Characterization of E-Rosette Forming Leukemia Cell Lines

In 1971, Minowada et al. [19] established a set of E-rosette forming leukemia cell lines, MOLT 1–4, from the peripheral blood of a 19 year old male with ALL. E-rosette formation of MOLT 1–4 was unexpected but was identified as a normal T cell membrane marker rather than a leukemia marker. An extensive study on MOLT 1–4 led to a number of significant contributions not only to leukemia research but also to immunobiological research. From these findings, CCRF-CEM and CCRF-HSB-2 were identified positively as T-cell leukemia cell lines.

2.1 MOLT 1–4 Cell Lines

Besides E rosette formation, the study of MOLT 1–4 demonstrated that T-cells lack immunoglobulin and that the absence of HLA-DR (class II MHC antigen) correlates with the absence of mixed leukocyte reaction [22]. This series of new observations resulting from the study of MOLT 1–4 is in part responsible for subsequent advances in cellular immunobiology research.

2.2 T-, B- and Non T-, Non B Cell Leukemias

Renewed enthusiasm stemming from the study of MOLT 1–4 cell lines led to the establishment of various leukemia cell lines [23]. The search for a leukemia specific antigen has thus far not been successful. Instead of this, all of the antigens detected on the leukemia-lymphoma cells to date are found to be normal gene products. Miyoshi et al. [24] reported a B-cell leukemia line, BALL-1 and Rosenfeld et al. [25] reported, for the first time, a non T- non B cell leukemia line, Reh, from a patient with ALL. As described above, Burkitt lymphoma cell lines are of B-cell type and RPMI 8226 reported by Moore et al. [14] is a myeloma cell line, an other B-cell type. Nilsson et al. [26] also reported a B-cell type lymphosarcoma cell line, U-698-M. The discovery of the pre-B cell ALL by Vogler et al. [27] in 1975 made us re-examine all pre-existing non T-non B cell leukemia cell lines. As expected, some of them were found to be the pre-B cell leukemia lines. Minowada et al. [28] were able to report a Ph^1chromosome prositive pre-B cell leukemia line, NALM-I, derived from a patient with chronic myelogeneous leukemia in its blastic phase. Similar to the discovery of EBV from the Burkitt lymphoma cell line by Epstein et al. [6], Reitz et al. [29] utilizing the established T cell leukemia and lymphoma lines found, for the first time, a retrovirus designated as human T-lymphotropic virus (HTLV type I). The HTLV I is known to be associated with some of the T cell leukemias and lymphomas, notably in the adult T cell leukemia in Japan [30]. It should be noted also that discovery and progress of the research of another retrovirus (HTLV-III, LAV, ARV or HIV) which is associated etiologically with AIDS were greatly helped by the use of in vitro virus-infected cell lines [32].

2.3 Panel of Leukemia-Lymphoma Cell Lines

During the last two decades, we were able to establish and collect a total of over 200 human hematopoietic cell lines. While all of them are growth-factor independent cell lines, a total of over 100 cell lines are the proven leukemia-lymphoma cell lines. For our convenience, they are divided into several groups of categories. Table 1 summarizes those leukemia-lymphoma cell lines including a few lines of in vitro retrovirus-infected transformant of normal T-cells.

2.4 Immunologic Diagnosis of Leukemias

The significance of immunologic membrane phenotyping of leukemias [33] has been increasingly appreciated not only for clinical management of leukemia patients but also for providing a model for immunobiology research [34]. The study of leukemia-lymphoma cell lines continues to play a major role in this area. The heterogeneity of clinical response in the management of leukemic patients has been known, and the immunologic parameters provide some rationale to account for it [35]. Figure 1 is a work-up chart we are currently using for the immunologic diagnosis of leukemia

DX	Immunologic subtype		E	EA	EAC	B: SmIg	B: CyIg	B: B-Ag	cALL	T: Pan-T	T: I/H	T: S/C	T: Thy	M: Pan-M	M: Gr	M: Mo	Ia	TdT	%	FAB
ALL	Common		-	-	-	-	-	-	+	-	-	-	-	-	-	-	+	+	55	L-1,2
	Pre-B		-	-	-	-	+	+/-	+	-	-	-	-	-	-	-	+	+	22	L-1,2
	T	I	-	-/+	-	-	-	-	+	+	-/+	-/+	-	-	-	-	+	+		
		II	+/-	-	-/+	-	-	-	+	+	+	+	+	-	-	-	-	+		
		III	+/-	-	-/+	-	-	-	-	+	+	+	+	-	-	-	-	+	15	L-1,2
		IV	+/-	-	-/+	-	-	-	-	+	+/-	-/+	+/-	-	-	-	-	-		
		V	+/-	-	-/+	-	-	-	-	+	+/-	-/+	-	-	-	-	+/-	-		
	B		-	+/-	+/-	+	+	+	-	-	-	-	-	-	-	-	+	-	3	L-3
	"Null"		-	-/+	-	-	-	-	-	-	-	-	-	-	-	-	-/+	-/+	5	L-1,2 M-1
ANLL	ANLL	I	-	-	-	-	-	-	-	-	-	-	-	-	-	-	+	-	10	M-1
		II	-	-	-	-	-	-	-	-	-	-	-	+	+	+/-	+	-	40	L-1,2 M-1,2,
		III	-	+	+/-	-	-	-	-	-	-	-	-	+	+/-	+/-	-/+	-	50	3,4,5
CML-BP	"Lymphoid"		-	-	-	-	-	-	+	-	-	-	-	-	-	-	+	+	20	L-1,2
	"Mixed"		-	-/+	-/+	-	-	-	+	-	-	-	-	+	+	+/-	+	-/+	30	M-1,
	"Myeloid"		-	+/-	+/-	-	-	-	-	-	-	-	-	+	+	+/-	+/-	-	50	2,3
CLL	B		-	+	+	+	+	+	-	-	-	-	-	-	-	-	+	-	95	
	T		-	-	-	-	-	-	-	+	+/-	-/+	-	-	-	-	-	-	5	

Fig. 1. Marker profile of leukemia. By means of multiple marker analysis, four major forms of leukemia (shown in the first column) are subtyped into immunologic subtypes, as shown in the second column. Combining all positive markers identifies the particular immunologic subtypes of leukemia in question. An appropriate combination of monoclonal antibody reagents may be substituted for or added to each marker

Abbreviations: DX = diagnosis
ALL = acute lymphoblastic leukemia
ANNL = acute non-lymphocytic leukemia
CML-BP = chronic myelogenous leukemia in blastic phase
CLL = chronic lymphocytic leukemia
E = E rosette
EA = EA-rosette (Fc receptor)
EAC = EAC-rosette (C3 receptor)
B = B-cell markers
SmIg = surface membrane immunoglobulin
B-Ag = B-cell associated antigens
cALL = antigen associated with non-T, non-B common ALL (CALLA)
T = T-cell antigens
Pan-T = antigens common for all T cells
I/H = antigen associated with inducer or helper T-cell subsets
S/C = antigens associated with suppressor or cytotoxic T-cell subsets
Thy = thymocyte antigens
Pan-M = antigens common for all myeloid-monocytes
Gr = antigens restricted to myelogranulocytes
Mo = antigens restricted to monocytes
Ia = HLA-DR (gp 28, 33) antigens
TdT = terminal deoxynucleotidyl transferase
% = percent estimates of incidence for each immunologic subtype within each diagnostic category
FAB = relative correlation of morphologic leukemia subtypes by FAB classification to immunologic subtypes of leukemia

Table 1. Human leukemia-lymphoma cell lines (N = 116: Jan. '87)

No.	Cell line	Origin[a]	No.	Cell line	Origin[a]
T-cell			49	LAZ-221**	ALL
1	P30/OHKUBO	ALL	50	KOPN 1-8**	ALL
2	MOLT-10	ALL	51	HPB-Null**	ALL
3	RPMI 8402	ALL	52	KLM-2**	AMOL
4	CCRF-CEM	ALL	53	BALM 1, 2	ALL
5	DND-41	ALL	54	U-698-M	LS
6	HPB-ALL	ALL	55	TANAKA	LB
7	HPB-MLT	LY	56	MANACA	LB
8	HD-Mar-2	HD?	57	JOK-1	HCL
9	MOLT-12	ALL	58	SKW-4	LB
10	MOLT 13, 14	ALL	59	BALM 3–5	LB
11	JM	ALL	60	SU-DHL-4	LY
12	TALL-1	ALL	61	SCOTT	LB
13	MOLT-3	ALL	62	MLB-1084	LB
14	MOLT-4	ALL	63	ABL-2	BL?
15	P12/Ichi.	ALL	64	EB-3	BL
16	KOPT-K1	LS	65	RAJI	BL
17	MOLT-11	ALL	66	HR1K	BL
18	MOLT 16, 17	ALL	67	B35M	BL
19	Jurkat	ALL	68	KOBK-101	BL
20	Jurkat/HTLV III***	ALL	69	DAUDI	BL
21	CCRF-HSB-2	ALL	70	NAMALVA	BL
22	PEER	ALL	71	RAMOS	BL
23	KE-37	ALL	72	BJAB	BL
24	SKW-3	CLL	73	DG-75	BL
25	A3/Kawa.	LY	74	CHEVALLIER	BL
26	MOLT-15	AMOL	75	DND-39	BL
27	ALL-Sil	ALL	76	NK-9	BL
28	MOLT-4/LAV***	ALL	77	B46M	BL
29	MOLT-4/ARV***	ALL	78	OGUN	BL
30	CCRF-CEM/HTLV III***	ALL	79	AL-1	BL
31	MT-4/HTLV III***	NT	80	SL-1	BL
32	MT-1	ATL	81	RIVA	LB
33	HUT-78	SS	82	BALL-1	ALL
34	HUT-102	MF	83	RPMI 8226	MM
35	CTCL-2	SS	84	U-266	MM
36	C5/MJ***	NT	85	ARH-77	MM
37	MT-2***	NT	86	Hs	MM
38	MT-2/HTLV III***	NT	87	Nak-Pc2	MM
39	C91/PL***	NT	88	Hair-M	HCL
40	C82/TK***	NT	89	JC-1	HCL
41	KH-2	ATL	90	Hair-MTI	HCL
42	H9	ALL			
43	H9/HTLV III***	ALL			
44	TOM-JIM***	NT	*Non-T, Non-B cell*		
45	467***	NT			
			91	REH	ALL
B-cell			92	KM-3	ALL
			93	NALL-1	ALL
46	NALM-1**	CML	94	NALM-16	ALL
47	NALM 6-15**	ALL	95	KOPN-N	ALL
48	NALM 17, 18**	ALL	96	BV-173	CML

(Continued)

Table 1 (continued)

No.	Cell line	Origin[a]	No.	Cell line	Origin[a]
Myelomonocyte-cell			108	KCL-22	CML
			109	U-937	LY
97	KG-1	AML			
98	ML 1–3	AML			
99	HL-60	APL	*Non-L, Non-M-cell*		
100	PL-21	APL			
101	JOSK-I	AMML	110	K-562	CML
102	CTV 1, 2	AMOL	111	HEL	EL
103	THP-1	AMOL	112	SPI 801, 802	ALL
104	JOSK-K	AMOL	113	SU-DHL-1	LY
105	JOSK-S	AMOL	114	HDLM 1–3	HD
106	JOSK-M	CML	115	L-428	HD
107	GDM-1	CML	116	KAN***	NT

[a] Abbreviations used: ALL = acute lymphoblastic leukemia; AML = acute myeloblastic leukemia; AMML = acute myelomonocytic leukemia; AMOL = acute monocytic leukemia; APL = acute promyelocytic leukemia; ATL = "adult T-cell leukemia"; BL = Burkitt lymphoma; CLL = chronic lymphocytic leukemia; CML = chronic myelocytic leukemia; EL = erythroleukemia; HCL = hairy cell leukemia; HD = Hodgin's disease; LB = B-cell lymphoma; LS = lymphosarcoma; LY = lymphoma; MF = Mycosis fungoides; MM = multiple myeloma; NT = normal transformed T cells; SS = Sezary syndrome; ** = Pre-B leukemia; *** = in vitro retrovirus infected.

relative to the FAB classification. The basis of the subtype classification of leukemia was initially developed in our laboratory with a panel of polyclonal antibody reagents [36]. Availability of increasing number of monoclonal antibody reagents has opened up the possibility that the immunologic diagnosis of leukemia can readily be done by anyone as one of the essential diagnostic tests [37] which include the FAB morphologic classification [1] and cytogenetic marker identification [38].

2.5 Multiple Marker Analysis of Leukemia Diagnosis

Despite the essential role of immunologic membrane phenotyping for the diagnosis of leukemia today, it does not replace or substitute for other parameters used in the diagnosis of leukemia. Each parameter complements the others. Thus, we emphasize that the multiple marker analysis of leukemia is the way by which the diagnosis and management of leukemia can be best accomplished [35,36]. The FAB morphologic diagnosis and the immunologic diagnosis of leukemia can be done on the first day of the patient's visit to clinic. Due to the technical limitation, cytogenetic test results can only be available a few days later. Nevertheless, these three parameters are the essential elements of the multiple marker analysis of leukemia diagnosis today. Additional markers including virologic [9,29], isoenzymatic [39], biochemical [40], functional [41] and molecular biologic [42] tests are all important, and are helpful in the clinical management of leukemia patients.

3. Model for Human Hematopoietic Cell Differentiation

The search for leukemia-specific antigens has thus far been disappointing, though there have been such claims in the past. All antigens and other markers in leukemia cells, except certain leukemia- and lymphoma-form specific cytogenetic markers, were of normal gene products. Therefore the marker profiles in the leukemia cells were found to reflect that of their normal counterparts [43-45]. In addition, most of the leukemias-lymphomas tested to date represented a single clonal cell population with arrested differentiation in their respective lineage.

3.1 Hypothetical Scheme for Normal Hematopoietic Cell Differentiation

These two characteristics, namely clonal origin and arrested cellular differentiation, of leukemia cells provided a basis by which a hypothetical scheme of the normal hematopoietic cell differentiation could be drawn. Based on our study both in cell lines and in uncultured fresh leukemias, Fig. 2 illustrates a scheme of hypothetical normal hematopoietic cell differentiation. With further functional, molecular and

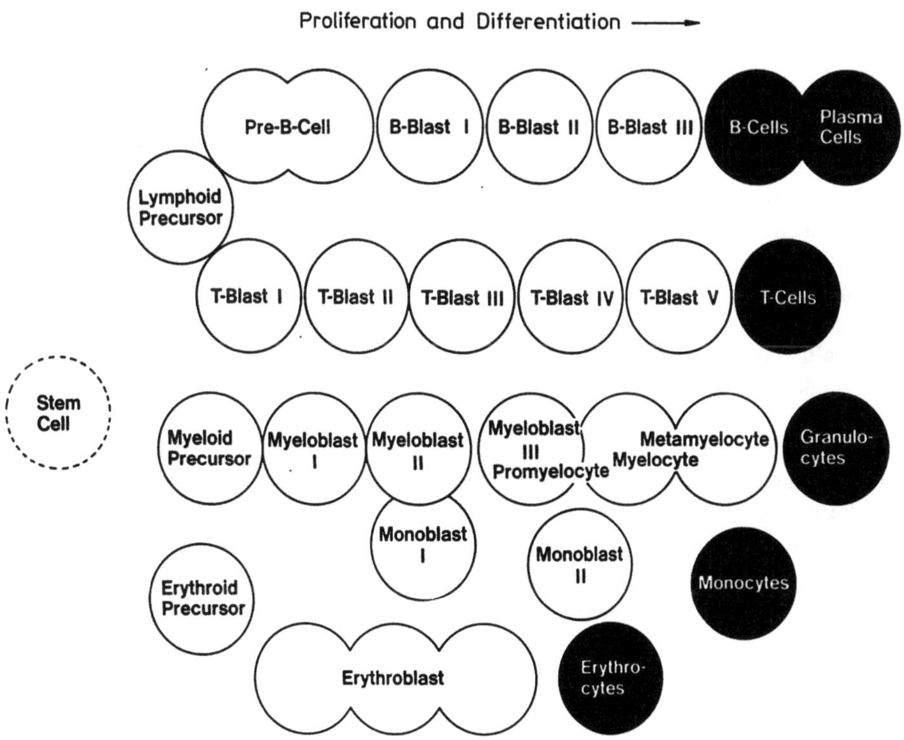

Fig. 2. Hypothetical differentiation scheme of normal hematopoietic cells based on the multiple marker analysis of leukemias and lymphomas. Assuming a single compartment of pluripotential stem cells, each lineage of hematopoietic cell type except megakaryocyte-platelet lineage is illustrated

other studies, it is possible to define further each compartment into discrete subset populations. Various detected antigens are, therefore, classified into three main groups, namely lineage restricted, differentiation associated and function-related antigens.

It should be mentioned, however, that unusual gene expression appears to exist in certain leukemia cases [46]. Accordingly some investigators [47] see the infidelity of marker expression in leukemias. Nevertheless, it is a remarkable fact that the majority of leukemias demonstrate what appears to be the normal marker profiles [48].

4. Utility of Hematopoietic Cell Lines

Besides the proven utility of the leukemia-lymphoma cell lines for leukemia research, the utility of hematopoietic cell lines has steadily increased in various biomedical research disciplines.

4.1 Lymphokine Research

Lymphokine research had been greatly hampered in the past because of its inconsistent materials since they were prepared from peripheral blood lymphocytes stimulated by various inducers [49].

Because of the extremely small quantity and the imprecise nature of the biological assay for its measurement, the crude material from the stimulated lymphocyte cultures is inconsistent in its content of the lymphokine in question and is usually a mixture of numerous other lymphokines. In 1970, Granger et al. [50] reported the successful detection of a lymphotoxic factor from a continuous human lymphoid cell line. From time to time since then, hematopoietic cell lines have been used as a source of the lymphokine production [51-57]. Advantages in the use of these cell lines are multifold. The cell line is quite stable in its characteristics. Only a few lymphokines or a single lymphokine from a single cell line is produced consistently for an unlimited period of time and amount. Certain kinds of the lymphokines can only be produced from cell lines, since the limitation in recombinant DNA technology available at the present time prevents production for some of these lymphokines.

Alternatively, these cell lines can now be used as an expression host for the lymphokine production after virus infection or other procedures [58,59]. Human monoclonal antibody is also successfully produced by utilizing these cell lines [60].

Table 2 summarizes our results of lymphokine production in the cell line panel.

4.2 Virus Research

I have reviewed the discovery of the Epstein-Barr virus from the established Burkitt lymphoma cell lines [6]. Similarly, in 1981 Reitz et al. [29] were able to identify the presence of the retrovirus HTLV type I from the established cell lines of T-cell leukemia

Table 2. Lymphokine production in human hematopoietic cell lines. In addition to 116 leukemia-lymphoma cell lines listed in Table 1, 16 EBV-transformed "normal" B lymphoblastoid cell lines as listed "Normal B cell" are tested for productivity of 5 lymphokines (DIF, IFN, TNF, IL-1 and IL-2) with or without various inducers. The inducers used include PHA, LPS, HVJ or MLC. The results denote either positive (+) or negative (−) in the unconcentrated culture supernatant. Blanks in the results are not tested.

No.	Cell line	DIF	IFN	TNF	IL-1	IL-2
T-cell						
1	P30/OHKUBO	+	+	−		−
2	MOLT-10	−				
3	RPMI 8402	+	+		−	−
4	CCRF-CEM	−	+	+	−	−
5	DND-41	−	+	+	−	
6	HPB-ALL	+	−		−	−
7	HPB-MLT	−			−	−
8	HD-Mar-2				−	−
9	MOLT-12					
10	MOLT 13, 14	−			−	+
11	JM	+			−	−
12	TALL-1	+	+	+	−	−
13	MOLT-3	−		+	−	+
14	MOLT-4	−		+	−	+
15	P12/Ichi.	−		+	−	−
16	KOPT-K1					
17	MOLT-11	−				
18	MOLT 16, 17		−		−	+
19	Jurkat		−	+		+
20	Jurkat/HTLV III***		−	−		−
21	CCRF-HSB-2	+		+	−	+
22	PEER	−	+	+	−	−
23	KE-37	−	+	+	−	−
24	SKW-3	−		+	−	−
25	A3/Kawa.	−			−	−
26	MOLT-15	−			−	−
27	ALL-Sil		−	−		
28	MOLT-4/LAV***		−	+		−
29	MOLT-4/ARV***		−	−		−
30	CCRF-CEM/HTLV III***		−	+		−
31	MT-4/HTLV III***		−	−		−
32	MT-1					
33	HUT-78	−			−	
34	HUT-102	+		+		+
35	CTCL-2		−			
36	C5/MJ***	+	+	+	+	−
37	MT-2***	−		+		
38	MT-2/HTLV III***		+	+		−
39	C91/PL****	+			−	
40	C82/TK***	+		+		
41	KH-2	−				
42	H9		−	−		+
43	H9/HTLV III***		−	−		+
44	TOM-JIM***		−	−		−
45	467***		−	−		−
B-cell						
46	NALM-1**	−	+	+	−	−
47	NALM 6–15**	+	−		−	−
48	NALM 17, 18**	−		+	−	−
49	LAZ-221**		−		−	−
50	KOPN 1–8**	−			−	−
51	HPB-Null**	−			−	−
52	KLM-2**		+	+	−	−
53	BALM 1, 2			+	−	−
54	U-698-M				−	−
55	TANAKA				−	−
56	MANACA		+		−	−
57	JOK-1			+	−	−
58	SKW-4		−	+	−	−
59	BALM 3–5	−			−	−
60	SU-DHL-4					
61	SCOTT		−		−	−
62	MLB-1084		−	−		
63	ABL-2		+	−		
64	EB-3		+	+	−	−
65	RAJI		−		−	−
66	HR1K					
67	B35M	−		+	−	−
68	KOBK-101					
69	DAUDI		−		−	−
70	NAMALVA	−	+		−	−
71	RAMOS	−	−	+	−	−
72	BJAB					
73	DG-75	−	+		−	−
74	CHEVALLIER					
75	DND-39				−	−
76	NK-9					
77	B46M		+	+	−	−
78	OGUN		+	+	+	−
79	AL-1				−	−
80	SL-1		+	+	−	−
81	RIVA				−	

(Continued)

Table 2 (continued)

No.	Cell line	DIF	IFN	TNF	IL-1	IL-2
82	BALL-1	−	+	+	+	−
83	RPMI 8226	−	−		−	−
84	U-266	−				
85	ARH-77	−		+	−	−
86	Hs					
87	Nak-Pc2		+	−		
88	Hair-M					
89	JC-1					
90	Hair-MTI					
	Non-T, Non-B cell					
91	REH	−	−	+	−	−
92	KM-3		−		−	
93	NALL-1		−	+	−	
94	NALM-16		+	+	−	−
95	KOPN-N					
96	BV-173		+	+	−	
	Myelomonocyte-cell					
97	KG-1	+	+	+	−	−
98	ML 1–3	−	+	+	+	−
99	HL-60	+	+		−	−
100	PL-21		−			
101	JOSK-I				+	
102	CTV 1, 2		−			
103	THP-1	+	+	+	+	−
104	JOSK-K				+	
105	JOSK-S				−	
106	JOSK-M				−	

No.	Cell line	DIF	IFN	TNF	IL-1	IL-2
107	GDM-1		−	−		
108	KCL-22		−	−		
109	U-937	−	+	+	+	−
	Non-L, Non-M-cell					
110	K-562		−	+	−	−
111	HEL	−	−			
112	SPI 801, 802			+	−	−
113	SU-DHL-1	−	+	+	−	−
114	HDLM 1–3	+	−	+	−	−
115	L-428	+				
116	KAN***		−	+		−
	Normal B cell					
	BCH 10		+	+	−	−
	BCH 11		+	+	−	−
	BCH 21		+	+	−	−
	BCH 23		−	+	−	−
	BCH 24		+		−	−
	BCH 25		+	+	−	−
	BCH 26		+	+	−	−
	BCH 27		+	+	−	−
	HUSP-3L	−		+	−	−
	HUSP-5L	+	+	+	−	−
	RPMI 1788	+	+	+	−	−
	RPMI 8382		+	+	−	−
	RPMI 8392	−	+	+	−	−
	BES-1	+				
	ASPC3L	−		+		
	B85	+		+		

Abbraviations used: DIF = Differentiation inducting factor; IFN = Interferon; TNF = Tumor necrosis factor; IL-1 = Interleukin 1; IL-2 = Interleukin 2; PHA = phytohemagglutinin; LPS = lipopolysaccharide; HVT = hemagglutinating virus of Japan; MLC = mixed leukocyte culture.

and lymphomas. Subsequent advances and progress in the study of these viruses were largely attributed to the availability of cell lines. This proved to be the case in the study of AIDS associated virus, HTLV type III [31,32]. Further characterization of the virus and research on the problem associated with this virus are highly dependent on the availability of these cell lines.

4.3 Molecular Biology of Lymphocytes

Croce et al. [61] have used the human leukemia-lymphoma cell lines to identify chromosomal location of human immunoglobulin genes by the procedure of human-mouse cell fusion. Similarly, utilization of leukemia cell lines contributed greatly to the advancement of oncogene studies in the current years [62-64]. Recently, Mak and his associates [65] utilizing a T-cell leukemia line were able to clone the DNA responsible for the long-awaited T-cell receptor genes. The subsequent burst in the research of T cell receptor genes is also attributable to the availability of these characterized T cell leukemia-lymphoma cell lines. Indeed, important knowledge associated with immunoglobulin genes, T-cell receptor genes and numerous proto-oncogenes are accumulating to provide insights into hematological malignancy at the molecular level [66-67]. It is particularly significant that a consistent cytogenetic abnormality common in the majority of cultured Burkitt lymphoma cells is related to the immunoglobulin heavy chain genes and c-myc genes which are localized on the break point of the chromosomes No. 14 and No. 8, respectively [68-70]. Some of the T-cell leukemias and their established cell lines showed a consistent translocation between chromosomes No. 14 and No. 8 where the break point on the chromosome No. 14 is the same region of α-chain of the T-cell receptor genes [71]. The break point on the chromosome No. 8 is in the vicinity of c-myc gene location. It is thus postulated that the c-myc gene is associated with certain types of both T-cell and B-cell malignancies [72]. While a number of studies pointing to a possibility of the oncogene involvement in human neoplasia are accumulating, Duesberg [73] raises, however, a question whether activated proto-oncogenes are sufficient or necessary for human cancer.

4.4 Differentiation and Functional Studies

Because of the fact that leukemia cell lines are of monoclonal population with an arrested differentiation profile, cell lines are effectively used to study in vitro cellular differentiation by various chemical and lymphokine inducers [74]. Obviously, the in vitro induced cellular differentiation of leukemia cell lines also provide a system to study the mechanism of cellular differentiation as well as to look for new growth or differentiation inducing factors [75]. The leukemia cell lines with or without differentiation inducers can now be used to look into functional characteristics of hematopoietic cells [76-78]. These findings are taken further to substantiate and supplement the validity of the hypothetical hematopoietic cell differentiation scheme as described earlier.

4.5 Applied Research and Usage

The panel of hematopoietic cell lines is useful for many other research purposes of seemingly unlimited potential. These cell lines were used to determine their susceptibility to certain anti-leukemic drugs [79] and radiation [80]. This approach has already proved to be clinically relevant when one tries to develop a new treatment protocol for certain forms of leukemias and lymphomas.

Beside various usages in research, the cell lines are utilized for large scale production of certain lymphokines in industry [81]. Because recombinant DNA technology for complex lymphokine expression in bacterial and yeast hosts is limited, mass production of hematopoietic cells is successfully used for industrial scale production of lymphokines [82]. Comparative study of efficacy between recombinant and natural lymphokines in their biological effects is yet unknown, though it is conceivable that human tumors are heterogeneous in their biological properties and therefore each should have a respective target tumor spectrum. As already discussed, modern biotechnology should undoubtedly enhance further utility of these hematopoietic cell lines in the future.

5. Conclusion

A brief historical background of the leukemia-lymphoma cell lines was described. The findings from the study of leukemia-lymphoma cell lines have provided not only an essential test procedure for the multiple marker analysis in the diagnosis and management of leukemia and lymphoma today but also unlimited possibility in the use of cell lines for a wide range of research and applied industry.

6. Abbreviations

ALL = acute lymphoblastic leukemia; EBV = Epstein-Barr virus; EBNA = Epstein-Barr virus associated-nuclear antigen; AML = acute myeloblastic leukemia; MHC = major histocompatibility complex; HTLV = human T-lymphotropic virus; LAV = lymphadenopathy-associated virus; ARV = AIDS-associated retrovirus; HIV = human immunodeficiency virus; AIDS = aquired immunodeficiency syndrome.

7. Acknowledgements

The author gratefully acknowledges numerous colleagues, technical and clerical staff at Roswell Park Memorial Institute, New York State Department of Health, Buffalo, N.Y., Hines V. A. Medical Center, Hines, IL, Loyola University of Chicago, Stritch School of Medicine, Maywood, IL, U.S.A. and Fujisaki Cell Center, Hayashibara Biochemical Labs, Okayama, Japan for their help and contribution. For generous gifts of a number of leukemia-lymphoma cell lines and the monoclonal antibodies by colleagues to this study, the author also gratefully acknowledges.

8. References

1. Bennett, J. M., Catovsky, D., Daniel, M. T., Flandrin, G., Galton, D. A. G., Gralnick, H. R., Sultan, C.: Brit. J. Haematol. *33*, 451 (1976)
2. Minowada, J.: Immunology of leukemic cells, in: Leukemia (Gunz, F. W., Henderson, E. S., eds.) p. 119, New York, Grune and Stratton 1982[4]
3. Pulfertaft, R. J.: Lancet *1*, 238, (1964)
4. Epstein, M. A., Barr, Y. M.: J. Natl. Cancer Inst. *34*, 231 (1965)
5. Minowada, J., Klein, G., Glifford, P., Klein, E., Moore, G. E.: Cancer *20*, 1430 (1967)
6. Epstein, M. A., Achong, B. G., Barr, Y. M.: Lancet *1*, 702 (1964)
7. Tanigaki, N., Yagi, Y., Moore, G. E., Pressman, D.: J. Immunol. *97*, 634 (1966)

8. Epstein, A. L., Kaiser-McCaw, B., Hecht, F., Kaplan, H. S.: Functional and cytogenetic characterization of established human malignant lymphoma cell lines, in: Human Lymphocyte Differentiation: Its application to Cancer, INSERM Symposium No. 8 (Serrou, B., Rosenfeld, C., eds.) p. 327, Amsterdam, Elsevier/North-Holland Biochemical Press 1978
9. Reedman, M., Klein, G.: Int. J. Cancer *2*, 499 (1973)
10. Zur Hausen, H., Schulte-Holthausen, H.: Nature *227*, 245 (1970)
11. Iwakata, S., Grace, J. T. Jr.: N.Y. State J. Med. *64*, 2279 (1964)
12. Moore, G. E.: J. Surg. Oncol. *4*, 320 (1972)
13. Moore G. E., Gerner, R. E., Franklin, H. A.: J. Amer. Med. Ass. *199*, 519 (1967)
14. Moore, G. E., Kitamura, H.: N.Y. State J. Med. *68*, 2054 (1968)
15. Foley, G. E., Lazarus, H., Farber, S., Uzman, B. G., Boone, B.A., McCarthy, R. E.: Cancer *18*, 522 (1965)
16. Adams, R. A., Foley, G. E., Uzman, B. G., Farber, S., Lazarus, H., Kleinman, L.: Cancer Res. *27*, 772 (1967)
17. Gerber, P., Hoyer, B. H.: Nature *231*, 46 (1971)
18. Henle, G., Henle, W., Diehl, V.: Proc. Natl. Acad. Sci. USA *59*, 94 (1968)
19. Minowada, J., Ohnuma, T., Moore, G. E.: J. Natl. Cancer Inst. *49*, 891 (1972)
20. Jondal, M., Holm, G., Wigzell, H.: J. Exp. Med. *136*, 207 (1972)
21. Borella, L., Sen, L.: J. Immunol. *111*, 1257 (1973)
22. Han, T., Minowada, J.: Clin. Exp. Immunol. *15*, 535 (1973)
23. Minowada, J., Sagawa, K.. Trowbridge, I. S., Kung, P. D., Goldstein, G.: Marker profiles of 55 human leukemia-lymphoma cell lines, in: Advances in Malignant Lymphomas: Etiology, Immunology, Pathology, Treatment (Rosenberg, S. A., Kaplan, H. S., eds.) p. 53, New York, Academic Press 1982
24. Miyoshi, I., Hiraki, S., Tsubota, T., Kubonishi, I., Mitsuda, Y., Nakayama, T., Kishimoto, H., Kimura, I., Masuji, H.: Nature *267*, 843 (1977)
25. Rosenfeld, C., Goutner, A., Choquet, C., Venuat, A. M., Kayibanda, B., Pico, J. L., Greaves, M. F.: Nature *267*, 841 (1977)
26. Nilsson, K., Sundström, C.: Int. J. Cancer *13*, 808 (1974)
27. Vogler, L. B., Crist, W. M., Bockman, D. E., Pearl, E. R., Lawton, A. R., Cooper, M. D.: New Engl. J. Med. *298*, 872 (1978)
28. Minowada, J., Koshiba, H., Janossy, G., Greaves, M. F., Bollum, F. J.: Leukemia Res. *3*, 261 (1979)
29. Reitz, M. S., Poiesz, B. J., Ruscetti, F. W., Gallo, R. C.: Proc. Natl. Acad. Sci. USA *78*, 1887 (1981)
30. Yoshida, M., Miyoshi, I., Hinuma, Y.: Proc. Natl. Acad. Sci. USA *79*, 2031, (1982)
31. Gallo, R. C., Sarin, P. S., Gelmann, E. P., Robert-Guroff, M., Richardson, E., Kalyanaraman, V. S., Mann, D., Sidhu, G., Stahl, R. E., Zolla-Pazner, S., Leibowitch, J., Popovic, M.: Science *220*, 865 (1983)
32. Popovic, M., Sarngadharan, M. G., Read, E., Gallo, R. C.: Science *224*, 497 (1984)
33. Janossy, G., Campana, D., Bollum, F. J.: Immunofluorescense studies in leukemia diagnosis, in: Methods in Haematology: Monoclonal Antibodies (Beverley, P., ed.) p. 97, London, Churchill Livingstone 1985
34. Greaves, M., Janossy, G.: Biochim. Biophys. Acta. *516*, 193 (1978)
35. Minowada, J.: Laboratory Med. *16*, 305 (1985)
36. Minowada, J.: Immunology of leukemia and lymphoma, in: Contemporary Issues in Clinical Oncology: Leukemias and Lymphomas (Wiernik, P., ed.) p. 183, New York, Edinburgh, London and Melbourne, Churchill-Livingstone 1985
37. Greaves, M.: Monoclonal antibodies as probes for leukemic heterogeneity and haemopoietic differentiation, in: Leukemia Markers (Knapp, W., ed.) p. 19, London, Academic Press 1981
38. Rowley, J. D.: Science *216*, 749 (1982)
39. Drexler, H. G., Gaedicke, G., Minowada, J.: J. Natl. Cancer Inst. *72*, 1283 (1984)
40. Bollum, F. J.: Blood *54*, 1203 (1979)
41. Han, T., Minowada, J.: Immunol. *35*, 333 (1978)
42. Korsmeyer, S. J., Arnold, A., Bakhshi, A., Ravetch, J. V., Siebenlist, U., Hieter, P. A.,

Sharrow, S. O., LeBien, T. W., Kersey, J. H., Poplack, D. G., Leder, P., Waldmann, T. A.: J. Clin. Invest. *71*, 301 (1983)
43. Minowada, J.: Markers of human leukaemia-lymphoma cell lines reflect haematopoietic cell differentiation, in: Human Lymphocyte Differentiation: Its Application to Cancer INSERM Symposium No. 8 (Serrou, B., Rosenfeld, C., eds.) p. 337, Amsterdam, Elsevier/North Holland Biomedical Press 1978
44. Foon, K. A., Billing, R. J., Terasaki, P. I., Cline, M. J.: Blood *56*, 1120 (1980)
45. Greaves, M. F., Mizutani, S., Furley, A. J. W., Sutherland, D. R., Chan, L. C., Ford, A. M., Molgaard, H. V.: Clinics in Haematology *15*, 621 (1986)
46. Roberts, G. T., Badawi, S. B. E., Sackey, K., Spence, D., Sheth, K. V., Aur, R. J.: Cancer *68*, 1473 (1986)
47. Smith, L. J., Curtis, J. E., Messner, H. A., Senn, J. S., Furthmayr, H., McCulloch, E. A.: Blood, *61*, 1138 (1983)
48. Greaves, M. F., Chan, L. C., Furley, A. J. W., Watt, S. M., Molgaard, H. V.: Blood *67*, 1 (1986)
49. DeWeck, A. L., Kristensen, F., Landy, M.: Biochemical Characterization of Lymphokines, New York, Academic Press 1980
50. Delvin, J. J., Klostergaard, J., Orr, S. L., Yamamoto, R. S., Masunaka, I. K., Plunkett, J. M., Granger, G. A.: Lymphotoxis: After fifteen years of research, in: Lymphokines, Vol. 9 (Pick, E., ed.) p. 313, New York, Academic Press 1984
51. Yoshida, T., Kuratsuji, T., Takada, A., Takada, Y., Minowada, J., Cohen, S.: J. Immunol. *117*, 548 (1976)
52. Gillis, S., Watson, J.: J. Exp. Med. *152*, 1709 (1980)
53. Papermaster, B. W., Gilliand, C. D., McEntire, J. E., Rodes, N. E., Dunn, P. A.: In vivo biological studies with lymphoblastoid lymphokines, in: Lymphokines and Thymic Hormones: Their Potential Utilization in Cancer Therapeutics (Goldstein, A. L., Chirigos, M. A., eds.) p. 289, New York, Raven Press 1981
54. Schook, L. B., Otz, U., Lazary, S., deWeck, A. L., Minowada, J., Odavic, R., Kniep, E. M., Edy, B.: Lymphokine and monokine acitivities in supernatants from human lymphoid and myeloid cell lines, in: Lymphokines Vol. 2 (Pick, E., Landy, M., eds.) p. 1, New York, Academic Press 1981
55. Rocklin, R. E., Blidy, A., Butler, J., Minowada, J.: J. Immunol. *127*, 534 (1981)
56. Minowada, J.: Production of cytokines by human hematopoietic cells, in: Mammalian Cell Culture Technology (Shikita, M., Yamane, I., eds.) p. 141, Tokyo, Soft Science 1985
57. Santoli, D., Tweardy, D. J., Ferrero, D., Kreider, B. L., Rovera, G.: J. Exp. Med. *163*, 18 (1986)
58. Gordon, J., Ley, S. C., Melamed, M. D., English, L. S., Hughes-Jones, N. C.: Nature *310*, 145 (1984)
59. Salahuddin, S. Z., Markham, P. D., Lindner, S. G., Gootenberg, J., Popovic, M., Hemmi, H., Sarin, P. S., Gallo, R. C.: Science *223*, 703 (1984)
60. Edwards, P. A. W., O'Hare, M. J.: The production of human monoclonal antibodies, in: Methods in Hematology, Vol. 13, Monoclonal Antibodies (Beverley, P. C., ed:) p. 60, Edinburgh, London, Melbourne, New York, Churchill Livingstone 1986
61. Croce, C. M., Shander, M., Martinis, J., Cicurel, L., D'Aneona, G. G., Dolby, T. W., Koprowski, H.: Proc. Natl. Acad. Sci. USA *76*, 3416 (1979)
62. Croce, C. M.: Cancer Res. *46*, 6019 (1986)
63. Dalla-Favera, R., Bregni, M., Erikson, J., Patterson, D., Gallo, R., Croce, C. M.: Proc. Natl. Acad. Sci. USA *79*, 7284 (1982)
64. Ryan, J., Barker, P. E., Shimizu, K., Wigler, M., Ruddle, F. H.: Proc. Natl. Acad. Sci. USA *80*, 4460 (1983)
65. Yanagi, Y., Yoshikai, Y., Leggett, K., Clark, S. P., Aleksander, I., Mak, T. W.: Nature *308*, 145 (1984)
66. Klein, G., Klein, E.: Nature *315*, 190 (1985)
67. Greaves, M. F.: Science *234*, 697 (1986)
68. Taub, R., Kirsch, I., Morton, C., Lenoir, G., Swan, D., Tronick, S., Aaronson, S., Leder, P.: Proc. Natl. Acad. Sci. USA *79*, 7837 (1982)
69. Rowley, J. D.: Nature *301*, 290 (1983)

70. Croce, C. M., Nowell, P. C.: Blood *65*, 1 (1985)
71. Shima, E. A., LeBeau, M. M., McKeithan, T. W., Minowada, J., Showe, L. W., Mak, T. W., Minden, M. D., Rowley, J. D., Diaz, M. O.: Proc. Natl. Acad. Sci USA *83*, 3439 (1986)
72. Finger, L. R., Harvey, R. C., Moore, R. C. A., Showe, L. C., Croce, C. M.: Science *234*, 982 (1986)
73. Duesberg, P. H.: Science *228*, 669 (1985)
74. Hozumi, M.: CRC Critical Reviews in Oncology/Hematology *3*, 235 (1985)
75. Olsson, I. L., Breitman, T. R., Gallo, R. C.: Cancer Res. *42*, 3928 (1982)
76. Metcalf, D.: Blood *67*, 257 (1986)
77. Takeda, K., Minowada, J., Bloch, A.: Cancer Res. *42*, 5152 (1982)
78. Drexler, H. G., Gaedicke, G., Komiyama, A., Minowada, J.: Amer. J. Hematol. *21*, 337 (1986)
79. Ohnuma, T., Arkin, H., Minowada, J., Holland, J. F.: J. Natl. Cancer Inst. *60*, 749 (1978)
80. Nakazawa, S., Minowada, J., Tsubota, T., Sinks, L. F.: Cancer Res. *38*, 1661 (1978)
81. Finter, N. B., Fantes, K. H.: The purity and safety of interferons prepared for clinical use: The case of lymphoblastoid interferon, in: Interferon 2 (Gresser, I., ed.) p. 65, London, Academic Press 1980
82. Shikita, M., Yamane, I.: Mammalian Cell Culture Technology, Tokyo, Soft Science Publications 1985

1.3 Scale-up of Oxygen Transfer in Animal Cell Cultures

Daniel I. C. Wang and Jonathan M. Goldstein

Department of Chemical Engineering and Department of Applied Biological Sciences, M.I.T., Cambridge, MA 02139, USA

Abstract

Animal cells are beginning to be considered for the production of recombinant proteins. The growth of animal cells can be achieved using either suspension or anchorage-dependent cells. One common feature using either system is the medium component, serum. The presence of serum introduces a foaming problem specific to animal cell cultures. Due to the fragility of animal cells, antifoams can not be readily used. Furthermore, direct sparging in animal cell cultures can have additional adverse effects due to the presence of the gas-liquid interfaces. Therefore the problem of oxygen transfer during scale-up of animal cell cultures must be addressed if the commercial development of this technology is to be realized. One method to oxygenate animal cell bioreactors without direct sparging is through the use of highly oxygen permeable membranes made of silicon. Furthermore the use of surface aeration can also be explored in combination. In this paper our research on the scale-up of oxygen transfer will be presented. The relationship of the operating parameters on the mass transfer coefficients using silicon tubing in combination with surface aeration will be presented for 0.5 liter and 10 liter fermentors. The studies were performed using anchorage-dependent, recombinant Chinese Hamster Ovary cells for the production of human plasminogen activator.

Bioproducts and Bioprocesses
Editors: Fiechter, Okada, Tanner

1.4 Stoichiometric Considerations of Mammalian Cell Metabolism in Bioreactors*

W. S. Hu and V. B. Himes

Department of Chemical Engineering and Materials Science, University of Minnesota, Minneapolis, MN 55455, USA

Contents

For mammalian cells in culture the primary energy sources include glucose and glutamine. Under normal culture conditions, both glucose and glutamine are essential for cell growth. At a high glucose concentration, most of the glucose consumed is converted to lactate. This high conversion of glucose to lactate can be reduced by controlling the glucose concentration at a low level via programmed feeding of glucose. With decreasing glucose concentration, the rate of glucose conversion to lactate is decreased and the rate of glutamine consumption is increased. When serving as energy sources, glucose and glutamine are substitutable substrates. However, besides serving as an energy source, glucose also provides cells with the ribose necessary for the synthesis of nucleic acids. Similarly, glutamine may be utilized for the synthesis of nucleic acids and non-essential amino acids. Thus, the two substrates can also be complementary. Under steady state conditions in a continuous culture, the carbon from glucose and glutamine results in primarily three places: assimilated into cell mass, converted to lactate and oxidized to carbon dioxide. Thus, the carbon balance on glucose and glutamine results in six fluxes. Among the six fluxes of carbon conversion, the two associated with assimilation are complementary in nature, whereas the other two pairs are substitutable in principle. Specific consumption or production rates of glucose, glutamine, lactate and ammonium, carbon dioxide, as well as oxygen can all be estimated from the measurement of the corresponding species. Furthermore, the stoichiometric coefficients may be estimated from the metabolic pathways. This analysis indicates that the state of mammalian cell energy metabolism can be estimated from the measurement of six species of substrates/metabolites in the bioreactor.

1. Introduction

Mammalian cell culture is recognized as an important method for the production of various biologicals such as viral vaccines for human and veterinary use, interferon, plasminogen activator and prourokinase [1,2,3,4,5]. Recombinant DNA technology has facilitated the production of complex proteins through mammalian cell culture.

* Presented at the US-Japan Joint Biotechnology Conference, Lake Biwa, Japan 1986.

Bioproducts and Bioprocesses
Editors: Fiechter, Okada, Tanner

Many of these proteins require post-translational modifications that can only be performed in mammalian cell systems. Such complex proteins currently being produced include Hepatitis B virus surface antigen [6], glycoprotein D of the Herpes Simplex virus [7], factor IX [8], antithrombin III [8] and tissue-type plasminogen activator [9]. Another example of complex proteins being produced by genetically manipulated mammalian cells are monoclonal antibodies synthesized by hybridoma cells.

Much progress has been made in mammalian cell culture technology in recent years. Various types of bioreactors for mammalian cell cultivation have been developed commercially recently and are reviewed elsewhere [10]. These cultivation methods include perfusion, cell recycle and microcarrier cultures all of which are being investigated for scale-up in order to achieve a cell concentration much greater than that achieved under conventional batch culture conditions. The selection of the appropriate cultivation method is dependent on the kinetic behavior of the cell type used. However, understanding of the kinetic behavior of cultured mammalian cells is still limited.

We have investigated the kinetics of mammalian cell growth and metabolism in batch, fed-batch and continuous culture systems using Chinese Hamster Ovary (CHO) and hybridoma HB8178 cells as model systems. In this communication we will present some of our results and give our thoughts on the perspective of biochemical engineering research in animal cell technology.

2. Kinetics of Batch Culture

The kinetics of a typical batch culture of hydridoma cells in a stirred bioreactor are shown in Fig. 1. Cells grow exponentially until reaching a concentration of approximately 1.0×10^6 cells per ml, at which point cell growth ceases and the viable cell concentration decreases rapidly. During the cultivation period, the glucose concentration decreases from an initial concentration of approximately 4.0 g l^{-1} to approximately 0.010 g l^{-1}. Typically the final lactate concentration is approximately 3.0 to 3.5 g l^{-1}. The antibody concentration increases during the growth period and continues to increase slightly after the viable cell concentration begins to decrease.

Another cell line, CHO cells, can be cultivated either in suspension or attached to a compatible surface. The growth kinetics of CHO cells in suspension in a batch culture are shown in Fig. 2. The growth kinetics in suspension are very similar to those of hybridoma cells except that the rapid transition to death phase after reaching the maximum cell concentration is not observed. Under the cultivation conditions used, a cell concentration of approximately 1.0 to 1.5×10^6 cells per ml can be achieved at the end of the cultivation period. As observed for the hybridoma cells, the residual glucose concentration decreases from an initial concentration of approximately 4.0 g l^{-1} to a final concentration of approximately 0.010 g l^{-1}.

The cellular yield of mammalian cells on glucose under conventional batch conditions is relatively low at approximately 2.5 to 3.5×10^5 cells per mg glucose cunsumed. In addition, a large fraction of the glucose consumed is converted directly to lactate.

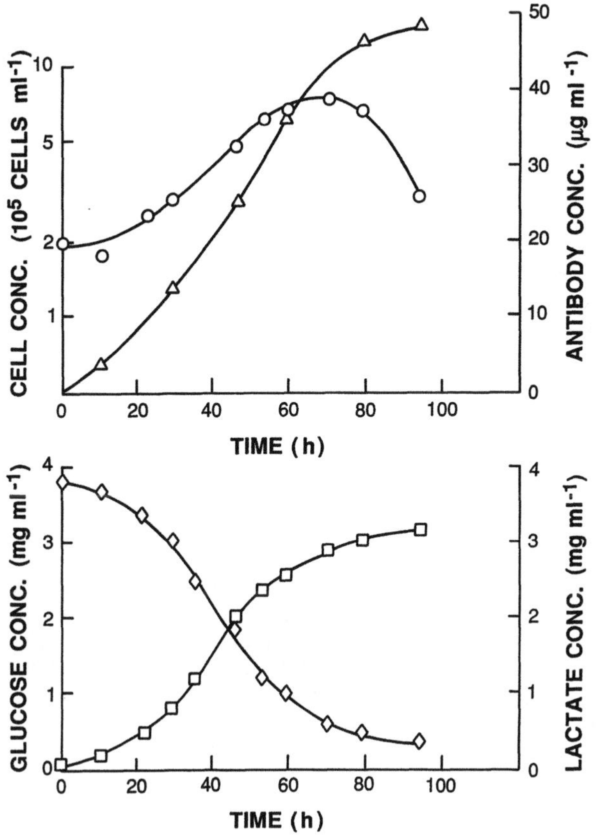

Fig. 1. Growth kinetics of hybridoma HB8178 cells in suspension. The growth medium used was DME supplemented with 10% horse serum. Symbols: cell concentration (○); antibody concentration (△); residual glucose concentration (◇); and lactate concentration (□)

The ratio of the amount of lactate produced to that of glucose consumed on a weight basis ranges from 0.6 to 1.0. Large accumulation of lactate in the growth environment causes a significant drop in pH which may become inhibitory to cell growth. Consequently, it is desirable to minimize the amount and rate of lactate produced during the culture period.

Besides lactate, ammonium accumulates in the growth environment primarily from the utilization of glutamine as a second energy source by mammalian cells. A combination of high lactate and ammonium concentrations in the growth environment can be growth inhibitory [11] as shown in Table 1. The data were obtained by determining the initial growth rates of hybridoma cells suspended in medium containing various concentrations of ammonium and lactate. With increasing concentrations of ammonium and lactate, a decrease in initial growth rate was observed. The inhibitory effect of lactate and ammonium has also been reported by other researchers, although the concentration levels which cause inhibition varies with cell types and cell lines [12].

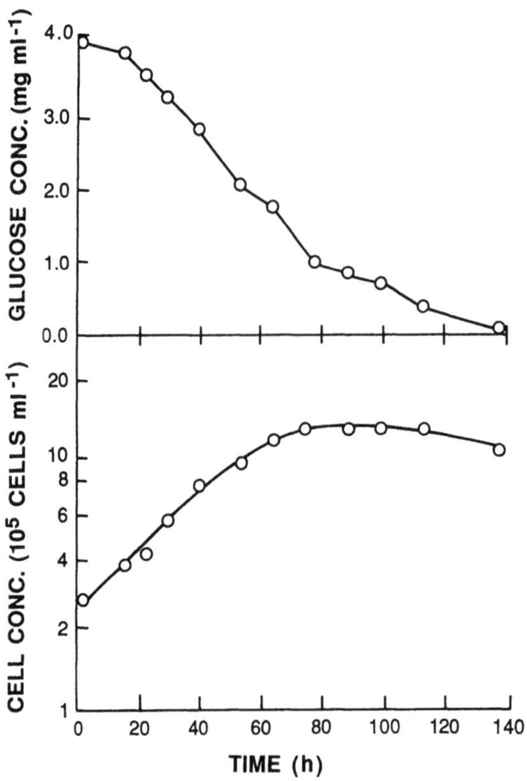

Fig. 2. Growth kinetics of CHO cells in suspension. Cells were cultivated in a spinner with 100 ml DME supplemented with 5% fetal bovine serum (FBS)

Table 1. Effect of ammonium and lactate

Ammonium conc. (mM)	Lactate conc. (mg ml^{-1})	Growth extent (10^5 cells per ml)	Initial growth rate (h^{-1})
0.0	0.0	6.0	0.044
3.0	0.0	2.9	0.017
0.0	3.0	5.0	0.038
6.0	0.0	0.50	0.0037
0.0	6.0	2.3	0.020
3.0	3.0	4.4	0.030
3.0	6.0	1.3	0.020
6.0	3.0	4.8	0.013
6.0	6.0	0.69	−0.0067

The cell concentration achieved in conventional stirred vessels is almost two orders of magnitude lower than that observed in vivo (10^6 cells per ml as opposed to 10^8 to 10^9 cells per ml). In the case of monoclonal antibody production from hybridoma cells as shown in Fig. 1, the maximum concentration achieved is also approximately two orders of magnitude lower than that obtained in vivo. To increase the productivity of conventional culturing methods, it is necessary to increase the cell and product concentrations. Examination of the batch kinetic information reveals that a few fold increase in cell concentration using conventional batch cultures is unlikely because the corresponding amount of lactate and ammonium produced would become growth inhibitory. To prevent the depletion of nutrients, and the accumulation of inhibitory metabolites, cultures are often operated in a fed-batch mode or in a continuous mode. For suspension cultures, such periodic or continuous medium replenishment inevitably removes the cells from the bioreactor along with the medium being replaced, thus preventing the culture from reaching a higher cell concentration. Furthermore, because of the lack of understanding in the kinetic behavior of cells in culture, the scheme of medium replenishment (e.g. flow rate or frequency of medium replenishment) is determined arbitrarily and is probably suboptimal. Too frequent medium replenishment results in costly waste of unused components of the growth medium such as supplemental growth factors and serum. Too infrequent medium replenishment may result in accumulation of potentially growth inhibitory metabolites and adversely affect the production of the desired biological.

A possible approach to increase cell cencentration is to change the reactor operating mode from a batch to a continuous flow system with a mechanism of retaining cells within the reactor while medium is continuously perfused through it. [13] However, unless the reaction kinetics of the system are changed, the higher cell concentration is achieved at the expense of a proportional increase in the amount of medium used. In order to improve the medium utilization efficiency, the reaction kinetics of the system have to be improved. In order to alter the reaction kinetics of the system, a better knowledge of cellular metabolism and a means for manipulation of the metabolism are necessary.

3. Kinetic Study of Glucose and Glutamine Metabolism in Continuous Culture

For mammalian cells in culture, the primary energy sources are glucose and glutamine [14]. When cells are cultivated in a typical cell culture medium such as Dulbecco's Modified Eagle's medium (DME) supplemented with serum, both glucose and glutamine are essential for cell growth. At a high glucose concentration, most of the glucose consumed is converted to lactate as illustrated in Figs. 1 and 2. With decreasing glucose concentrations, the rate of glucose conversion to lactate is decreased and the rate of glutamine consumption is increased [14].

For both hybridoma and CHO cells, the utilization of nutrients appears to be more efficient in a continuous culture than in a batch culture. A typical cellular yield on glucose that we have observed in continuous culture is approximately 2.0×10^6 cells per mg glucose consumed. The growth kinetics of a continuous culture of CHO cells are shown in Fig. 3. The medium used was DME supplemented with 5% fetal calf serum. The culture was first operated in a batch mode during the first 44 h before it was changed to a continuous mode. The glucose concentration in the feed was $1.0\ g\,l^{-1}$. At a dilution rate of $0.024\ h^{-1}$ a steady state was achieved as judged by the concentrations of the cells, glucose, glutamine and lactate.

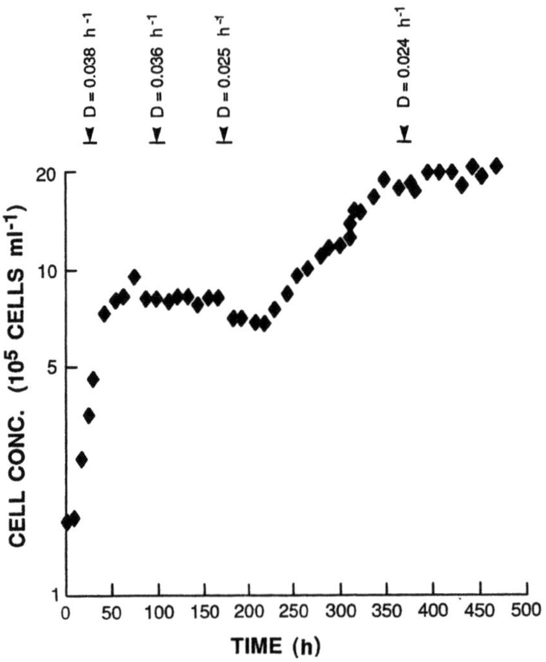

Fig. 3. Growth kinetics of CHO cells in continuous culture. Feed medium is DME containing 1.0 g l^{-1} glucose supplemented with 5% FBS. The dilution rates are shown in the figure

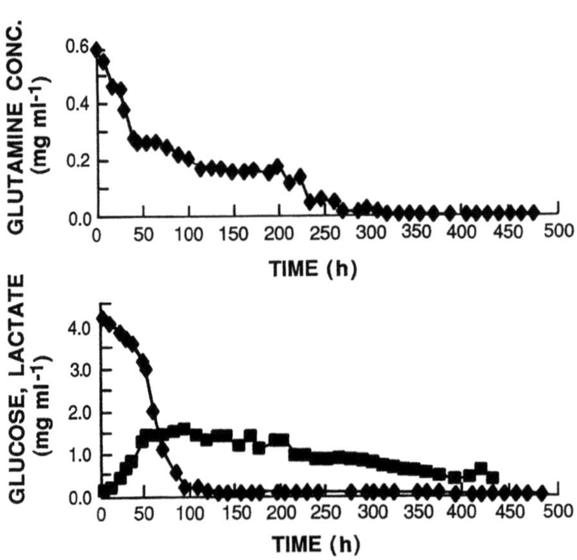

Fig. 4. Residual glutamine concentration and residual glucose (◆) and lactate (■) concentrations for continuous culture of CHO cells. The cultivation conditions are the same as those shown in Fig. 3

The glucose and glutamine consumed as well as the lactate produced in the first 470 h are shown in Fig. 4. The ratio of lactate produced to glucose consumed appears to be lower when the culture was operated in a continuous mode than during the batch mode. This was accompanied by an increase in glutamine consumption. It was noted that the decreased conversion to lactate and increased consumption of glutamine coincided with a low residual glucose concentration (in the range of 0.01 to 0.1 g l^{-1}).

4. Glucose Feeding

It has been observed that glucose consumption by mammalian cells in culture is affected by the glucose concentration. By daily feeding of glucose to maintain a glucose concentration of 0.14–0.22 μg l^{-1}, the glucose consumed by human fibroblasts cultured in T-25 cm^2 tissue flasks was greatly reduced as compared to the culture in which the initial glucose concentration was 31 μg l^{-1} and not subsequently replenished [15]. Such daily feeding of a small amount of glucose is feasible in the study cited above only because of the low cell number utilized and, consequently, low overall consumption rate. With the relatively high overall consumption rate observed in a bioreactor due to a greater number of cells per unit volume, the feeding has to be

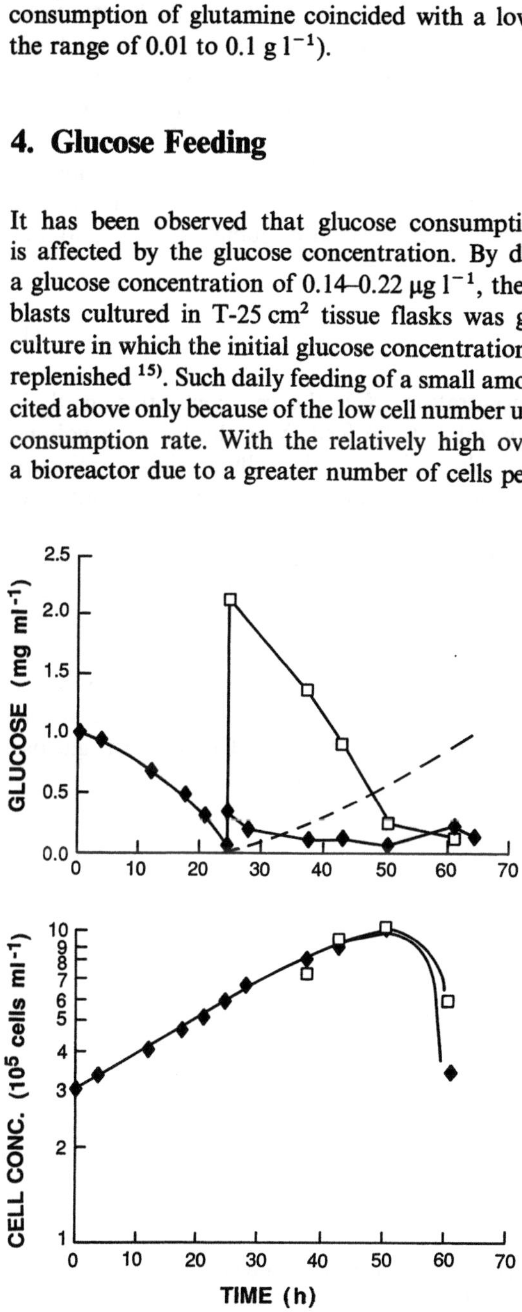

Fig. 5. Cultivation of HB8178 hybridoma cells in batch cultures with step feeding (□) and programmed feeding (◆) of glucose. *Symbols:* residual glucose concentration for step feeding (□) and programmed feeding (◆). Cumulative glucose addition to the programmed feeding vessel (– – –)

at a more frequent interval. Because of the transient nature of the process in which the cell concentration is increasing, the feeding also has to be dynamic as opposed to static. Obviously, too rapid a feeding rate may result in an increase in the residual concentration of glucose and, therefore, result in a higher consumption rate. On the other hand, too low a feeding rate can possibly result in the depletion of glucose and cause cell death as in the case of the cultivation of hybridoma cells. [16] We have attempted to maintain the residual glucose concentration at a desired level via programmed feeding in the cultivation of CHO and hybridoma HB8178 cells.

An experiment was carried out using an initial glucose concentration of 1.0 g l^{-1}. Hybridoma HB8178 cells grew exponentially after inoculation. The glucose concentration was allowed to decrease to 0.08 g l^{-1} before the culture was divided into two. To one culture, 2.0 g l^{-1} of glucose was added. To the other, glucose was fed exponentially with a microprocessor controlled pump according to the following equation:

$$\frac{dS}{dt} = \frac{FS_0}{V} - \frac{\mu X}{Y} \tag{1}$$

where S is the residual glucose concentration, X is the cell concentration, S_0 is the feed concentration, F is the flow rate of the feed, V is the working volume of the culture, μ is the growth rate of the cells, Y is the cellular yield on glucose and t has the dimension of time. Both cultures grew at the same rate until the maximum cell concentration was reached (Fig. 5). In the culture with step feeding, the glucose concentration did not decrease below 1.0 g l^{-1} until the end of the cultivation period. In the case of programmed feeding, the glucose concentration remained below 0.1 g l^{-1} during the exponential growth period. In the case of programmed feeding the increase in residual glucose concentration at the end of cultivation was due to the onset of cell death where the feeding rate was no longer equal to the consumption rate of the cells. The glucose consumed and the lactate produced by the two cultures are compared in Fig. 6. The rate of glucose

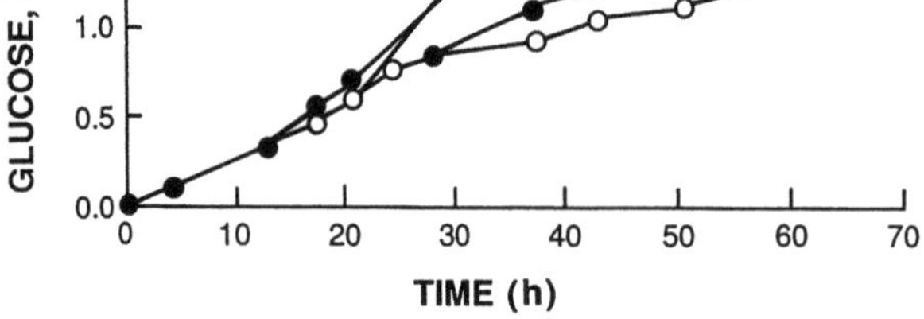

Fig. 6. Cumulative glucose consumption *(solid symbols)* and lactate production *(open symbols)* of step feeding (■, □) and programmed feeding (●, ○)

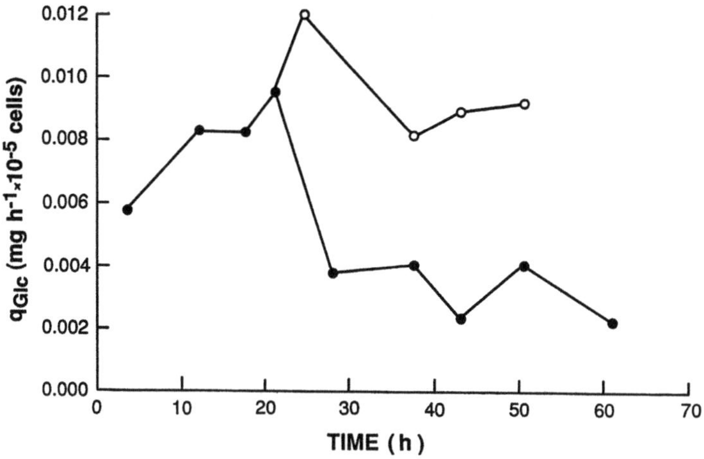

Fig. 7. Specific glucose consumption rate for step feeding (○) and programmed feeding (●)

consumption was lower in the case of programmed feeding than that of step feeding. The amount of lactate produced is also reduced with the programmed feeding of glucose. Figure 7 shows that the specific glucose consumption rate first increases during the early stage of exponential growth. A step feeding of glucose appears to raise the specific consumption rate to a maximum. This is followed by a gradual return to its normal level after 15 h. In the case of programmed feeding, the specific glucose consumption rate remains at a relatively low value throughout the cultivation period.

Similar experiments with programmed feeding of glucose and glutamine employing CHO cells have also been carried out. Again, the glucose consumption rate with programmed feeding was decreased as compared to step feeding with a high concentration of glucose at the initiation of the cultures.

The results presented above illustrate the feasibility of manipulating cell metabolism in a bioreactor. However, without a kinetic model, such programmed feeding is feasible only when both the yield coefficient and the specific growth rate are constant. Obviously the kinetic relationship between the metabolic state of cells and the concentration of major substrates should be established for any programmed feeding scheme.

5. Stoichiometric Considerations

The two primary energy sources for mammalian cells in culture, glucose and glutamine, are present in the growth medium at much higher concentrations than the other components. Reitzer et al. showed that glutamine oxidation by mammalian cells can provide as much as 50% of the cellular energy when cultured in the presence of glucose [17]. Glucose is catabolized via the glycolytic pathway yielding two moles of ATP per mole of glucose consumed. However, complete oxidation of glucose to carbon dioxide via the tricarboxylic acid (TCA) cycle yields a total of 36 moles of ATP although this occurs to a lesser extent by mammalian cells cultivated in vitro[15,17,18]. Under culture conditions with high glucose concentrations, most of the glucose consumed is converted to lactate. With decreasing residual glucose concentrations, the rate of glucose conversion to lactate is decreased and the rate of glutamine consumption is increased.

Because glucose and glutamine can both serve as energy sources for mammalian cells, they can be classified as substitutable for one another insofar as energy demand is concerned. However, glucose and glutamine fulfill other non-energetic needs of the cells as well. Glucose is catabolized via the pentose monophosphate shunt to form ribose moieties which are necessary for nucleotide synthesis. Glutamine can be utilized for the synthesis of non-essential amino acids, proteins and nucleosides. Clearly, in the case of glucose and glutamine utilization for non-energetic purposes, one nutrient can not substitute for the other. They are complementary in nature for non-energetic processes. Therefore, the two compounds, when serving as energy sources are substitutable substrates and when consumed for non-energetic purposes arc complementary substrates[19,20]. This complementary and substitutable nature of glucose and glutamine cannot be ignored in the kinetic analysis and model development for mammalian cell metabolism.

The state of understanding of mammalian cellular kinetic behavior in culture is minimal due to a lack of reliable kinetic data. Although it is important to monitor the specific rates of substrate consumption and metabolite accumulation in the growth environment, it is also important to identify the flux of the substrate through the primary metabolic pathways available to the cell. Because glucose and glutamine can be substitutable in providing cellular energy, various combinations of the fluxes of the two substrates through energy yielding metabolic pathways may fulfill the cell's needs. Different combinations of these fluxes result in different metabolic states, or different degrees of utilization efficiency. That is, the fractions of glucose and glutamine being completely oxidized to carbon dioxide versus conversion to lactate may vary according to cultivation conditions.

A material balance of quantities derived from glucose and glutamine consumption by mammalian cells in a continuous culture can be shown as follows:

$$\frac{dX}{dt} = \mu X - DX \tag{2}$$

$$\frac{dGlc}{dt} = D(Glc_0\text{-}Glc) - q_{Glc}X \tag{3}$$

$$\frac{dGln}{dt} = D(Gln_0\text{-}Gln) - k(Gln)_{DEG} - q_{Gln}X \tag{4}$$

$$\frac{dNH_3}{dt} = D(NH_{30}\text{-}NH_3) + k(Gln)_{DEG} + q_{NH_3}X \tag{5}$$

$$\frac{dLac}{dt} = D(Lac_0\text{-}Lac) + q_{Lac}X \tag{6}$$

$$\frac{dCO_2}{dt} = D(CO_{2_0}\text{-}CO_2) - CTR + q_{CO_2}X \tag{7}$$

$$\frac{dO_2}{dt} = D(O_{2_0}\text{-}O_2) + OTR - q_{O_2}X \tag{8}$$

where X is the viable cell concentration, D is the dilution rate, μ is the growth rate and t has the dimension of time. We define Glc and Gln to be the glucose and glutamine concentrations respectively, NH_3 and Lac to be the residual ammonia and lactate concentration respectively, and CO_2 and O_2 to be the residual carbon dioxide and oxygen concentrations respectively. Glc_0, Gln_0, NH_{3_0}, Lac_0, CO_{2_0}, and O_{2_0} are the respective concentrations in the feed stream. The specific consumption or production rate is given by q_s where s denotes the appropriate substrate or metabolite. The first order rate constant for spontaneous degradation of glutamine is k and the oxygen and carbon dioxide transfer rates to the culture medium for oxygen and carbon dioxide is given by OTR and CTR respectively.

The concentration of the seven species as well as the rate of spontaneous glutamine degradation and oxygen and carbon dioxide transfer rates can all be measured in the bioreactor. Knowing the inlet concentrations of the species, the specific rates, q_s, can be determined. Assuming glucose and glutamine to be the primary energy sources, their specific consumption rates and the corresponding metabolite production rates can theoretically be determined. If the assumption is made that glucose and glutamine utilization by the cell is restricted to either the provision of energy or as carbon sources incorporating all the carbon, then the carbon from glucose and glutamine can be converted to: cell mass, lactate and carbon dioxide. Consequently, there are six potential fluxes associated with carbon utilization from the two substrates. Two fluxes are associated with assimilation into cell mass and are complementary in nature. The other four fluxes are associated with energy production and are substitutable in nature. The six consumption/production rates describing the six fluxes are:

$$q_{Glc} = (Glc)_{assm} + (Glc)_{Lac} + (Glc)_{CO_2} \tag{9}$$

$$q_{Gln} = (Gln)_{assm} + (Gln)_{Lac} + (Gln)_{CO_2} \tag{10}$$

$$q_{NH_3} = \alpha(Gln)_{assm} + \beta(Gln)_{CO_2} + \gamma(Gln)_{Lac} \tag{11}$$

$$q_{Lac} = \sigma(Glc)_{Lac} + \varepsilon(Gln)_{Lac} \tag{12}$$

$$q_{CO_2} = \xi(Glc)_{CO_2} + \delta(Gln)_{CO_2} \tag{13}$$

$$q_{O_2} = \psi(Glc)_{CO_2} + \omega(Gln)_{CO_2} \tag{14}$$

The subscripts assm, Lac and CO_2 represent the rates of glucose or glutamine being assimilated into cell mass, converted to lactate and oxidized to carbon dioxide respectively. The greek letters signify the appropriate stoichiometric coefficients.

If the ammonia production due to the metabolism of other amino acids and transamination reactions during glutamine consumption are minimal, then the specific production rate of ammonia can be expressed as Eq. (11). With similar assumptions of negligible contribution to metabolite production and oxygen consumption from other amino acids, lactate and carbon dioxide production and oxygen consumption can be written as Eqs. (12), (13) and (14).

The overall specific consumption and production rates represent six equations with six unknowns being the individual fluxes of glucose and glutamine through the three different metabolic pathways. With the six overall specific rates determined from the mass balances in Eqs. (3)–(8), and with the necessary stoichiometric coefficients, the six unknown fluxes can be determined for any given steady state residual glucose and glutamine concentration in the growth environment. Determination of the fractional conversion of glucose and glutamine to lactate and carbon dioxide allows one to monitor the energy utilization efficiency of a given culture system. Such an approach may provide us with a means of monitoring the metabolic state of cells in the bioreactors and facilitate the control of animal cell metabolism for bioprocessing.

6. Conclusion

With the increasing importance of mammalian cell culture for production of biologicals, there exists a great need for extensive kinetic analysis of cellular growth and metabolism. Of particular concern is the utilization of the energy sources available to the cells. It is desirable to be able to culture mammalian cells in an environment that allows for efficient utilization of the energy sources, glucose and glutamine. This will minimize the accumulation of potentially growth inhibitory metabolites in the cell environment while maximizing the culture productivity.

We have shown through continuous cultures of CHO and hybridoma cells the ability of some mammalian cells to utilize glucose and glutamine more efficiently than observed under conventional batch conditions. The more efficient energy metabolism of the cells can be correlated with low residual glucose concentrations in the cell environment. For this reason, we have also investigated programmed feeding of the energy sources in order to improve substrate utilization in a batch system. By maintaining a residual glucose concentration below 0.1 $g\,l^{-1}$ in the programmed feeding fed-batch cultures, the rate of glucose consumption was decreased with a concomitant increase in glutamine consumption and decrease in the rate of lactate production during exponential cell growth.

By taking advantage of the substitutable and complementary nature of the two energy sources, glucose and glutamine, six stoichiometric equations can be established that describe six fluxes of glucose and glutamine through available metabolic pathways. This allows for a quantitative description of glucose and glutamine utilization for any given environmental growth condition. The kinetics of cellular growth and metabolism can then be described through model development based on the six stoichiometric equations in order to operate mammalian cell culture systems under optimal and efficient conditions.

7. Acknowledgement

This work was supported in part by grants from Ecolab, Inc., St. Paul, Minnesota, the National Science Foundation (ECE-8512427 and ECE-8552670), and the Graduate School of the University of Minnesota.

8. List of Symbols

s	Substrate concentration (mg ml)
t	Time (h)
F	Flow rate (ml h^{-1})
S_0	Substrate concentration in feed (mg ml^{-1})
V	Working volume (ml)
μ	Specific growth rate (h^{-1})
X	Viable cell concentration (cells per ml)
Y	Yield coefficient (cells per mg substrate consumed)
d/dt	Time derivative
D	Dilution rate (h^{-1})
Glc	Glucose concentration (mg ml^{-1})
Gln	Glutamine concentration (mg ml^{-1})
NH_3	Ammonia concentration (mg ml^{-1})
Lac	Lactate concentration (mg ml^{-1})
O_2	Oxygen concentration (mg ml^{-1})
Glc_0	Glucose concentration in feed (mg ml^{-1})
Gln_0	Glutamine concentration in feed (mg ml^{-1})
NH_{30}	Ammonia concentration in feed (mg ml^{-1})
Lac_0	Lactate concentration in feed (mg ml^{-1})
CO_{20}	Carbon dioxide concentration in feed (mg ml^{-1})
O_{20}	Oxygen concentration in feed (mg ml^{-1})
q	Specific consumption/production rate (mg per cell per h)
k	First order rate constant for spontaneous glutamine degradation (h^{-1})
CTR	Carbon dioxide transfer rate (mg ml^{-1} h^{-1})
OTR	Oxygen transfer rate (mg ml^{-1} h^{-1})

9. References

1. Spier, R. E. and Whiteside, J. P. W.: Biotech. Bioeng. 18: 659 (1976)
2. van Wezel, A. L., van Herwaarden, J. A. M. and van de Heuvel-de Rijk, E. W.: Dev. Biol. Stand. 42: 65 (1979)
3. Giard, D. J., Fleischaker, R. J., Sinskey, A. J. and Wang, D. I. C.: Develop. Indust. Microbiol. 22: 299 (1981)
4. van Wezel, A. L.: Develop. Biol. Stand. 55: 3 (1984)
5. Tao, T. Y., Bohn, M. A., Ji, G. Y., Einsele, A. and Hu, W. S.: J. Biotechnol. 6: 205 (1987)
6. Michel, M. L., Sobczak, E., Malpiece, Y., Tiollais, P., and Streeck, R.: Bio/technology 3:561 (1985)

7. Lasky, L. A., Dowbenko, D., Simonsen, C. C. and Berman P. W.: Bio/technology 2: 527 (1984)
8. Kaufman, R. J., Dorner, A. J., Wasley, L. C., Murtha, P., Gure, M. and Raney, P.: "The Potentials and Limitations of High Level Expression of Secreted Proteins in Mammalian Cells" ACS abstract, 194th ACS National Meeting, New Orleans, La. 1987
9. Kaufman, R. J., Wasley, L. C., Spiliotes, A. J., Gossels, S. D., Latt, S. A., Larsen, G. R. and Kay, R. M.: Mol. Cell. Biol. 5: 1750 (1985)
10. Hu, W. S. and Dodge, T. C.: Biotechnol. Progr. 1: 209 (1985)
11. Hu, W. S., Dodge, T. C., Frame, K. K. and Himes, V. B.: Develop. Biol. Standard. 66: 279 (1987)
12. Butler, M. and Spier, R. E.: J. Biotechnol. 1: 187 (1984)
13. Kitano, K.: Appl. Microbial. Biotechnol. 24, 282–286 (1986)
14. Zielke, H. R., Zielke, C. L. and Ozand, P. T.: Fed. Proc. 43: 121 (1984)
15. Zielke, H. R., Ozand, P. T., Tildon, J. T., Sevdalian, D. A. and Cornblath, M.: J. Cell. Physiol. 95: 41 (1978)
16. Dodge, T. C., Ji, G. Y. and Hu, W. S.: Eng. Microb. Technol. 9: 607 (1987)
17. Reitzer, L. J., Wice, B. M. and Kennell, D.: J. Biol. Chem. 254(8): 2669 (1979)
18. Donnelly, M. and Schaffler, I. E.: J. Cell. Physiol. 89: 39 (1976)
19. Bader, F. G.: Biotech. Bioeng. 20: 183 (1978)
20. Baltzis, B. C. and Fredrickson, A. G.: Biotech. Bioeng., 31: 75 (1988)

2. Biocatalysts

2.1 Microbial and Enzymatic Processes for the Production of Biologically Useful Compounds

Hideaki Yamada

Department of Agricultural Chemistry, Kyoto University, Kyoto, Japan

Contents

Reactions catalyzed by enzymes and enzyme systems display far greater specificities than more conventional forms of organic reactions and, of all the reactions available, enzymatic synthesis and transformations have the greatest potential. We have recently been carrying out studies on the processes for production of various biologically useful coenzymes, vitamins, amino acids and amides, using microbial enzymes as catalysts.

1. Production of Coenzymes

Synthesis of coenzyme A and related compounds: Coenzyme A (CoA) is biosynthesized from pantothenic acid, cysteine and AMP. We found that *Brevibacterium ammoniagenes* is a good enzyme source for the synthesis of CoA. Yields from phosphorylated substrates are about 10 times greater than from pantothenic acid.

D-Pantoyl lactone, a starting material for production of D-pantothenic acid, is prepared by asymmetric reduction of ketopantoyl lactone using microbial enzyme. When cells of *Candida parapsilosis* are incubated with ketopantoyl lactone and glucose, D-pantoyl lactone in yields of nearly 100% can be produced.

Production of S-adenosylmethionine and S-adenosylhomocysteine: We have established efficient biological processes for practical preparation of these nucleosides. For the production of S-adenosylmethionine (AdoMet), *Saccharomyces sake* is cultivated in a medium containing methionine as the precursor of AdoMet. S-Adenosylhomocysteine (AdoHcy) is synthesized from adenosine and homocysteine with AdoHcy hydrolase from *Alcaligenes faecalis* as catalyst.

2. Synthesis of Optically Active Amino Acids

Synthesis of D-phenylglycine and related D-amino acids: Dihydropyrimidinase is an enzyme which catalyzes the ring-opening reaction of dihydropyrimidines. We found that the enzyme also catalyzed the hydrolysis of a variety of 5-monosubstituted

Bioproducts and Bioprocesses
Editors: Fiechter, Okada, Tanner

hydantoins. The hydantoins are hydrolyzed to the corresponding D-forms of the N-carbamoyl glycine derivatives. These N-carbamoyl amino acids are quantitatively transformed to D-amino acids by treatment with $NaNO_2$ under acidic conditions.

Synthesis of aromatic and sulfur-containing amino acids: β-Tyrosinase, tryptophanase, cysteine desulfhydrase and β-chloro-D-alanine chloride lyase have been found to synthesize several aromatic and sulfur-containing amino acids through β-replacement and the reverse of α,β-elimination reactions.

3. Synthesis of Amides by Enzymatic Hydration of Nitriles

Acrylamide and methacrylamide are industrially produced. We have suggested that a microbial enzyme, nitrile hydratase, may be applicable to the practical production of these amides. When cells of *Pseudomonas chlororaphis* are mixed with nitriles, stoichiometric conversion of nitriles to amides is possible.

In this seminar, one or two examples will be selected from our studies and discussed in some detail.

2.2 Inducible Recombinant Cell Cultures and Bioreactors

Wolf R. Vieth

Department of Chemical and Biochemical Engineering, Rutgers University,
Piscataway, New Jersey 08855-0909, USA

Contents

This investigation, in common with another recent one, examined the kinetics of a recombinant *E. coli* strain in conventional chemostats and immobilized recombinant cell (IMRC) bioreactor sytems. Both studies showed similar findings: i.e., a significant decrease in the selection rate between CSTR and IMRC systems. However, the link between the observed transient response of the bioreactor and longer cell doubling times in the immobilized state awaited our study. These results are in agreement with those obtained in an earlier study of selection between recombinant and wild-type cell populations, in which wall growth (i.e., *self-immobilization*) effects played a significant role in the long-term bioreactor response. Similarly, in our work, the rate of selection was found to decrease significantly in the presence of a fixed (immobilized) cell population. Using rigorous bioreactor modeling and parameter estimation methods, analysis of the data provides a clear picture of the relative importance of plasmid partitioning kinetics and the ratio of recombinant to free cell growth rates in the overall transient response of the bioreactor.

1. Introduction

In the work described here, the host was *E. coli* MBM7061 (z^-, y^+), harboring the plasmid pMLB1108 which is lac i^q z^+. Thus, while the host strain does contain the gene which codes for the permease (y^+), it is deficient in the z gene for expression of β-galactosidase, in the normal position on the chromosome. This deficiency is made up by the plasmid pMLB1108, which contains the proximal half of the lac operon; i.e., the repressor gene (i^q) and the gene coding for β-galactosidase (z^+). The q mutation causes the repressor protein to be over-

Bioproducts and Bioprocesses
Editors: Fiechter, Okada, Tanner

produced, allowing better control for the switching on and off of protein synthesis in chemostats and in immobilized cells. Thus, one has a very flexible genetic system to study strategies for maintaining plasmids in genetically engineered systems.

The second rather important aspect of this work is the comparative study of plasmid-bearing cells in free suspension cultures and in immobilized states. So-called "plasmid-shedding" occurs as a consequence of natural selection; i.e., competitive growth of wild-type and recombinant strains on the available substrate. If cell division is blocked (by immobilization), differential reproduction cannot take place. Thus, immobilized recombinant cell cultures are expected to be more stable [7,8]. However, weighing against this advantage is the possibility that immobilized cells may not over-produce the desired metabolite as actively. But that, too, is mitigated by the flexible metabolic switching policies afforded by the inducer/repressor combination which, at least in theory, could be applied at an optimal time in the growing cycle.

2. Strain Development

The strain employed in this investigation is a *lac* i^q strain, meaning that it over-produces the lac repressor protein which in turn blocks transcription of the lac structural gene. A *lac* i^q host can produce as much as ten times the amount of repressor present in a normal cell [1]. Overproduction of the lac repressor by the resulting re-combinant cells is necessary to provide a basis for inducible enzyme biosynthesis.

In the development of a recombinant strain to be used in subsequent bioreactor studies, it was necessary to have a host possessing the lac permease gene so that lac-tose could function as the inducer. The host therefore had to be *lac* z^-, y^+ so that the *lac* z^+ plasmid and host could be easily distinguished when grown on indicator plates. A plasmid/host system was chosen which meets these requirements. The plasmid, pMLB1108, was developed by Michael L. Berman at the National Cancer Institute. It contains *lac* i^q *o, p and z* genes. The plasmid repressor gene overproduces the lac repressor protein, thereby providing the ability to regulate protein biosynthesis by controlling the inducer (i.e., lactose or IPTG) concentration.

The host is a *lac* z, y^+ strain, MBM7061 [3]. The permease activity was obtained through lysogenic incorporation of the transducing phage λ p1048 in the *E. coli* host strain MC4100 [4]. The resulting recombinant strain was tested in small-scale culture and found to possess the desired characteristics. It was used for all subsequent bioreactor studies. Therefore, all experimental results were obtained using this recombinant strain.

3. IMRC Bioreactor

An IMRC bioreactor is, usually, a three-phase packed or fluidized bed reaction system, possessing solid, liquid and gas phases. An IMRC bioreactor may be operated in either mode. The disadvantage in using a packed bed reactor is that channeling

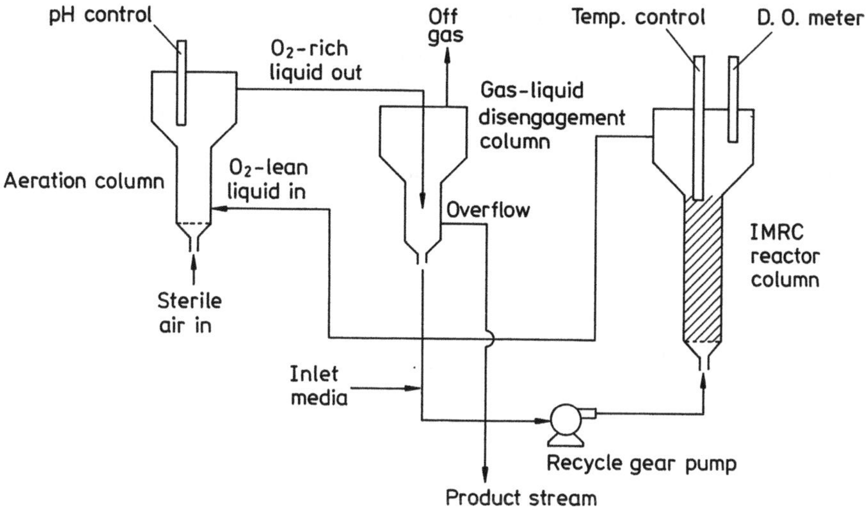

Fig. 1. IMRC reactor design

of air or medium through the bed can occur, leading to stagnant zones in the reactor with very poor local contact. Most often gas (air) bubbles are used to provide mixing of the solid and liquid phases. However, this type of three-phase system is difficult to scale up which in turn makes it difficult to obtain consistent results among reactor runs, due to differences in fluidization characteristics. Aeration of the reaction mixture in a separate vessel and rapid recycle of the oxygenated liquid to the IMC reactor column provides more gentle and predictable fluidization while obtaining excellent mass transfer characteristics [10]. As a result, scale-up of the IMC reactor becomes much more predictable.

The bioreactor [2] actually used for immobilized cell reactor studies is shown schematically in Fig. 1. The reactor is a three column system, composed of the IMRC reactor column (shown right), the liquid phase oxygenation column (shown left), and the gas-liquid disengagement column (located in the center). The main feature of this reactor system is that the liquid phase is oxygenated in a separate vessel from the column containing the gel particles. The advantages of this design are:

- The rates of fluidization and oxygenation can be controlled independently.
- The mixing pattern inside the IMRC reactor column is more thorough, yet more gentle due to lower shear forces, thereby reducing the rate of bead attrition.

The total system volume (liquid + gel particles) varied between 650 and 750 ml, depending on the liquid level in the IMRC reactor. The temperature was maintained at 37 °C. Approximately 200 ml of gel beads were used in each reactor run. The cell broth was recycled between the IMRC reactor column and the oxygenation unit; a liquid recycle rate of approximately 700 ml min^{-1} was maintained.

When using a phosphate buffered medium, it was unnecessary to control the pH in the reactor system. The phosphate salts had an ample buffering capacity. The

E. coli culture has a pH of approximately 6.80 when grown to stationary phase in a phosphate buffered medium. In contrast, when grown in tris buffered medium, a stationary phase *E. coli* culture has a pH of 4.20. Tris buffered medium had to be used with calcium alginate gel, because high phosphate concentrations dissolve the alginate material, by chelating the calcium ions. It was therefore necessary to control the system pH when using tris buffered media.

4. IMRC Bioreactor Studies

The IMRC bioreactor system was operated using tris-buffered minimal lactose media as the inlet feed, and calcium alginate as the gel matrix material. Steady state effluent concentrations of cell, lactose and β-galactosidase were obtained as functions of dilution rate, which is based on the total (liquid + gel phase) volume of the three column bioreactor system. A protracted experiment was performed twice, at one week intervals, using the same bioreactor system. Experimental results are shown in Figs. 2 and 3, respectively.

These results show that the IMRC bioreactor system can be operated at a dilution rate as high as 1.8 h^{-1} without incurring the cell washout problem observed in the free cell system at a dilution rate of 0.97 h^{-1}. In the second trial, higher values for maximum β-galactosidase concentration (5600 units g^{-1} cells) and reactor productivity (2900 units l^{-1} h^{-1}) were obtained than in the first one. This revealed the presence of a substrate limitation to cell growth other than carbon source (lactose) in the former. The tris-buffered growth medium has a phosphate concentration of 1.0 mM; however, calcium chloride has to be added to the growth medium

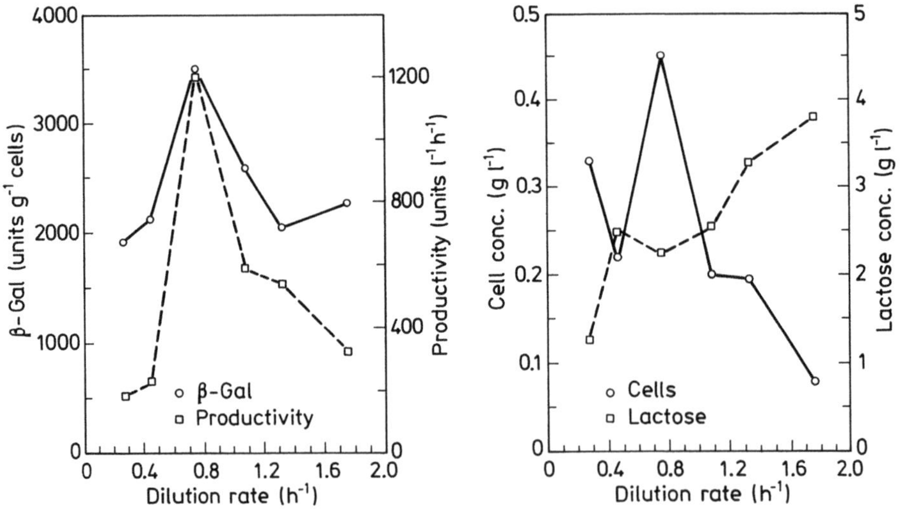

Fig. 2. Steady state IMRC bioreactor performance: lactose feed

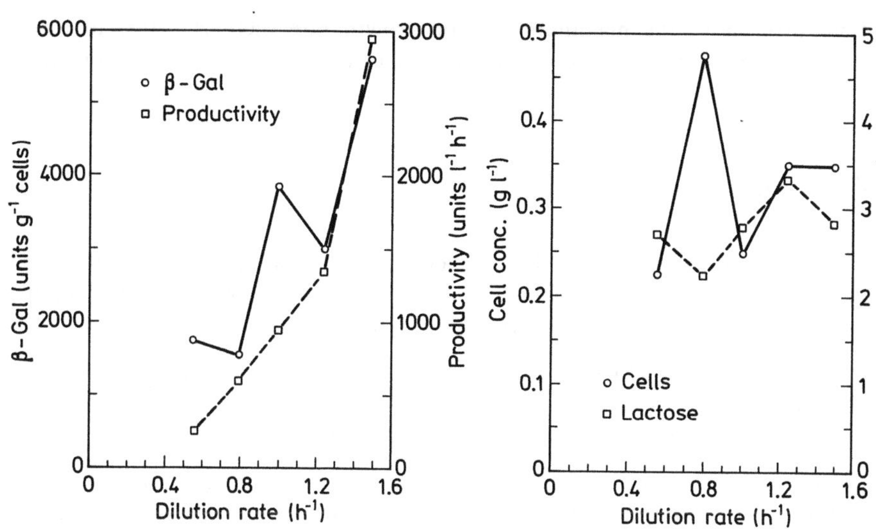

Fig. 3. Steady state IMRC bioreactor performance: lactose feed

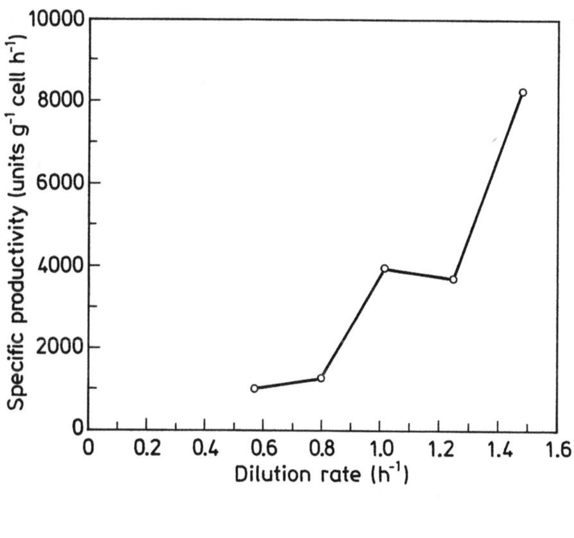

Fig. 4. Steady state IMRC bioreactor performance: lactose feed, specific productivity data

to prevent the gel particle beads from swelling and dissolving. But phosphate anion chelates calcium ions, the net result being that free phosphate is limited.

Nonetheless, in spite of the lack of smoothness in the IMRC bioreactor profiles, the IMRC bioreactor system did demonstrate that product concentrations and volumetric productivities similar to the free cell CSTR bioreactor system were obtainable; however, this might be vastly improved by increased cell density. The specific productivity data, shown in Fig. 4, indicate that the rate of enzyme biosynthesis in the immobilized cells is approximately equal to the corresponding rate of enzyme production in the CSTR bioreactor study at similar dilution rates; however, at dilution rates above the maximum specific growth rate there is a two to three-fold improvement for the IMRC bioreactor productivity over the CSTR bioreactor case.

5. Gene Expression System

Regulation of protein biosynthesis is crucial for a growing cell to regulate the concentration of catabolic and anabolic enzymes and ensure efficient use of available resources. Protein production involves two discrete steps:

1. Transcription of the structural gene coding for the protein to produce messenger RNA (mRNA).
2. Translation of mRNA into protein by ribosomal enzymes.

In most cases, control of protein biosynthesis occurs at the first step, by controlling the frequency of transcription initiation. Initiation may be controlled by the promoter gene which binds RNA polymerase, or by the operator gene, which, if bound by repressor protein, blocks transcription from the promoter site. A multitude of different promoter-operator genes exists, so the exact control mechanism is dependent upon which control region governs expression of the structural gene. The lac operon system is perhaps the best understood and was the system chosen in the studies conducted in our laboratory.

The lac operon is shown schematically in Fig. 5. Control of mRNA synthesis for β-galactosidase takes place at two independent sites.

1. The repressor gene (i) constitutively produces a protein (R) which, in the absence of inducer (lactose), binds to the operator gene (o), blocking transcription from the promoter.

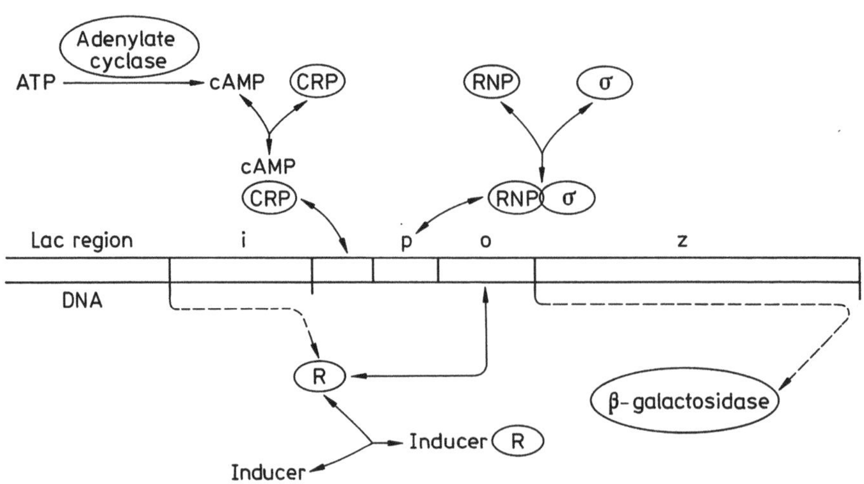

Fig. 5. Schematic of lac operon transcription control system

2. In the presence of high levels of cyclic adenosine monophosphate (cAMP), a complex forms between cAMP and the catabolite receptor protein (CRP) which then binds to the P_I region of the promotor gene, facilitating RNA polymerase binding at the P_{II} site [5].

Thus negative regulation takes place at the operator gene while positive regulation occurs at the promoter P_I site. A high lactose concentration is sufficient to prevent the repressor protein from binding to the operator gene, but in the presence of glucose, low cAMP levels prevent the CRP-cAMP complex from forming and binding to the promoter. As a result, extremely low levels of β-galactosidase are synthesized by the cell. There is evidence that the cAMP-CRP complex is not the sole mediator of catabolite repression [21]. This is confirmed by the fact that catabolite repression can take place in *crp* mutants, in the absence of a functional cAMP-CRP complex [9]. An intermediate in the metabolism of glucose is thought to bind free RNA polymerase (RNAP), inhibiting the binding of RNAP to the P_{II} site of the promoter gene [6]. This intermediate has been given the name "catabolite modulating factor".

The effects of catabolite modulating factor have been incorporated into a rigorous mathematical model of β-galactosidase biosynthesis in wildtype *Escherichia coli* [17]. The model provides a quantitative description of active inducer transport [11] and the role of PTS (phosphoenolpyruvate dependent sugar transport system) enzymes in the regulation of cAMP levels and in inducer exclusion.

Genetic engineering technique allows researchers to modify promotor and operator functions and to use them to govern expression of a foreign gene by cloning it into the structural gene region of the operon. The lac operon is used frequently in many plasmid constructions, either in its entirety or as a portion. Since its expression is so well characterized, it is convenient to use the lac operator-promoter region to regulate expression of cloned genes. The pUC plasmids have been developed for this purpose [20].

A structured model for lac promoter-operator function in wild-type and high copy number plasmids has been developed [13, 14]. The effect of plasmid copy number on protein production was modeled with results showing that the rate of transcription with respect to plasmid copy number deviates from linearity at high copy number, as one would expect.

6. Bioreactor Model

The mathematical model is, once again, written in terms of its component parts. The important variables, characterizing the state of the system, are the recombinant and plasmid-free cell concentrations in the bulk phase, bulk phase substrate concentration and the concentration of the enzyme β-galactosidase in the reactor effluent. As the immobilized phase makes a substantial contribution to the bulk phase values of these variables, it is necessary to evaluate the rates of cell division and substrate consumption in the immobilized phase as well. Assuming that there is no net accumulation of substrate (lactose) in the immobilized phase, the diffusion of substrate through the spherical gel particle must equal the rate at which it is consumed. This is expressed by Eq. (1).

6.1 Intraparticle Substrate Diffusion-reaction Balance Equations

$$D_{eff} \frac{d}{dr} [r^2 \, dS/dr] = r\,[S] \,. \tag{1}$$

This differential equation may be solved by using two of the following boundary conditions (the first being essential):
Surface condition:

$$D_{eff} \frac{dS}{dr} = k_s[S_F - S] \,, \qquad r = R \,. \tag{2}$$

Interior condition:

$$\frac{dS}{dr} = 0 \,, \qquad r = r^* \tag{3}$$

or:

$$S = 0 = \frac{dS}{dr} \,, \qquad r = r^* \,. \tag{4}$$

Equations (1)–(4) are useful when the carbon source within the gel particle is the limiting nutrient, controlling the rate of immobilized cell growth and division. To allow for the possibility that another nutrient limits cell growth (e.g., oxygen or phosphate) or in the absence of sufficiently detailed information regarding (carbon) substrate diffusion to the immobilized cells, it is advantageous to express the immobilized cell growth rate as a fraction of the free cell growth rate. The reaction term, r(S) may be evaluated using a simple substrate utilization model to relate it to the rate of cell production.

$$r(S) = \frac{1}{Y_x} \, r(X) \tag{5}$$

where r(X) is obtained from a modified Monod model.

$$r[X] = \frac{\mu_{max} S X}{S + K_S[1 + K_i X]} \,. \tag{6}$$

6.2 IMRC Reactor Balance Equations

Recombinant and plasmid-free cell concentrations may be obtained from *component* balances for each species in the following reactor model. The cell concentration in the immobilized phase is assumed to remain constant with respect to time and dilution rate. The idea here is that live cells grow to saturate a pellicular layer near the particle surface, accessible to substrate. Any additional replication is then released to the bulk. The liquid (free) phase concentrations of recombinant and wild-type cells may be calculated for a given set of operating conditions. The net rate of free cell production of each species is the sum of contributions from the liquid and immobilized phases, minus the rate at which cell removal occurs from the bioreactor. The following assumptions are made in the development of the subsequent *component* balance equations for recombinant and plasmid-free cells.

1. All cells reproduce by binary fission.
2. All recombinant cells contain the same number of plasmids corresponding to the mean plasmid copy number.
3. Plasmid is divided randomly between daughter cells at the time of cell division.
4. No plasmid-free cell or its progeny can regain the plasmid.

Recombinant Cell Balance:

$$\frac{dX_F^+}{dt} = [1 - \theta_I]\,\mu_I^+ X_I \varphi_I \frac{1-\varepsilon}{\varepsilon} + [1 - \theta_F]\,\mu_F^+ X_F^+ - \frac{D}{\varepsilon} X_F^+ . \tag{7}$$

The first term on the right hand side of Eq. (7) represents release of plasmid-bearing cells from the immobilized phase to the bulk phase; the other two terms are the customary ones, expressing plasmid-bearing cell growth and removal in the reactor effluent.

Plasmid-Free Cell Balance:

$$\frac{dX_F^-}{dt} = [\theta_I \mu_I^+ X_I \varphi_I + \mu_I^- X_I (1 - \varphi_I)] \frac{1-\varepsilon}{\varepsilon} + \theta_F \mu_F^+ X_F^+ + \mu_F^- X_F^- - \frac{D}{\varepsilon} X_F^- . \tag{8}$$

The composite term appearing first on the right hand side of Eq. (8) represents release of plasmid-free cells, while the second term represents cell reversion in the free suspension culture.

Substrate Balance:

$$[\varepsilon + \xi_I(1 - \varepsilon)] \frac{dS_F}{dt} = D[S_0 - S_F] - \frac{1}{Y_X} r[X] \tag{9}$$

where:

$$r[X] = [\varphi_I \mu_I^+ + [1 - \varphi_I]\,\mu_I^-]\,X_I[1 - \varepsilon] + \mu_F^+ X_F^+ \varepsilon + \mu_F^- X_F^- \varepsilon . \tag{10}$$

The immobilized cell growth rate of each species can be related to the corresponding free cell growth rate as follows:

$$\mu_I^+ = \eta_I \mu_F^+ \quad \text{and} \quad \mu_I^- = \eta_I \mu_F^- . \tag{11}$$

The terms η_I and ξ_I are bioreaction and substrate distribution effectiveness factors, respectively. These effectiveness factors may be evaluated using the following equations:

$$\eta_I = \frac{\int_{r^*}^{R} r[S]\, 4\pi r^2\, dr}{\left[\int_{r^*}^{R} 4\pi r^2\, dr\right] r[S_F]} \tag{12}$$

and

$$\xi_I = \frac{\int_{r^*}^{R} S 4\pi r^2 \, dr}{\left[\int_{r^*}^{R} 4\pi r^2 \, dr\right] S_F} . \tag{13}$$

Equations (12) and (13) are included here for the sake of completeness. In the evaluation of steady state bioreactor kinetics, any term containing ξ_I drops out of the analysis. In the IMRC bioreactor simulations, shown later, all of the experiments were essentially steady state with respect to substrate concentration, and η_I was treated as a parameter and evaluated by simulation to match the predicted cell concentration to that observed experimentally.

The fraction of plasmid-containing cells in the immobilized phase, φ_I, may be evaluated using the following relation developed by Bailey et al. [1]. The fraction of plasmid-containing cells in the immobilized phase is given by:

$$\frac{1}{\varphi_I[t]} = 1 + \frac{\theta[1-\omega]}{[1-\theta][2-\omega]}[1 - [2^a/2 - \theta]^m] \tag{14}$$

where $\omega = [2-\theta]^{1/a}$, $\theta = 2^{1-N_p}$, $a = \mu_I^-/\mu_I^+$, $m = t/G_I^+$.

The recombinant cell fraction in the bioreactor effluent is the same as the fraction of recombinant cells in the liquid phase. It may be calculated directly from the integrated values of X_F^+ and X_F^-.

The fraction of plasmid-containing cells in the liquid phase is given by:

$$\varphi_F = \frac{X_F^+}{X_F^+ + X_F^-} . \tag{15}$$

6.3 Gene Expression Kinetics

Once the recombinant cell concentration in the bulk phase is known, lac operon kinetics are used to evaluate the concentration of β-galactosidase in the effluent stream [16]. The concentration of β-galactosidase is dependent upon the concentration of mRNA which codes for the enzyme. The β-galactosidase mRNA concentration is dependent upon the rate of initiation of transcription.

β-galactosidase mRNA balance:

$$\frac{d[\text{mRNA } X_F^+]}{dt} = K_{+M} N_p F Q \mu_F^+ X_F^+ - [K_{-M} + D/\varepsilon][\text{mRNA } X_F^+] \tag{16}$$

where the *gene copy number*, N_p, appears explicitly. The concentration of β-galactosidase can readily be calculated from the mRNA concentration, based on the translation kinetics.

β-galactosidase enzyme balance:

$$\frac{d(EX_F^+)}{dt} = K_{+E}[\text{mRNA } X_F^+] - K_{-E}[EX_F^+] - \frac{D}{\varepsilon}[EX_F^+] . \tag{17}$$

Other equations are used to evaluate the fraction of unbound operator gene, Q, and the binding efficiency, F, of RNA polymerase at the promoter site [17].
Operator index:

$$Q = \frac{1 + b_1 S_{in}}{1 + b_1 S_{in} + b_2 [R]_t} = \frac{1 + b_1 S_{in}}{1 + b_1 S_{in} + b_2} . \tag{18}$$

The intracellular lactose concentration, S_{in}, is typically a multiple of the lactose level in the extracellular environment. It may be evaluated with the aid of the following equation.
Intracellular lactose concentration:

$$S_{in} = \frac{A'' S_F}{B + S_F} . \tag{19}$$

Catabolic repression index:

$$F = \frac{K_1}{K_2 + [M]} . \tag{20}$$

The catabolite modulating factor concentration ([M]) has a negative effect on the RNAP binding efficiency. Although its exact chemical structure and mechanism of action are not yet known, it is responsible for the existence of an optimal dilution rate at which the specific enzyme concentration reaches a maximum. This has been modeled by assuming that a critical growth rate exists such that the concentration of catabolite modulating factor is negligible at growth rates below the critical value, and increases hyperbolically at growth rates above the critical value [15]. Catabolite modulating factor:

$$[M] = \frac{k_m[\mu_F^+ - \mu_{crit}]}{Y_X[k_{-m} + \mu_F^+ - \mu_{crit})]} . \tag{21}$$

Lastly, the growth rate of cells in the bulk phase may be evaluated using the following growth model.
Cell growth rate:

$$\mu = \frac{\mu_m S_F}{S_F + K_s[1 + K_i X]} = \frac{\mu_m S_F}{S_F + K_s[1 + K_p(S_0 - S_F)]} . \tag{22}$$

6.4 Steady State IMRC Bioreactor Dynamics with Selection Pressure

When a sufficient quantity of antibiotic is present in the reactor feed, the growth rate and concentration of plasmid-free cells in the bioreactor is negligible. For steady state reactor operation, the IMRC reactor balance equations simplify, allowing for an analytical solution of substrate, recombinant cell and enzyme concentrations as functions of dilution rate. The bulk substrate concentration can be obtained by solving the following

$$[\nu\delta - \beta] S_F^2 + [\beta S_0 - \varkappa - \nu\delta S_0 - \nu\lambda] S_F + \varkappa S_0 = 0 \tag{23}$$

where

$$\nu = \frac{\mu_m}{D}$$

$$\beta = 1 - K_s K_p$$

$$\varkappa = K_s[1 + K_p S_0]$$

$$\delta = \varepsilon[1 - \theta_F]$$

$$\lambda = \frac{\eta_I[1-\varepsilon]}{Y_X} \frac{[1-\theta_I]}{[1-\theta_F]} X_I \, .$$

The substrate concentration is obtained using the quadratic formula; the negative root is the meaningful one.

The free cell growth rate is evaluated using Eq. (22). The free cell concentration is evaluated using the following equation:

$$X_F^+ = Y_X[1 - \theta_F][S_0 - S_F] \, . \tag{24}$$

The specific activity of β-galactosidase (units g^{-1} cells) can be solved directly using the following equation:

$$[E] = \frac{K_{+E} K_{+M} F N_p Q \mu_F^+}{[K_{-E} + D/\varepsilon][K_{-M} + D/\varepsilon]} \, . \tag{25}$$

7. Analysis and Discussion

The IMRC bioreactor model was used to evaluate the experimental results obtained from CSTR and IMRC bioreactor experiments. The bioreactor model was applied to shorter-term steady state bioreactor kinetics using selection pressure (ampicillin), and longer-term transient bioreactor kinetics, in which stability experiments were conducted, using a glucose minimal medium. The CSTR bioreactor model is a limiting case of the more general IMRC bioreactor model, and is obtained by setting the liquid fraction (ε) equal to one, and simplifying the component balance equations. The steady state bioreactor simulations are another limiting case, and were obtained by setting the time derivatives equal to zero and solving the component balances to obtain cell, substrate and product concentrations.

The parameter values estimated in the following bioreactor simulations were obtained by adjusting them to minimize the sum of squared deviations between the mathematical model and experimental results. This was accomplished through the use of NONLIN, the nonlinear statistical regression package.

Mathematical modeling results describing CSTR bioreactor dynamics are shown in Figs. 6–8, along with the experimental CSTR bioreactor data. Parameter estimates for the biomass yield factor, (Y_x), and the gene expression parameter, b_2, were obtained using NONLIN to give the best possible fit between the model and experimental data. The estimate for plasmid copy number N_p was obtained from

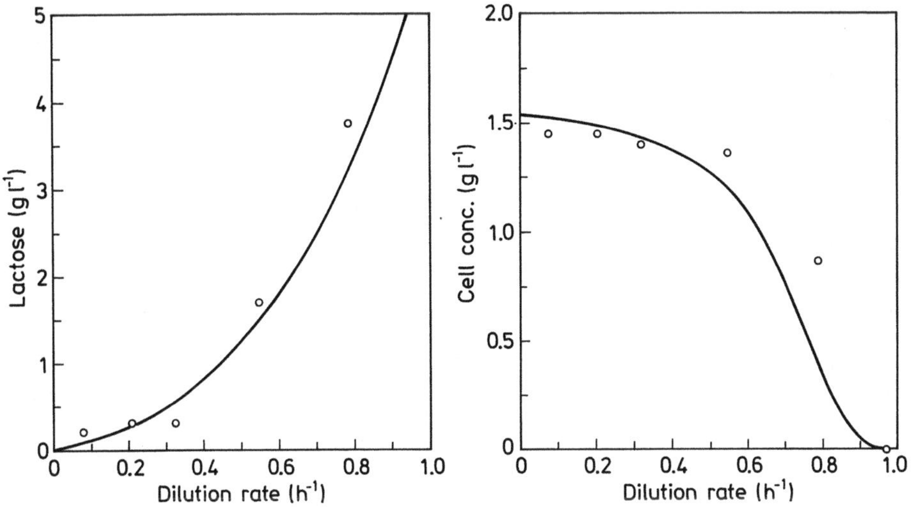

Fig. 6. Steady state CSTR modeling results: cell and lactose concentration profiles

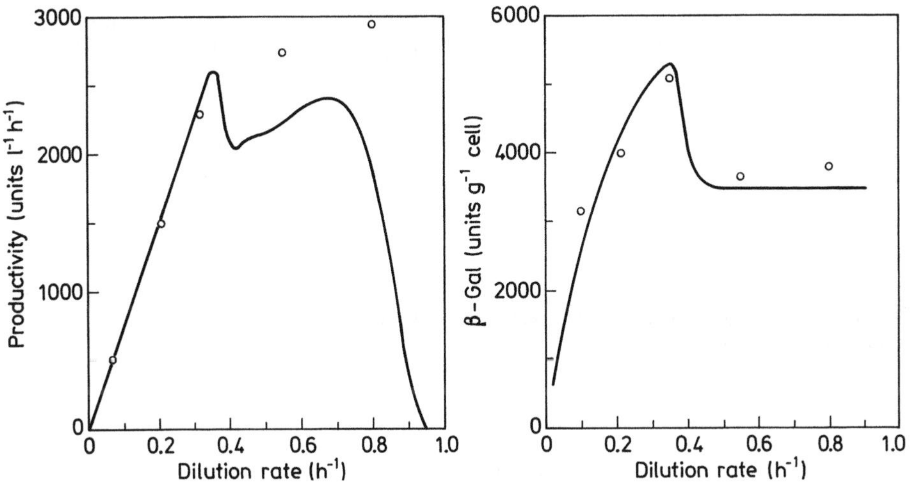

Fig. 7. Steady state CSTR modeling results: enzyme concentration and productivity profiles

fitting the unsteady state modeling and experimental results, shown in Fig. 10. The parameter value for μ_{crit} was the dilution rate at which the β-galactosidase specific activity displays its maximum value. The remaining parameter values were taken from previous mathematical modeling studies of wild-type *E. coli*, carried out in this laboratory [15]. The parameter values used in the simulation are shown in Tables 1 and 2.

The fit between the mathematical model and experimental data is good and can be attributed in part to the accuracy of the cell growth and lac operon gene

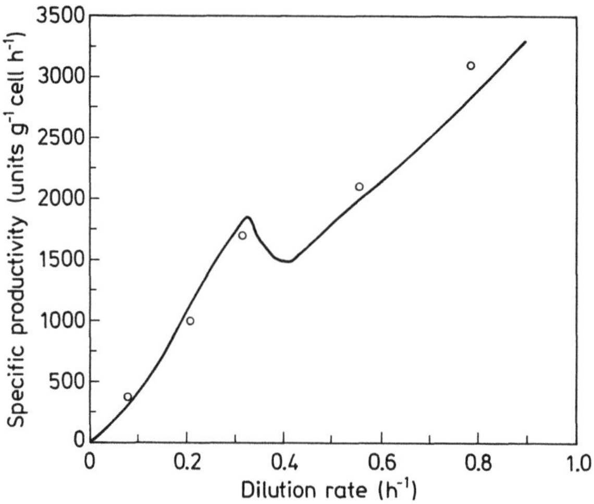

Fig. 8. Steady state CSTR modeling results: specific productivity profile

Table 1. Parameter values used in steady state bioreactor model

Parameters	Values	Parameters	Values
μ_m	0.92 h^{-1}	N_p	7.2 plasmids per cell
μ_{crit}	0.35 h^{-1}	K_{+E}	1000 units mg^{-1} h^{-1}
Y_x	0.32	K_{-E}	0.0 h^{-1}
K_s	0.04 g l^{-1}	b_1	5.0 l g^{-1}
K_p	10.0 l g^{-1}	b_2	245 dimensionless
K_1	4.82 mg l^{-1}	A	60.0 g l^{-1}
K_2	4.89 mg l^{-1}	B	0.40 g l^{-1}
K_{+M}	50.0 mg g^{-1}	k_{+m}	1.48 mg l^{-1}
K_{-M}	27.6 h^{-1}	k_{-m}	0.05 h^{-1}

Table 2. Operational parameter values used in steady state bioreactor simulations

CSTR		IMRC	
S_0	5.0 g l^{-1}	S_0	5.0 g l^{-1}
ε	1.0	ε	0.714
		X_I	3.25 g l^{-1}
		η_I	0.22

expression parameters obtained from previous studies [17], and also to the accuracy with which the plasmid copy number could be estimated from the CSTR stability experiment, shown in Fig. 10. Estimation of one lac operon gene expression parameter value, b_2, was necessary because the plasmid used in this study has a *lac i^q* repressor gene which overproduces the lac repressor protein. The additional level of repression is manifested by increasing the value of this dimensionless parameter, which has an estimated value of 245, as compared to a normal *lac i* gene, which has a value of 100.

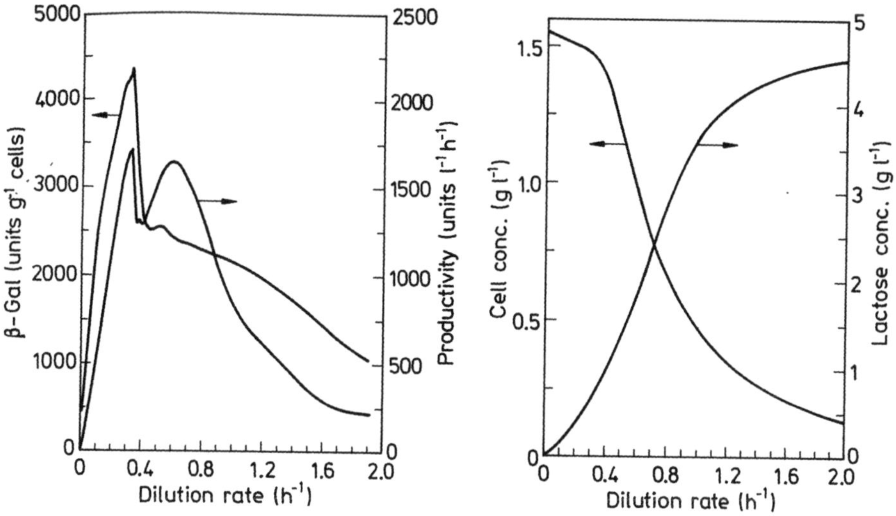

Fig. 9. Steady state IMRC bioreactor simulation results

A simulation of steady state IMRC bioreactor kinetics (Fig. 9) was performed using the parameter values shown in Tables 1 [17] and 2 (this work). When compared against the results of the IMRC bioreactor experiments, the model predicts higher cell concentrations at low dilution rates than those measured experimentally. At high dilution rates, the experimental cell concentrations are slightly higher. The lack of conformity of the experimental data to the model is due to the phosphate limitations discussed earlier. When compared against the CSTR bioreactor model, the IMRC bioreactor model predicts a maximum enzyme volumetric productivity at a lower dilution rate than that predicted by the CSTR bioreactor model. The IMRC bioreactor model does predict substantial enzyme productivities at dilution rates above the maximum specific growth rate, thus demonstrating the enhanced operational stability of the IMRC bioreactor at high dilution rate. With higher immobilized cell loading or a protein secretion system, the IMRC productivity values could be increased dramatically.

8. Plasmid Stability Studies

The mathematical modeling of plasmid segregational instability kinetics requires consideration of the cell growth kinetics only. The gene expression kinetics do not enter the calculations, because very little product was produced when the recombinant cells were grown on glucose. The phosphate limitation was obviated by switching to κ-carageenan as the carrier. The focus of this phase of the modeling and experimental study was to analyze the loss of the recombinant cell population from the bioreactor, and the mechanisms involved. There are two main processes which govern the rise in the fraction of wild-type (plasmid-free) cells in the bioreactor. The first is the rate at which recombinant cells divide to give one

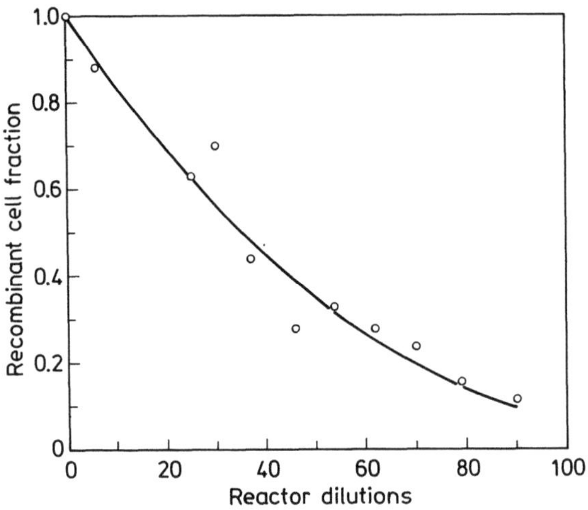

Fig. 10. CSTR bioreactor stability modeling results

Table 3. List of operational and estimated parameter values used in bioreactor stability simulations

CSTR		IMRC	
$\mu_F^+ = D$	0.67 h^{-1}	D	1.08 h^{-1}
ε	1.0	ε	0.71
$X_F^+ (t = 0)$	1.57 $g\,l^{-1}$	X_I	0.35 $g\,l^{-1}$
$\varphi_F (t = 0)$	1.0	N_p	7.2 plasmids per cell
a	1.02	$X_F^+ (t = 0)$	0.35 $g\,l^{-1}$
N_p	7.2 plasmids per cell	μ_F^-	0.98 h^{-1}
		a	1.10
		η_I	0.22

recombinant and one plasmid-free cell. This process of unequal plasmid partitioning gives an exponential decline in the recombinant cell fraction. The second process involves the ratio of growth rates between the wild-type and recombinant cells, which causes the fraction of recombinant cells to decrease as a result of having a longer doubling time. This growth-rate difference manifests itself more slowly than the plasmid partitioning process, in that it becomes more noticeable as the fraction of wild-type cells increases.

The parameter values used in the simulation, shown in Fig. 10, are listed in Table 3. The parameters estimated in this simulation are the plasmid copy number (N_p), and the ratio of growth rates between the wild-type and recombinant cells (*a*). The remaining parameter values are operational constants used in the bioreactor experiment.

The estimated value of the mean plasmid copy number, N_p, was 7.2 copies per cell, which is less than the value of 11 obtained by isolation and quantification of plasmid from shake-flask experimentation. The statistical estimate is likely to be the more accurate of the two methods, due to the inaccuracies of gel electrophoresis and cell plating quantification procedures. The estimated value of *a* was 1.02 which indicates that the plasmid-free cell growth rate is 2% greater than the

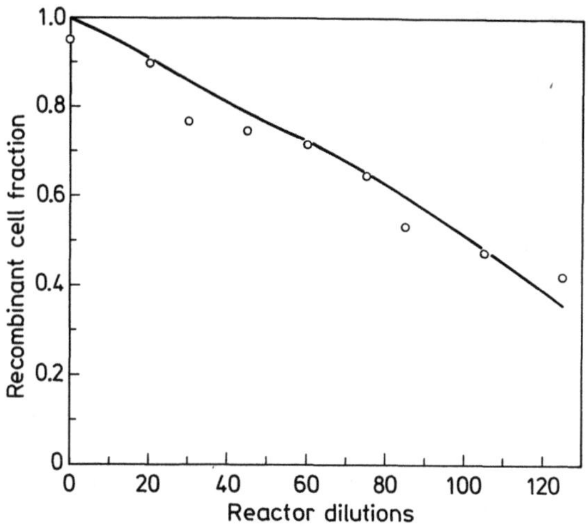

Fig. 11. IMRC bioreactor stability modeling results

recombinant cell growth rate, which is a very small difference. Our results for N_p and a are in good agreement with those of Seo and Bailey [19] and Kim et al. [12]. Had a parameter value of 1.0 been used for a, the estimated plasmid copy number would have been 6.5 copies per cell. Therefore, in this experiment, the main event governing the rate of decline in the recombinant cell fraction is unequal plasmid partitioning.

The parameters used in the IMRC bioreactor stability simulation (Fig. 11) are shown in Table 3. The plasmid copy number estimate from the free cell bioreactor stability experiment was used. The value of η_I was estimated by simulation, using the IMRC bioreactor equations, to obtain the observed free cell concentration at the initial conditions. The specific growth rate of wild-type cells (μ_F^-) was obtained by solving the IMRC bioreactor model equations at the dilution rate used in the experiment. The value obtained is the same as the maximum specific growth rate, as a result of the high dilution rate providing ample substrate to the free cells. The ratio of growth rates between wild-type and recombinant cells is very sensitive to the actual growth rate. It has been shown that this ratio increases with decreasing growth rate [19]. The statistical estimate of a obtained in this simulation was 1.10, which is a significant increase over the value of 1.02 obtained in the free cell bioreactor simulation, shown in Fig. 10.

The difference in growth rates between recombinant and wild-type cells becomes very important in an immobilized cell bioreactor system. Despite the fact that this ratio increases the rate of loss of plasmid-containing cells, there is a marked reduction in the overall rate due to the slower growth rate of immobilized cells. It is this net improvement in the time profile for the recombinant cell fraction that suggests that immobilized cell bioreactor systems be tested widely for protein product biosynthesis.

A total of five parameters were estimated using nonlinear regression. The parameter estimates, standard deviations and 95% confidence intervals are shown in Table 4. It is apparent from the values of a obtained in each bioreactor system and the confidence intervals associated with each that a clear difference exists in the ratio

Table 4. Parameter estimates and regression statistics

Parameter	Estimate	σ	95% CI
Y_x	0.32	0.04	0.24–0.40
N_p	7.2	0.38	6.35–8.07
b_2	244.7	22.7	181–308
α_{CSTR}	1.02	0.01	0.99–1.05
α_{IMRC}	1.10	0.007	1.09–1.12

of wild-type to recombinant cell growth rates between the two bioreactor systems, and this ratio becomes more important in the process kinetics in the slower-growing immobilized cell system.

Mathematical analysis of IMRC and CSTR bioreactor kinetics is an important tool in developing new bioreactor design and operation strategies, as well as in allowing one to evaluate observed experimental results with greater understanding and insight. The most important outcome of the bioreactor simulation efforts was the finding that growth rate differences between wild-type and recombinant cells are magnified in the IMRC bioreactor system, in comparison to the free cell bioreactor system. In addition, application of the plasmid copy number estimate obtained in the free cell bioreactor stability study to steady state gene expression kinetics provided good agreement with the observed levels of enzyme biosynthesis.

9. Conclusions

The single most important conclusion that can be deduced from the experimental results is that IMRC bioreactor systems demonstrate dramatically improved operational stability over continuous free cell bioreactor processes in the absence of selection pressure. Using the plasmid containing cell fraction as the indicator, there is a five- to ten-fold improvement (i.e., decrease) in the level of plasmid segregational instability in going from the CSTR to the IMRC bioreactor system. This appears to have great potential when applied to a continuous large scale protein production process, where trace quantities of antibiotic in the final product would violate good manufacturing practice (GMP) guidelines, set forth by the Food and Drug Administration.

10. Nomenclature

CSTR	continuous stirred tank reactor
D_{eff}	intraparticle substrate diffusion coefficient, $cm^2\ s^{-1}$
[E]	β-galactosidase specific activity, units g^{-1} cells
$[EX_F^+]$	β-galactosidase concentration, units l^{-1}
G_I	immobilized cell doubling time, h
IMRC	immobilized recombinant cell

m	number of generations
[mRNA X_F^+]	mRNA concentration, mg l^{-1}
N_p	plasmid copy number
r	radial distance from particle center, cm
S	substrate concentration, g l^{-1}
S_F	liquid phase substrate concentration in the microenvironment of a cell, g l^{-1}
S_{in}	intracellular lactose concentration, g l^{-1}
S_0	inlet substrate concentration, g l^{-1}
X	cell concentration, g l^{-1}
Y_x	biomass yield coefficient

Subscripts

F	free (liquid) phase
I	immobilized phase

Superscripts

+	recombinant (plasmid-containing)
—	wild-type (plasmid-free)

Greek Symbols

a	ratio of wild-type to recombinant cell growth rates
ε	liquid fraction
η_I	ratio of IMC to free cell growth rate $[\mu_I/\mu_F]$
θ	probability of obtaining plasmid-free segregant
μ	specific growth rate, h^{-1}
μ_m	maximum specific growth rate, h^{-1}
σ	standard deviation
ξ_I	substrate diffusion effectiveness factor
φ	recombinant cell fraction

11. References

1. Bailey, J. E., M. Hjortso and F. Srienc: Ann. N.Y. Acad. Sci. Biochem. Eng. III **413**, 71 (1983)
2. Bailey, K.: Ph. D. Thesis in Chemical and Biochemical Engineering, Rutgers U. 1986
3. Berman, M. L. and D. E. Jackson: J. Bacteriol. **159**, 750 (1984)
4. Casadaban, M. J.: J. Mol. Biol. **104**, 541 (1976)
5. De Crombrugghe, B., S. Busby and H. Buc: Science **224**, 831 (1984)
6. Dessein, A., F. Tillier and A. Ullmann: Molec. Gen. Genet. **162**, 89 (1978)
7. De Taxis du Poet, P., P. Dhulster, J. N. Barbotin and D. Thomas: J. Bacteriol. **165**, 871 (1986)
8. Dykhuizen, D. E. and D. L. Hartl: Microbiol. Rev. **47**, 150 (1983)
9. Guidi-Rontani, C., A. Danchin and A. Ullmann: Proc. Natl. Acad. Sci. USA **77**, 5799 (1980)
10. Karkare, S. B., R. C. Dean and K. Venkatasubramanian: Biotechnology **3**, 247 (1985)
11. Kaushik, K. R., W. R. Vieth and K. Venkatasubramanian: J. Molec. Catalysis **30**, 39 (1985)
12. Kim, B. G., T. A. Good, M. M. Atasi and M. L. Shuler: Ann. N.Y. Acad. Sci., Biochem. Eng. V, 384 (1987)
13. Lee, S. B. and J. E. Bailey: Biotech. Bioeng. **26**, 1372 (1984a)
14. Lee, S. B. and J. E. Bailey, Biotech. Bioeng. **26**, 1383 (1984b)

15. Ray, N. G.: Ph. D. Thesis in Chemical and Biochemical Engineering, Rutgers U. 1985
16. Ray, N. G., W. R. Vieth and K. Venkatasubramanian: Ann. N.Y. Acad. Sci. Biochem. Eng. *IV*, **469**, 212 (1986)
17. Ray, N. G., W. R. Vieth and K. Vankatasubramanian: Biotech. Bioeng. **29**, 1003 (1987)
18. Scott, J. R.: Microbiol. Rev. **48**, 1 (1984)
19. Seo, J. H. and J. E. Bailey: Biotech. Bioeng. **27**, 156 (1985)
20. Vieira, J. and J. Messing: Gene **19**, 259 (1982)
21. Wanner, B. L., R. Kodaira and F. C. Neidhardt: J. Bacteriol. **136**, 947 (1978)

2.3 Intradiffusion Effect on Reactivity of Immobilized Microorganisms

Shintaro Furusaki

Department of Chemical Engineering, University of Tokyo
7-3-1, Hongo, Bunkyo-ku, Tokyo 113, Japan

Contents

Intraparticle diffusion resistance played an important role in the behavior of immobilized microorganisms. Intraparticle diffusivity did not depend on the gel concentration up to 5%, however it depended on the cell concentration. The latter dependence is discussed by the use of the random pore theory of Wakao and Smith originally proposed for solid catalyst particles. The overall reaction rates of viable immobilized *Saccharomyces cerevisiae* and *Zymomonas mobilis* were measured and analyzed with consideration of the intraparticle diffusivity, the cell concentration in the particle and reaction rate of the native microorganisms. Behavior of the immobilized *S. cerevisiae* was well described by the analysis. In case of *Z. mobilis*, reaction rate in the immobilized state appeared to be about 20% smaller than that in the native condition. Overall reaction rate of the immobilized microorganisms became slower at high cell concentration and at low substrate concentration.

1. Introduction

Immobilization is applied more and more extensively in the field of the biocatalyst application. Engineering researches of immobilized biocatalyst have covered a wide range of aspects such as immobilizing materials and nature of biocatalysts. However, the behavior of immobilized biocatalysts from the viewpoint of mass transfer is also an important topic.

The effect of mass transfer on overall reactivity of solid phase catalyst has been considered by the use of the concept of so-called effectiveness factor [1,2], i.e.

Bioproducts and Bioprocesses
Editors: Fiechter, Okada, Tanner

conversions in solid phase catalyst reactors can be estimated [3] by the use of the factor. In the field of biochemical engineering the factor for the Michaelis-Menten type reactions is kown to be obtained by the theories of Bischoff [4] and Kobayashi et al. [5,7]. Yamane et al. [8] generalized their analyses taking into account the effect of film mass transfer resistance and partition coefficient.

Now, in the application of the effectiveness factor, knowledge of effective diffusivity in the gel is salient. Nakanishi et al. [9] studied about this property in the case of immobilized enzymes. Later a number of investigations of the diffusivity and its effect on reactivity have been carried out in connection with the behavior of immobilized enzyme reactors.

Several papers [10,12] have been published on the effect of changing particle size, cell concentration, etc. on internal diffusion using reactions by immobilized microorganisms. Klein et al. [13,14] clarified that effective diffusivity dependend on the cell concentration in the gel, i.e. the diffusivity decreased with the increase of the cell concentration. Other investigations are summarized in recent reviews by Radovich [15,16] and Karel [17].

In this paper, the effect of intraparticle diffusion effect on immobilized microorganisms is analyzed from a fundamental approach of diffusion and reaction kinetics. Dependence of the effective diffusivity on the cell concentration, and the behavior of the overall reaction rate are also discussed.

2. Materials and Methods

Saccharomyces cerevisiae from Oriental Yeast Co. and *Zymomonas mobilis*, Type culture NRRL B14023, were used as models of microorganisms. They were cultivated in standard media and collected at the final stage of the exponential growth period [18-20].

The immobilized microorganisms were prepared by the use of Ca-alginate gel. Microorganisms were suspended in Na-alginate solution (0.5–2%) and the suspension was extruded in the form of drops through a needle into $CaCl_2$ solution (1.5–13%). Thus the immobilized microorganism particles were prepared. Details of the procedures together with analytical methods are given elsewhere [18,20]. Distribution coefficient of the substrate glucose between the gel and liquid phases was close to unity and its influence will be ignored in later analyses.

3. Intraparticle Diffusion in Immobilized Microorganism Particles

3.1 Theoretical Consideration

Diffusion characteristics inside particles should be understood in the analysis of immobilized biocatalysts. About diffusion in gel particles, Nakanishi et al. [9] studied the behavior extensively. They showed diffusivities in the gel decreased with the gel concentration. There have been a number of studies since their investigation into diffusion in immobilization particles, such as polyacrylamide gel, κ-carrageenan gel and calcium alginate gel. It is generally recognized that diffusivity in the gels is 60–80% of that in water.

Diffusion resistance in immobilized microorganisms involve different aspects from those in immobilized enzymes. There are cell bodies to interfere with diffusion in the particle. Also diffusion inside the cell bodies may need to be accounted for. After considerating the latter resistance it was concluded that this resistance can be neglected, because the time-constant of the diffusion in cell bodies is in the order of 10^{-4} s. This value may be obtained from the result of Gurney and Lurie [21] on unsteady state thermal conduction. From their results, the above time-constant of diffusion in spherical particles was obtained from the relation $D_e t/R^2 = 0.1$. The cell radius may be considered to be about 10^{-6} m. The time-constant value, 10^{-4} s, is negligible with compared to the duration of the reactions.

From the above consideration diffusion through particles of immobilized microorganism may be considered similar to diffusion through solid catalyst carriers. If a substrate reaches cell surfaces of the microorganism, reaction will occur without further mass transfer limitations. Wakao and Smith [22] proposed a mass transfer model through solid catalysts shown in Fig. 1. This may be called as the random pore model. According to their model, effective diffusivity in the immobilized microorganism system is smaller than that in the entrapping gel by the factor of ε^2, i.e.

$$\frac{D_e}{D_g} = \varepsilon^2 . \tag{1}$$

Converting this ratio to more convenient form,

$$\frac{D_e}{D_0} = k_1(1 - k_2 c_c)^2 \tag{2}$$

where k_1 is the ratio of the diffusivities in the gel to that in water, and k_2 corresponds to the specific cell-volume.

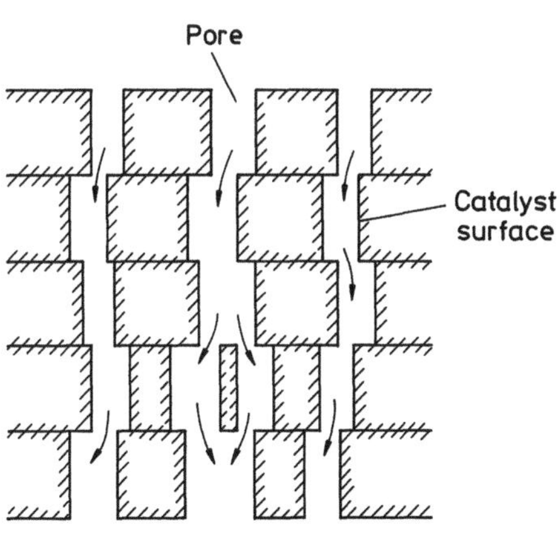

Fig. 1. Random pore model for solid catalysts

3.2 Experimental

Measurement of effective diffusivity can conveniently be carried out by the unsteady-state measurements [23,24]. Here, the stepwise elution method of Horowitz and Fenichel [25] was chosen for its simplicity of analysis. By their method, the diffusing material is initially included within the particles and the change of its concentration is measured against time. The fundamental diffusion equation is given below [26].

$$\frac{\partial u}{\partial t} = D_e \frac{\partial^2 u}{\partial r^2} \tag{3}$$

$$u = cr\,. \tag{4}$$

The boundary and initial condition is:

$$\text{at} \quad r = 0\,, \quad \text{then} \quad u = 0 \tag{5}$$

$$\text{at} \quad r = R\,, \quad \text{then} \quad u = 0 \tag{6}$$

$$\text{at} \quad t = 0\,, \quad \text{then} \quad u = r\bar{c}(0)\,. \tag{7}$$

The solution of the above equation is

$$\frac{\bar{c}(t)}{\bar{c}(0)} = \frac{6}{\pi^2} \sum_{n=1}^{\infty} \frac{1}{n^2} \exp\left(-n^2\pi^2 D_e t/R^2\right)\,. \tag{8}$$

The higher terms of the right side diminish quickly, so a plot $\bar{c}(t)/\bar{c}(0)$ vs. $\pi^2 t/R^2$ will give a straight line for sufficiently large t, whose slope gives D_e.

An important condition to carry out the diffusion experiment is to choose the diffusing material unconsumed by the viable cells. The use of inviable cells instead

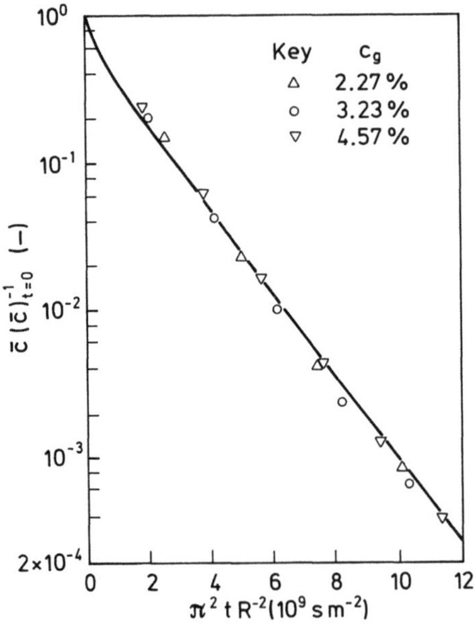

Fig. 2. Unsteady state measurement of effective diffusivity in gel. Dependence on gel concentration

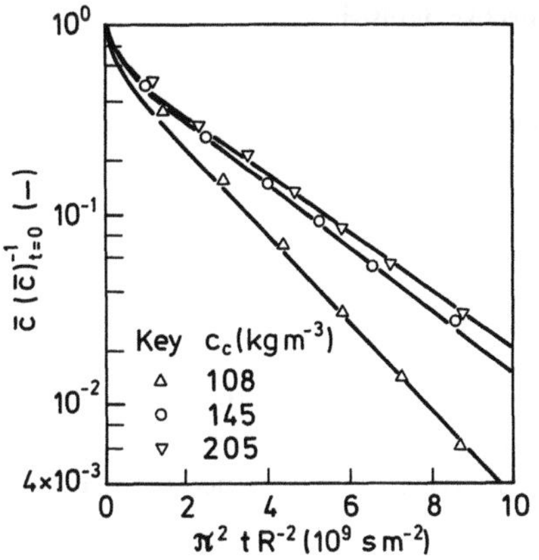

Fig. 3. Unsteady state measurement of effective diffusivity in gel. Dependence on cell concentration

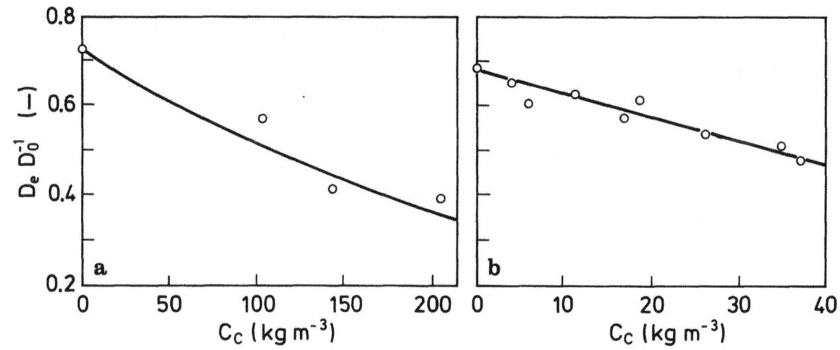

Fig. 4. Dependence of the effective diffusivity on cell concentration **a** *S. cerevisiae*; **b** *Z. mobilis*

of the above condition is not recommended because materials may diffuse through the cell body. From this reasoning xylose and galactose were chosen as the diffusing materials instead of glucose for the case of viable immobilized *S. cerevisiae* and *Z. mobilis*, respectively.

An example of the diffusion plot for the xylose-*S. cerevisiae* system is shown in Fig. 2. In this figure the gel concentration c_g was changed but found inaffective to diffusivity within the range of the experimental gel concentration. Here, the gel concentration is defined as concentration of gel material, i.e. calcium alginate, in total particle. Similarly the effect of the cell concentration, i.e. concentration of cells in the gel (in dry cells in gel, kg m^{-3}), was studied. In this case, the slope of the plot changed according to the cell concentration as shown in Fig. 3. The dependence on the cell concentration is summarized in Fig. 4 for the cases of *S. cerevisiae* and *Z. mobilis*. The obtained values of k_1 and k_2 are listed in Table 1.

The value of k_1 was found to be reasonable since the specific volume of *S. cerevisiae* was 1.78×10^{-3} m^3 kg^{-1} (volume of dry cell). The specific volume of *Z. mobilis* was not measured but that of bacteria is normally considered to be about 4×10^{-3} m^3 kg^{-1} (volume of dry cell). Thus the value of k_1 agrees with the

Table 1. Values of k_1 and k_2 in the correlation of the effective diffusivity in gel

Immobilized microorganism	*S. cerevisiae*	*Z. mobilis*
k_1 (—)	0.728	0.68
k_2 ($m^3 kg^{-1}$)	1.54×10^{-3}	4×10^{-3}

specific volume of cells. The small discrepancy may be due to the steric effect revealing that cells must grow in limited space in the gel. The value of k_2 is also in the agreeable range. The effective diffusivity of glucose can be obtained from above data by using the ratio of diffusivities in water.

4. Reaction Characteristics of Immobilized Microorganisms Producing Ethanol

Using the above data of effective diffusivity the overall reaction characteristics can be quantitatively analyzed. The reaction producing ethanol by *S. cerevisiae* and *Z. mobilis* was investigated.

4.1 Reaction Kinetics of Native Microorganisms

In order to accomplish the above investigation the reaction kinetics of the native cells must be known. The reaction rate was measured by batch experiments in our research. Cultivation media are shown elsewhere [18,20]. The preparation of cells to be used in the kinetic study was carefully made to obtain the same condition of the cell growth as that of the cells immobilized in gel. The collected cells are in the final stage of the exponential growth period. They are kept in sterilized water for about 12 h at room temperature.

Reaction media are shown in Table 2. The measurement was carefully programmed so that the cells are in the induction period of growth during the measurement. Thus the concentration of the cells was kept constant. The reaction rate of the cells thus prepared was assumed to be the same as the inherent reaction rate of immobilized cells in later analyses.

Table 2. Reaction media for native and immobilized cells

	S. cerevisiae	*Z. mobilis*
Glucose	1–100 kg	0.8–180 kg
$MgSO_4 \cdot 7 H_2O$	0.5 kg	0.5 kg
KH_2PO_4	1.0 kg	1.0 kg
$CaCl_2 \cdot 2 H_2O$	7.35 kg	2.5 kg
H_2O	1 m^3	1 m^3

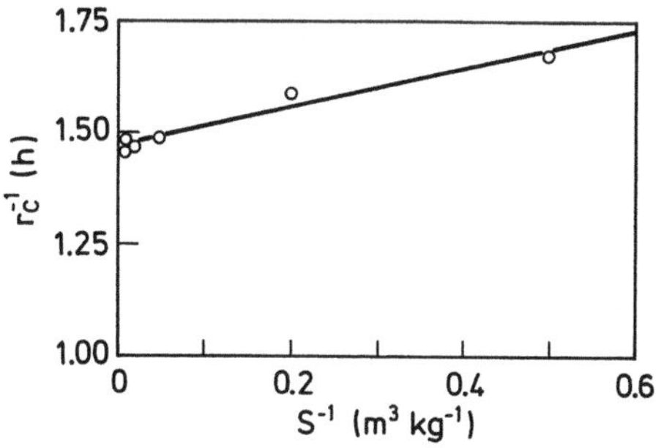

Fig. 5. Initial reaction rate of ethanol production by *S. cerevisiae*

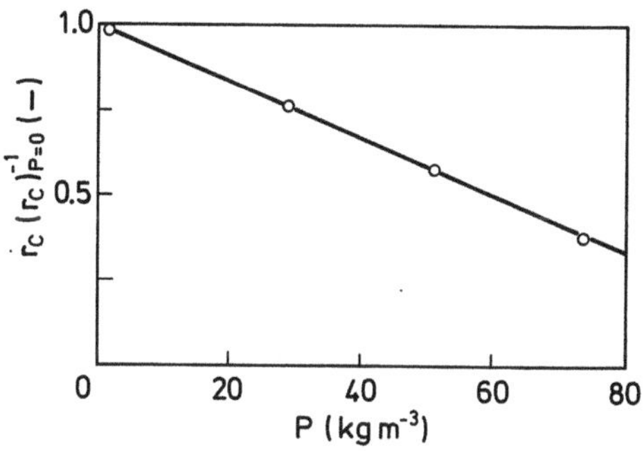

Fig. 6. Ethanol inhibition on reaction rate of native *S. cerevisiae*

Reaction rate of *S. cerevisiae* was investigated by means of measurement of initial reaction rate and ethanol inhibition. Initial reaction rate is shown in Fig. 5, which suggest the applicability of the Michaelis-Menten (M-M) type kinetics. Ethanol inhibition is measured by adding ethanol to the reaction medium. The ratio of the reaction rates with and without ethanol is plotted in Fig. 6. Combining the above results the reaction rate of ethanol production by *S. cerevisiae* is expressed by Eq. (9).

$$r_c = \frac{V_{max}S}{K_m + S}(1 - aP)\,. \tag{9}$$

Values of the parameters are given in Table 3. The Michaelis-Menten type initial reaction is similar to the results by previous investigations [27-31]. The ethanol inhibition is the same as the Ghose-Tyagi relationship [32].

Reaction kinetics of *Z. mobilis* were also studied. In this case the initial reaction rate was investigated because the analysis of the immobilized microbe was done under

Table 3. Parameters in the reaction rate of *S. cerevisiae* and *Z. mobilis*

	S. cerevisiae	*Z. mobilis*
V_{max} (h^{-1})	0.678	5.0
K_m (kg m^{-3})	0.229	1.0
a (m^3 kg^{-1})	8.40×10^{-3}	—

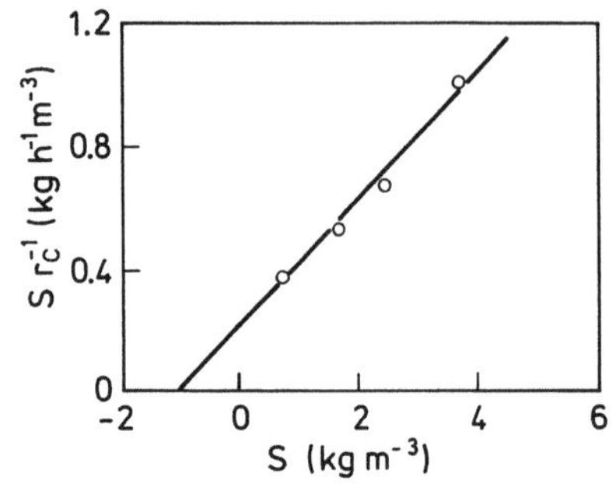

Fig. 7. Initial reaction rate of ethanol production by native *Z. mobilis*

the condition that ethanol produced was sufficiently dilute. The reaction medium is again shown in Table 2. The rate is expressed in Fig. 7 showing that the kinetics follow the Michaelis-Menten equation. Values of K_m and V_{max} are summarized in Table 3 above. It must be noted that this reaction rate relationship is valid at negligible product concentration.

4.2 Overall Reaction Rate of Immobilized Resting Microorganisms

One of the application methods of immobilized viable cells is the immobilized resting microorganism. In this case the microorganisms are viable but do not grow. Thus, the cell concentration in immobilizing particles is homogeneous and about the same as that just after immobilization. Reaction characteristics of immobilized resting *S. cerevisiae* and *Z. mobilis* in Ca-alginate gel are described in this section.

4.2.1 Behavior of Immobilized Resting *Saccharomyces cerevisiae*

The immobilized yeast pellets were prepared by the method described in the previous section. By dissolving the gel we could count the number of viable yeast cells by dying with methylene blue. Doing this, the fraction of viable cells was found to be over 99%. The reaction medium was the same as that used in the measurement of the reaction rate of native yeast. The film mass transfer resistance was found to be negligible from experiments where the rotation rate was altered.

The overall initial reaction rate of immobilized resting *S. cerevisiae* thus obtained is plotted in Fig. 8 in the form of the Lineweaver-Burk (L-B) plot. Solid lines in the figure are the calculated lines according to the method of Kobayashi et al. [5-7]. The lines show curvature due to the intraparticle diffusion resistance as suggested by Hamilton et al. [32]. The above L-B plot shows the following interesting features. Firstly, the value of V_{max}, 0.538 kg h^{-1} kg^{-1} (ethanol per dry cell), is almost identical with that of free cells, 0.554 kg h^{-1} kg^{-1} (ethanol per dry cell). This means that the immobilized yeast seems to be as active as free cells if it is in large cell concentration. Secondly, the apparent K_m value is large for the case of high cell concentration. Thirdly, the reaction rate is larger for the case of lower cell concentration if the substrate concentration is low. This is an interesting effect of the mass transfer limitation.

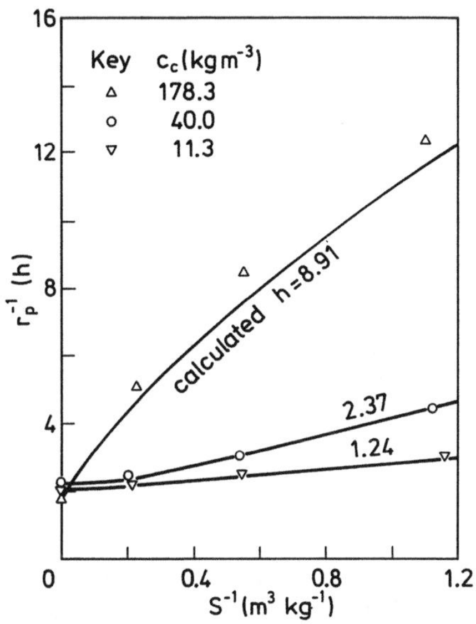

Fig. 8. Lineweaver-Burk plot of the overall reaction rate by immobilized resting *S. cerevisiae*

The detailed theoretical calculation procedures [5-7,18] for the analysis of the overall reaction rate are not shown here, however, the results of the study are presented. The parameter, h, in Fig. 8 is the one derived from Bischoff's analysis [4] on generalized Thiele modulus. It is a generalized Thiele modulus [33] for zero order reactions using K_m for bulk concentration. The calculations were carried out by using the initial reaction rate of native yeast and the effective diffusivity of glucose described above. The calculated results coincide well with the experimental data. Thus, the inherent activity of the immobilized yeast is found to be about the same as that of the native yeast. This nature can be used to calculate the reactor analysis of immobilized *S. cerevisiae.*

4.2.2 Behavior of Immobilized Resting *Zymomonas mobilis*

The study of immobilized resting microorganism was carried out for the case of *Z. mobilis* as well. The microorganism was cultivated in a programmed way so that cells in the final stage of the exponential growth period were always obtained. The immobilization procedure was similar to that for *S. cerevisiae.*

Two different sizes of cell beads were prepared, i.e. 2.4–3.0 mm and 3.4–4.0 mm in diameter. Each bead has a variety of cell concentrations, i.e. dry cell in gel 0.25–40 kg m^{-3}. The overall initial reaction rate of the immobilized *Z. mobilis* was again measured by batch reactions tracing the change of ethanol concentration.

The results can again be analyzed by the use of the Kobayashi relation since the reaction kinetics of native cells follow the M-M equation. A comparison between the experimental data and the calculation is made in Fig. 9 in terms of the effectiveness factor. The effectiveness factor for the M-M kinetics should come between those for the zero and first order kinetics. Looking at the figure we can find the data behave quite similarly to the calculated values. However, the data values are

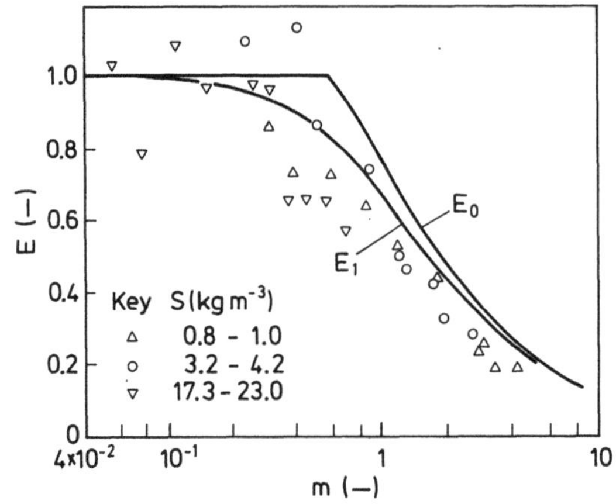

Fig. 9. Effective factor analysis of the overall reaction rate by immobilized resting *Z. mobilis*

slightly lower than expected. This appears to indicate that the inherent reaction rate is about 20% lower than that of the native cells. This may be due to the environmental interference caused by the immobilization and/or the aftereffect of osmotic shock at the time of immobilization.

Thus, using the effectiveness factor is useful in the investigation of the reaction rates of immobilized microorganisms if we carefully adopt the effective diffusivity in gel.

4.3 Overall Reaction Rate of Immobilized Growing Microorganisms

Behavior of the immobilized growing microorganism is another important topic to be studied. In this case, the cells grow while they metabolize substrates. Thus the cell concentration is not homogeneous, and a heterogeneous concentration distribution is observed. Usually a higher cell concentration is observed near the surface of particles since the substrate concentration is high in this region.

Here, behavior of the immobilized growing *S. cerevisiae* was studied. The immobilized yeast particles were prepared and placed in the medium shown in Table 4 at 303 K. After 42 h growth the cell concentration profile in the particles was found to be stationary. Therefore, the particles were collected after the 42 h growth period and used for the overall reaction rate measurement. The reaction

Table 4. Cultivation medium for immobilized growing *S. cerevisiae*

Glucose	100 kg
$MgSO_4 \cdot 7\,H_2O$	0.5 kg
KH_2PO_4	1 kg
$CaCl_2 \cdot 2\,H_2O$	7.35 kg
$(NH_4)_2SO_4$	5 kg
Yeast extract	1.5 kg
H_2O	1 m^3

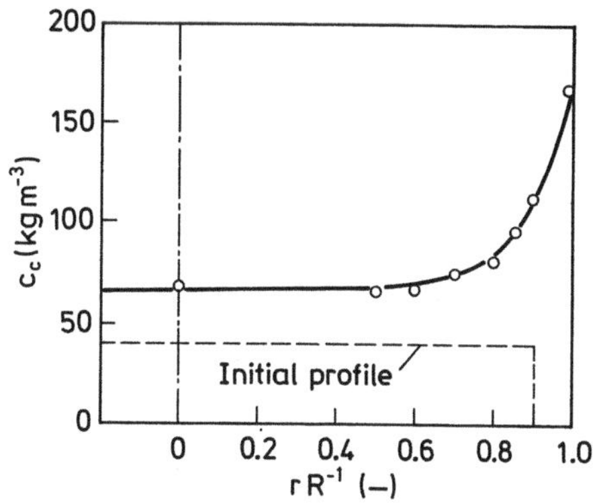

Fig. 10. Concentration profile of immobilized growing *S. cerevisiae* in Ca-alginate gel

rate was measured in the same substrate solution as that used in the measurement of reaction rate of native yeast.

The cell concentration profile was measured by combining the data from electron microscopic studies and from experiments using gel dissolution in 10% Na polyphosphate. From the latter method, the average cell concentration could be obtained. The profile obtained is shown in Fig. 10. It was found that the cell concentration was very high near the surface. The concentration is almost homogeneous and low in the core region ($rR^{-1} < 0.7$).

The overall reaction rate was measured by the same procedure. It is shown in the form of the L-B plot in Fig. 11 by closed circles. The solid line in the figure shows the result by numerical calculation [19] using the previous effective diffusivity relationship, the cell concentration profile and the reaction rate of native cells. The experimental data again fits the calculated data well. For comparison the overall reaction rate of the resting cells (before the growth) is shown in the figure by open circles together with the calculated data shown by the dotted line. The difference between the solid and dotted curves represents the contribution of cell growth. For small substrate concentration the overall reaction rate becomes smaller when the cells grow. This is probably caused by the effect of the diffusion resistance due to cell body.

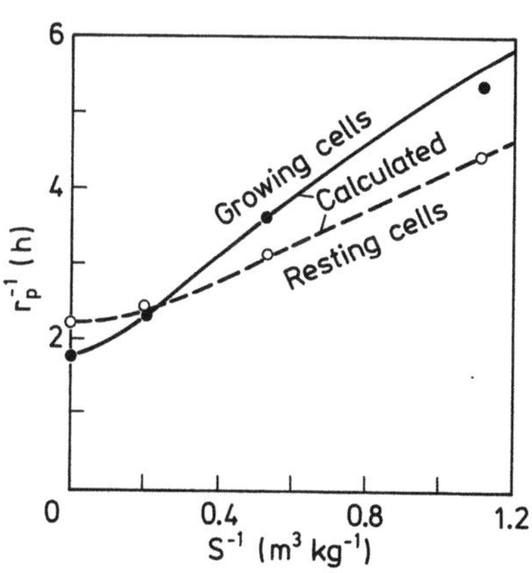

Fig. 11. Lineweaver-Burk plot of the overall reaction rate by immobilized growing S. *cerevisiae*

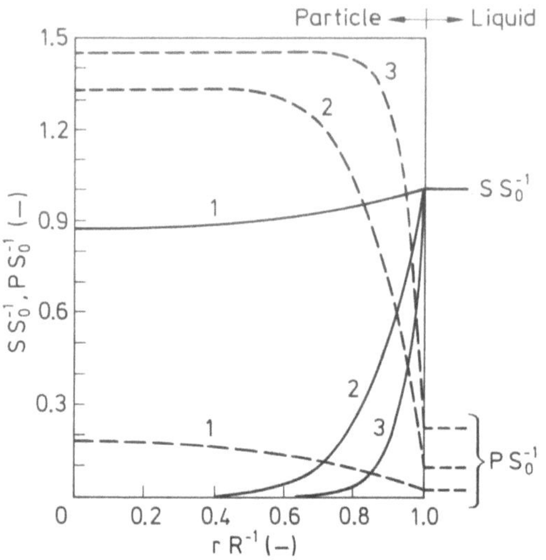

Fig. 12. Concentration profiles of the substrate and product in the immobilized growing yeast particle. *1* S_0 = 98.0 kg m^{-3}; *2* S_0 = 4.77 kg m^{-3}; *3* S_0 = 0.89 kg m^{-3}. *Solid lines* denote reduced substrate concentrations, and *broken lines* denote reduced product concentrations

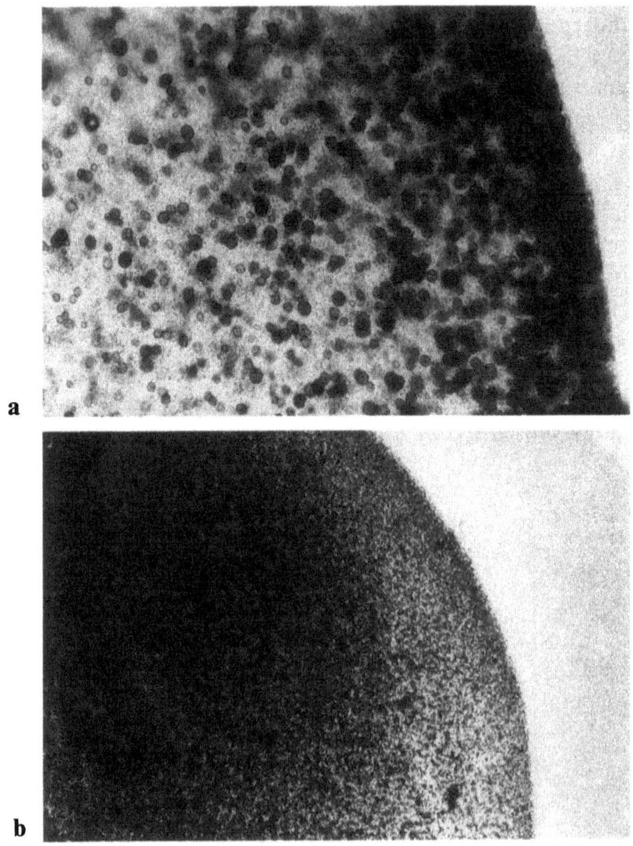

Fig. 13a, b. Cell concentration distributions of the growing *Z. mobilis* after the 43 h growth at dilution rate of 0.5 h^{-1}. **a** Normal profile, S_0 = 52 kg m^{-3}. **b** The cells are populated in the center, S_0 = 239 kg m^{-3}

The calculation was also tried [34] using a stepwise bimodal cell concentration profile deduced from Fig. 10 i.e. the populated region near the surface and the sparce core region. The result was almost identical to the numerical calculation. Therefore, this approach may become one of simpler analysis on the immobilized growing cells. Examples of concentration profiles of the substrate and product are shown in Fig. 12 as a result of the numerical calculation [19].

5. Importance of the Study of Intraparticle Diffusion

The above results successfully show the usefulness of the reaction engineering analysis on the effect of intraparticle diffusion on immobilized microorganisms. The analysis is useful not only to estimate the conversion or selectivity but also to study metabolic activity of the immobilized microorganism. For the case of *S. cerevisiae*, the activity did not change by immobilization, but the activity of *Z. mobilis* decreased due to immobilization. This kind of effect includes salient features of immobilization, which would be worthwhile investigating further.

The cell concentration profile is another interesting feature which could be studied further. There are several causes affecting the profile of cell distribution. Figure 13(a) is an example of normal distribution. The cell concentration is high near the surface and becomes lower toward the center. Figure 13(b) shows an unusual profile. The cell concentration is high at the central core region due to the high substrate concentration causing inhibition of cell growth.

6. Conclusion

Reaction behavior of immobilized microorganisms was investigated by the use of *Saccharomyces cerevisiae* and *Zymomonas mobilis* immobilized by Ca-alginate gel. Both immobilized resting and growing microorganisms were studied. Intraparticle diffusivity was expressed by a function of the cell concentration using the random pore model of Wakao and Smith.

The overall reaction rate was estimated using the effective diffusivity relation, the cell concentration profile and the reaction rate of native microorganisms. The inherent reaction rate of immobilized *S. cerevisiae* was about the same as that of native *S. cerevisiae*. However, the inherent reaction rate of *Z. mobilis* was about 20% lower than that of the native cell. The analysis of the activity of immobilized microorganisms from the metabolic viewpoint will be important in the development of application of immobilized microorganism. For this purpose, the reaction engineering approach will play an important role in the prediction of behavior of immobilized biocatalysts.

7. Nomenclature

a	$m^3\ kg^{-1}$	parameter in the ethanol inhibition relationship
c	$kg\ m^{-3}$	concentration of the diffusing material
$\bar{c}$	$kg\ m^{-3}$	average concentration of the diffusing material

c_c	$kg\,m^{-3}$	cell concentration in gel particles, dry basis
c_g	%	gel concentration
D_0	$m^2\,s^{-1}$	diffusivity in water
D_e	$m^2\,s^{-1}$	effective diffusivity in the cell-containing gel
D_g	$m^2\,s^{-1}$	effective diffusivity in the cell-free gel
E	—	effectiveness factor
E_0	—	effectiveness factor for zero order reaction
E_1	—	effectiveness factor for first order reaction
h	—	modulus, $(R/3)\,(V_{max}/2D_eK_m)^{1/2}$
K_m	$kg\,m^{-3}$	Michaelis constant
k_1	—	ratio of diffusivity in gel to that in water
k_2	$m^3\,kg^{-1}$	constant in Eq. (2), dry cell basis
m	—	generalized modulus, $hS(K_m + S)^{-1}\,[SK_m^{-1}\ln(1 + SK_m^{-1})]^{-0.5}$
P	$kg\,m^{-3}$	product concentration
R	m	particle radius
r	m	radial position
r_c	h^{-1}	reaction rate of native cell, weight of EtOH per h per unit weight of dry cells
r_p	h^{-1}	reaction rate of immobilized cell, based on the weight of the dry cell
S	$kg\,m^{-3}$	substrate concentration
S_0	$kg\,m^{-3}$	substrate concentration at particle surface
t	s	time
u	$kg\,m^{-2}$	cr (Eq. (4))
V_{max}	h^{-1}	maximum reaction rate in the Michaelis Menten equation, weight of product per h per unit weight of dry cells
ε	—	volume fraction of the gel unoccupied by cells

8. References

1. Thiele, E. W.: Ind. Eng. Chem. *31*, 916 (1939)
2. Wicke, E., Brotz, W.: Chem. Ing. Tech. *21*, 219 (1949)
3. Satterfield, C. N.: Mass Transfer in Heterogeneous Catalysis, Huntington, R. E. Krieger Publ. 1981
4. Bischoff, B.: AIChE J. *11*, 351 (1965)
5. Moo-Young, M., Kobayashi, T.: Can. J. Chem. Eng. *50*, 162 (1972)
6. Kobayashi, T., Moo-Young, M.: Biotech. Bioeng. *15*. 47 (1973)
7. Kobayashi, T., Ohmiya, K., Shimizu, S.: J. Ferment. Technol. *54*, 260 (1976)
8. Yamane, T., Araki, S., Sada, E.: J. Ferment. Technol. *59*, 367 (1981)
9. Nakanishi, K., Adachi, S., Yamamoto, S., Matsuno, R., Tanaka, A., Kamikubo, T.: Agric. Biol. Chem. *41*, 2455 (1977)
10. Hiemstra, H., Dijkhuizen, L., Harder, W.: Eur. J. Appl. Microbiol. Biotechnol. *18*, 189 (1983)
11. Onuma, M., Onuma, T., Umita, T., Aizawa, J.: Biotech. Bioeng., *27*, 1533 (1985)
12. Adlercreutz, P.: Biotech. Bioeng. *28*, 223 (1986)
13. Klein, J., Schara, P.: Appl. Biochem. Biotechnol. *6*, 91 (1981)
14. Klein, J., Manecke, G.: New Development in the Preparation and Characterization of Polymer-bound Biocatalysts, in: Enzyme Engineering 6 (Chibata, I., Fukui, S., Wingard, L. B. eds.) p. 181, New York, Plenum 1982
15. Radovich, J. M.: Biotechnol. Advan. *3*, 1 (1985)

16. Radovich, J. M.: Enzyme Microbiol. Technol. *7*, 2 (1985)
17. Karel, S. F.: Chem. Eng. Sci. *40*, 1321 (1985)
18. Furusaki, S., Seki, M.: J. Chem. Eng. Japan *18*, 389 (1985)
19. Seki, M., Furusaki, S.: J. Chem. Eng. Japan *18*, 461 (1985)
20. Sakaki, K., Nozawa, T., Furusaki, S.: Biotech. Bioeng., *31*, 603 (1988)
21. Gurney, H. P., Lurie, J.: Ind. Eng. Chem. *15*, 1170 (1923)
22. Wakao, N., Smith, J. M.: Chem. Eng. Sci. *17*, 825 (1962)
23. Matsuno, R., Nakanishi, K.: Kagaku Kogaku *43*, 283 (1979)
24. Hannoun, J. M., Stephanopoulos, G.: Biotech. Bioeng. *28*, 829 (1986)
25. Horowitz, S. B., Fenichel, I. R.: J. Phys. Chem. *68*, 3378 (1964)
26. Crank, J., Mathematics of Diffusion, vol. 2, p. 91, Clarendon Press 1975
27. Aiba, S., Shoda, M., Nagatani, M.: Biotech. Bioeng., *10*, 845 (1968)
28. Nagatani, M., Shoda, M., Aiba, S.: J. Ferment. Technol. *46*, 241 (1968)
29. Aiba, S., Shoda, M.: J. Ferment. Technol. *47*, 790 (1969)
30. Novak, M., Strehaiano, P., Moreno, M., Goma, G.: Biotech. Bioeng. *23*, 201 (1981)
31. Luong, J. H. T.: Biotech. Bioeng. *27*, 280 (1985)
32. Hamilton, B. K., Gardner, C. R., Colton, C. K.: AIChE J. *20*, 503 (1974)
33. Aris, R.: Chem. Eng. Sci. *6*, 262 (1957)
34. Seki, M.: M. S. Thesis, Univ. of Tokyo 1984

2.4 The Adsorption of *Trichoderma reesei* C30 Cellulase Protein to Protein Adsorbents: Application to Cellulase Recovery and Reuse *

Jonathan Woodward and Jamie C. Bales[1]

Chemical Technology Division, Oak Ridge National Laboratory,
P.O. Box 2008, Building 4500N, Oak Ridge, Tennessee 37831-6194, USA

Contents

The adsorption of *Trichoderma reesei* C30 cellulase protein to DEAE-, CM-, Cibacron Blue- and concanavalin A-Kieselguhr-based supports (protein adsorbents termed Macrosorb) has been measured. The DEAE- and CM-Macrosorb possessed the greatest capacity for cellulase protein that was unaffected by 0.5 M glucose and minimally affected by 10% (v/v) ethanol. Adsorption of protein to DEAE-Macrosorb was measured at pH 3.5, 5.0, 6.0, and 7.0. At pH 3.5 and 7.0, 25 and 76% of the protein contacted with the adsorbent was bound, respectively. Cellulase protein adsorbed on DEAE-Macrosorb at pH 7.0 was eluted by shaking the support in 10 mM sodium acetate buffer, pH 3.5. The eluted protein was used together with an immobilized β-glucosidase to hydrolyze microcrystalline cellulose (Avicel) to glucose. At the completion of hydrolysis, it was demonstrated that soluble protein could be readsorbed back onto the support and reused for further cellulose hydrolysis. Cellulase protein adsorbed on undigested cellulose could be partially desorbed by suspending the residue at pH 7.0 and recovered by addition of DEAE-Macrosorb.

* Research sponsored by the Office of Basic Energy Sciences, U.S. Department of Energy, under contract DE-AC05-84OR21400 with Martin Marietta Energy System, Inc.

1 Oak Ridge Associated Universities summer research participant from Hanover College, Indiana.

Bioproducts and Bioprocesses
Editors: Fiechter, Okada, Tanner

1. Introduction

The enzymatic hydrolysis of cellulose could become a more economical process if the enzyme could be recovered from the reaction mixture in active form and reused several times. This could be achieved if an immobilized cellulase were used to hydrolyze lignocellulosic substrates. However, the use of an immobilized enzyme to catalyze the hydrolysis of an insoluble substrate does not appear to be logical, because effective interaction between enzyme and substrate would be greatly impaired by the enzyme's immobility.

There are, however, some reports describing the use of immobilized cellulase to hydrolyze insoluble cellulose; for example, Karube et al. [1] immobilized cellulase in a collagen fibril matrix. Although details on the recovery of activity after immobilization were not reported, the immobilized enzyme (in a fluidized-bed reactor) reportedly hydrolyzed insoluble microcrystalline cellulose (Avicel, 0.33% w/v) circulating through the bed. Greater than 80% hydrolysis of the substrate was achieved after 160 h at 30 °C. Furthermore, no leakage of the enzyme was reported, which suggests that intimate association between the immobilized enzyme and insoluble cellulose had to occur somehow, even during constant circulation of substrate. Woodward and Zachry [2] immobilized the cellulase complex of *T. reesei* C 30 by covalent coupling to cyanogen bromide-activated Sepharose after aminoalkylation of the carbohydrate side chains of the enzyme with ethylenediamine. However, more than 90% of the Avicelase activity was lost when the cellulase was immobilized by this method. Similar results have been obtained by Fadda et al. [3]. Rogalski et al. [4] attached cellulase from *Aspergillus terreus* to controlled pore glass and claimed 72% retention of activity. Some of the data on immobilized cellulases provided by these authors are summarized in Table 1.

When using an immobilized cellulase, it is essential to ensure that leakage of enzyme from the support does not occur; otherwise, hydrolysis will obviously be caused by the soluble enzyme. This has been shown to be true for cellulase immobilized by adsorption onto concanavalin A attached to an inorganic Kieselguhr-based support [5]. In these experiments, the substrate, microcrystalline cellulose

Table 1. Immobilized cellulase preparations

Support	Substrate (% w/v)	Leakage	Recovery of activity (%)	Hydrolysis of substrate (%)	References
Collagen	Avicel (0.33)	None	n.d.[a]	>80 (160 h, 30 °C)	Karube et al., 1977
CNBr-Sepharose	Avicel (2)	None	8	~6 (28 h, 50 °C)	Woodward & Zachry, 1982
CNBr-Sepharose	Alfa-Alfa (3)	n.d.	20	~60 (6 d, optimal temperature)	Fadda et al., 1984
Controlled-pore glass	Avicel (1)	n.d.	72	~25 (72 h, 50 °C)	Rogalski et al., 1985

[a] n.d. = not determined

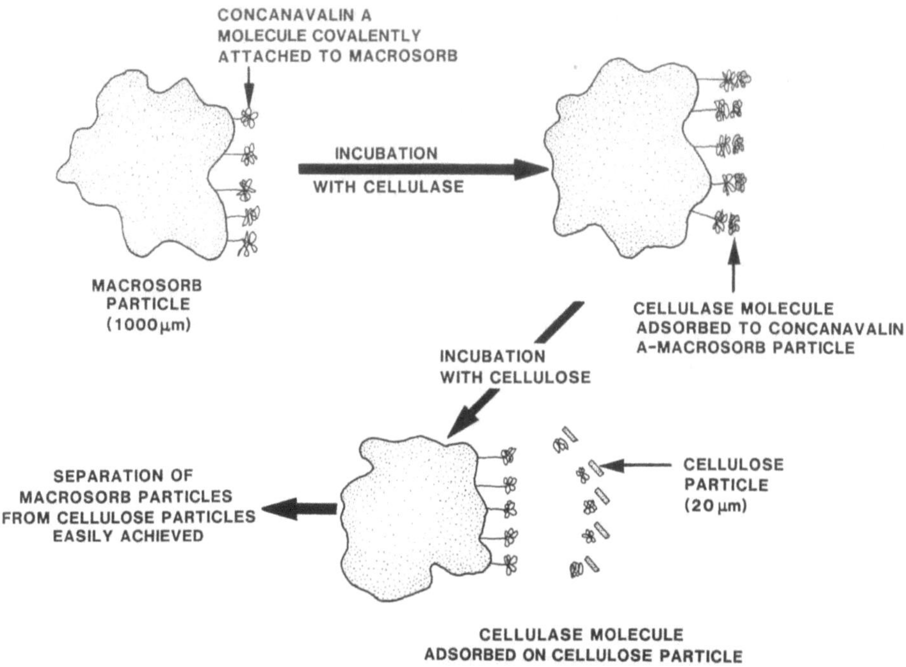

Fig. 1. Desorption of cellulase adsorbed on concanavalin A-Macrosorb in the presence of cellulose

(Avicel, PH 105, particle size 20 μm), was easily separated from the enzyme support whose particle size ranges from 160–1000 μm. Thus, when the immobilized cellulase was mixed with the substrate, hydrolysis of cellulose to glucose commenced and continued even after removal of the support from the reaction mixture. This phenomenon is illustrated in Fig. 1. In such a system containing an immobilized cellulase and cellulose, hydrolysis of the latter proceeds by virtue of the fact that desorption of the enzyme from the support occurs followed by adsorption on the substrate. During the hydrolysis, released enzyme could be readsorbed back onto the support and ultimately used again. However, it would not be necessary for the support to be present in the reaction mixture during hydrolysis. In order for this system to be practical, it is essential that readsorption of released enzyme would occur in the presence of the products of hydrolysis, as well as in the presence of bioproducts; for example, when saccharification and transformation are carried out simultaneously. In a practical system, however, it is unlikely that complete hydrolysis of a cellulosic substrate will occur, and sometimes as much as 50% of the original enzyme used in a saccharification will remain in the adsorbed state, even when there is a high percentage of conversion of the substrate. Elution of the cellulase bound to the undigested cellulose is necessary if it is also to be recovered and reused. Of the three major active components associated with the cellulase enzyme complex, β-glucosidase is relatively easy to recover because this enzyme is not adsorbed by cellulose to any great extent. In fact, β-glucosidase can be used in an immobilized form since its substrate, cellobiose, is soluble [6]. The elution of cellulase from undigested cellulose might be achieved using a protein adsorbent, providing that adsorbent has a higher affinity for the enzyme than does the residual cellulose or lignocellulose. However, recent studies [7,8] have shown that a high percentage of the cellobiohydrolase and glucanohydrolase components which are adsorbed to lignocellulosic residues after hydrolysis can be eluted by simple washing of the residues in phosphate buffer, pH 7.0.

In order to develop a method for cellulase recovery, the adsorption of cellulase protein to four macroporous Kieselguhr-based supports (Macrosorb) in the absence and presence of glucose and ethanol has been measured. It has been shown also that cellulase protein immobilized on DEAE-Macrosorb can be used to hydrolyze cellulose after desorption from this support, recovered by the same support, and reused. Such adsorbents may find use in cellulase recovery from reaction and media.

2. Materials and Methods

2.1 Enzyme and Enzyme Support Sources

Trichoderma reesei C30 cellulase was a generous gift from Dr. Mary Mandels, U.S. Army Natick R & D Center, Natick, Massachusetts, U.S.A. A commercial preparation of β-glucosidase (cellobiase 250L) was kindly provided by Novo Laboratories, Inc. (Wilton, Connecticut, U.S.A.). A variety of inorganic macroporous granules (Macrosorb) fabricated from Kieselguhr was kindly donated by Sterling Organics, Rensselaer, New York, U.S.A. These included: (1) Macrosorbs KAX · DEAE and KAX · CM ion exchangers in which cross-linked derivatized agarose is retained in the granules, (2) Macrosorb KA · Cibacron Blue on affinity dye coupled to agarose retained in the granules, and (3) Macrosorb K(GA) in which the granule surface is activated with glutaraldehyde.

These materials were chosen as supports for cellulase because of the ease by which they can be separated from the substrate used in these studies — microcrystalline cellulose (Avicel PH 105, FMC Corporation, Philadelphia, Pennsylvania, U.S.A.). As has been noted in the introduction, the Macrosorb particles, being much denser than the cellulosic particles, settle rapidly to the bottom of a reaction vessel and, therefore, simple decantation can be used to separate substrate and support. Separation of the latter two is essential to prove that desorption of enzyme occurs when the immobilized enzyme and substrate are mixed together. Also, after hydrolysis of cellulose is complete, the enzyme, having been readsorbed back on the support, would be easily recoverable.

2.2 Immobilization of Cellulase on the Macrosorb Supports

Macrosorbs KAX · DEAE, KAX · CM, and KA · Cibacron Blue were washed thoroughly with 50 mM sodium acetate buffer, pH 5.0. Each support (1 g) was shaken in a rotary shaking water bath at 30 °C for 30 min with 20 mg of cellulase

(≡27% protein) in 10 ml of 50 mM sodium acetate buffer, pH 5.0. The supernatant was assayed for protein and the amount of cellulase protein adsorbed to the substrate extrapolated. Macrosorb K(GA) was "activated" with concanavalin A prior to its use as an adsorbent for cellulase. Concanavalin A (200 mg) was dissolved in 200 ml of 0,1 M sodium phosphate buffer, pH 8.0, containing 1 mM $CaCl_2$, 1 mM $MgCl_2$, 0,15 M NaCl, and 0.2 M glucose and mixed (by gentle rolling) with ~10 g wet weight of Macrosorb K(GA) for 1 h at 23 °C. The supernatant was decanted, and the amount of concanavalin A covalently bound to the support was determined by measuring the absorbance at 280 nm. The amount of concanavalin A bound was 2–3 mg g^{-1} Macrosorb. Excess active groups were removed by washing the concanavalin A-Macrosorb (CAM) complex with 200 ml of 0,1 M glycine-NaOH buffer, pH 9.0, containing 0.72 g of mannitol and 0.23 g sodium cyanoborohydride for 30 min at 23 °C. The CAM complex was then washed thoroughly in 50 mM sodium acetate buffer, pH 5.0, containing 1 mM $MgCl_2$, 1 mM $MnCl_2$, and 1 mM $CaCl_2$. Adsorption of cellulase protein to CAM was carried out as described previously for the other Macrosorb supports.

2.3 Immobilization of β-Glucosidase onto CAM

Concanavalin A-Macrosorb (20 g) was mixed with 200 ml of β-glucosidase (original enzyme diluted by 50) in 50 mM sodium acetate buffer, pH 6.0, containing 1 mM $MgCl_2$, 1 mM $CaCl_2$, and 0.15 M NaCl, for 30 min at 23 °C. The supernatant was decanted, diluted four times with 50 mM sodium acetate buffer, pH 5.0, and assayed for activity at pH 5.0, 30 °C, with 10 mM cellobiose as the substrate [6]. The Macrosorb was washed thoroughly in 50 mM sodium acetate buffer, pH 5.0, containing 1 mM $MgCl_2$, 1 mM $MnCl_2$, and 1 mM $CaCl_2$, and its β-glucosidase activity was determined by incubating 1 g of Macrosorb together with 10 mM cellobiose in the pH 5.0 buffer (plus ions) and shaking the mixture at 30 °C for 30 min. Glucose production was monitored.

2.4 Saccharification of Microcrystalline Cellulose

Microcrystalline cellulose (Avicel, PH 105) from FMC Corporation, Philadelphia, Pennsylvania, was enzymatically hydrolyzed at 30 °C in a shaking water bath. The substrate (100 mg) was present in a 20.0 ml volume containing 1 g of immobilized β-glucosidase (~1.0 unit) and a known quantity of cellulase protein in soluble or immobilized form. The precise conditions of these experiments are described in the legends to the figures or tables. During the course of hydrolysis, glucose production was monitored, as was the amount of soluble protein in the medium.

2.5 Analytical Procedures

Protein was assayed using the Coomassie Blue reagent (Bio-Rad, Richmond, California). Glucose was measured using the hexokinase assay reagent (Sigma, St. Louis, Missouri). One unit of β-glucosidase activity is defined as that amount of enzyme required to generate 1 μmol of glucose (from cellobiose) per minute at 30 °C.

3. Results and Discussion

3.1 Immobilization of β-Glucosidase on CAM

The β-glucosidase was immobilized onto CAM in several experiments, and the range of activity obtained was 0.81–1.33 units g^{-1} CAM. Hydrolysis of cellulose (Avicel) was carried out in the presence and absence of the immobilized β-glucosidase, and the results indicated that the presence of the immobilized enzyme considerably enhanced the extent of hydrolysis (Fig. 2), which is in agreement with the work of others [9]. In this experiment, 1.25 mg of cellulase protein was incubated with the substrate, and after 72 h, 0.8 mg was determined to be soluble, while the remainder (0.45 mg) was still bound to the undigested cellulosic residue. In the concept for cellulase recovery, soluble enzyme would be recovered by adsorption onto an appropriate insoluble support that is easily separated from the reaction mixture. Ideally, such a support would also effect the desorption of protein bound to residual cellulose.

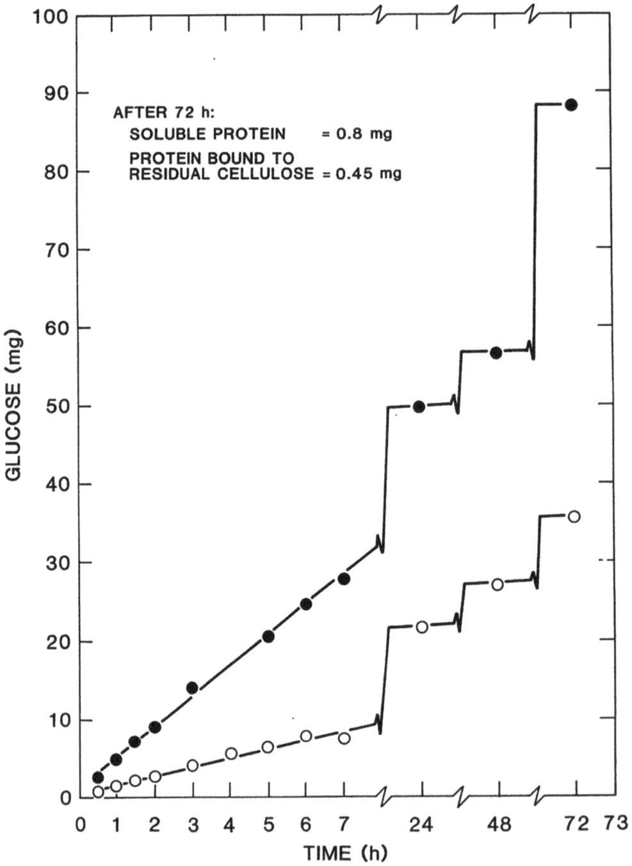

Fig. 2. Hydrolysis of Avicel by *T. reesei* C30 cellulase. The reaction mixture contained 100 mg Avicel and 1.25 mg of cellulase protein plus (●——●) or minus (○——○) 1 g of the Macrosorb-β-glucosidase complex, all in 10 ml, 50 mM sodium acetate buffer, pH 5.0, and was shaken at 30 °C

3.2 Immobilization of Cellulase Protein on Macrosorb Supports

Adsorption of cellulase protein to the four different inorganic Macrosorb supports (see Materials and Methods) was measured in the absence and presence of glucose (0.5 M) or thenaol (10% v/v). Clearly, it is important that recovery of cellulase protein occurs in the presence of such concentrations of metabolites typically encountered in a cellulose saccharification or ethanol production by yeast. The data in Table 2 show that 1 g of DEAE- or CM-Macrosorb possesses a greater capacity to adsorb cellulase protein than either Cibacron Blue or concanavalin A-Macrosorb. Furthermore, they have the same capacity for cellulase protein in the presence of 0.5 M glucose, and this capacity is only slightly reduced in the presence of 10% (v/v) ethanol.

Table 2. Adsorption of cellulase protein by four macrosorb preparations[a]

Additive	DEAE-macrosorb	CM-macrosorb	Cibacron Blue-macrosorb	Concanavalin A-macrosorb
None	46	47	28	29
Glucose (0.5 M)	48	47	17	21
Ethanol 10% (V/V)	39	41	10	14

[a] Values given are the percentages of cellulase protein (i.e., 5.4 mg) initially contacted with the support that is adsorbed to the macrosorb preparation. Adsorption conducted in 50 mM sodium acetate buffer, pH 5.0

3.3 Hydrolysis of Avicel by the Concanavalin A-Macrosorb (CAM) Complex

It has been shown previously [2)] that cellulase adsorbed on concanavalin A-Sepharose (CAS) is as active toward Avicel as is the soluble enzyme. Since CAS and Avicel are very difficult to separate, it could not be established that the hydrolysis of Avicel was actually being catalyzed by enzyme desorbed from the support rather than by a truly immobilized enzyme. This was confirmed by using CAM as a vehicle for cellulase immobilization. The CAM, to which cellulase was adsorbed, was mixed with Avicel and then easily separated from the substrate. Although the source of the enzyme was removed, hydrolysis of the substrate proceeded, indicating that desorption of cellulase protein from the support and adsorption onto Avicel had occurred during mixing (Fig. 3). Using the immobilized or soluble enzyme, the conversion of cellulose to glucose was 63% and 94%, respectively. The difference could be partially attributed to the fact that β-glucosidase is poorly adsorbed on concanavalin A at pH 5.0. Although it is tempting to speculate that Avicel has a greater affinity for cellulase than does CAM, the mechanism whereby the enzyme desorbs from the support in the presence of Avicel is unknown.

Since concanavalin A is the adsorbing species and also adsorbs glucose, it is likely that CAM would not prove useful as a means to recover cellulase from cellulosic hydrolysates because the clucose present would compete for the protein-binding sites

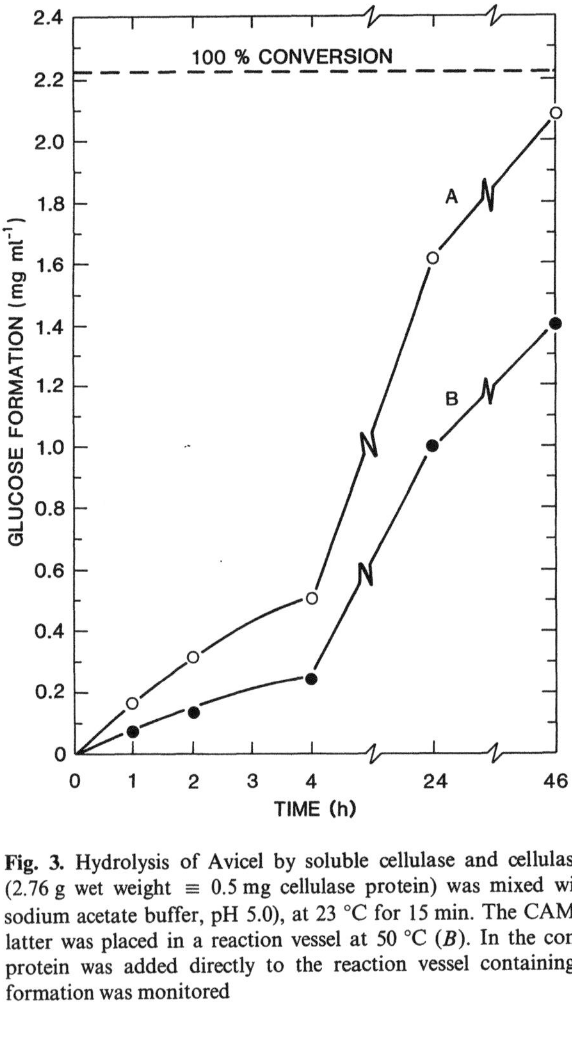

Fig. 3. Hydrolysis of Avicel by soluble cellulase and cellulase desorbed from CAM. The CAM (2.76 g wet weight ≡ 0.5 mg cellulase protein) was mixed with Avicel (10 mg in 5.0 ml, 50 mM sodium acetate buffer, pH 5.0), at 23 °C for 15 min. The CAM and Avicel were separated, and the latter was placed in a reaction vessel at 50 °C (*B*). In the control (*A*), 0.5 mg of soluble cellulase protein was added directly to the reaction vessel containing Avicel (10 mg) at 50 °C. Glucose formation was monitored

on concanavalin A. In fact, no evidence has been obtained to indicate that CAM will readsorb cellulase protein from a cellulosic hydrolysate.

3.4 Hydrolysis of Avicel by the DEAE-Macrosorb-Cellulase (DMC) Complex

When DEAE-Macrosorb, to which 3.7 mg of cellulase protein was adsorbed, was incubated with Avicel, virtually no hydrolysis of substrate occurred, compared with the results obtained when a similar amount of soluble protein was used (Fig. 4). It was assumed that desorption of cellulase protein from DEAE-Macrosorb did not occur in the presence of Avicel, unlike the case observed in the presence of CAM. To test this hypothesis, NaCl was added to each reaction mixture after 2 h to give a final concentration of 0.5 M. Whereas the addition of NaCl had no effect

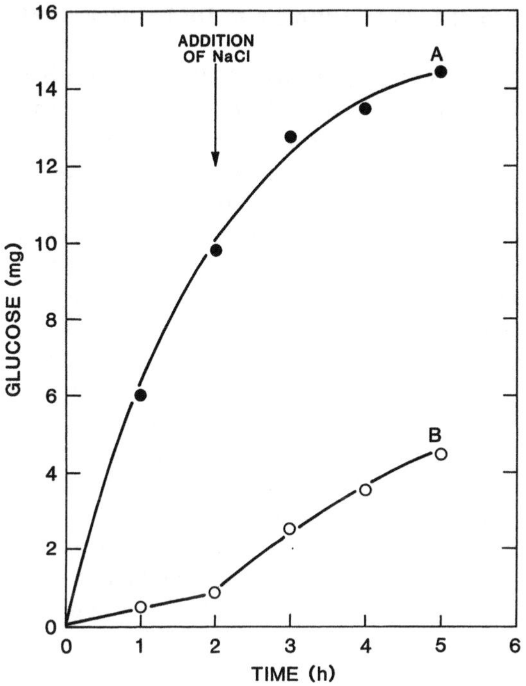

Fig. 4. Hydrolysis of Avicel by soluble cellulase and cellulase adsorbed on DEAE-Macrosorb. One gram of DEAE-Macrosorb containing 3.7 mg of cellulase protein was incubated with Avicel under the conditions described in Materials and Methods. In the control (*A*), 3.7 mg of soluble cellulase protein was added to the reactor. Glucose formation was monitored. The *arrow* indicates the time at which the reaction media were made 0.5 M by the addition of NaCl

on the rate of cellulose hydrolysis by soluble cellulase, there was an increase in the rate of glucose production in the reaction mixture containing the DEAE-Macrosorb-bound enzyme. It can be concluded from this observation that the increase in the rate of glucose production is caused by salt elution of cellulase protein from the support.

It can be seen from data in Fig. 4 that even after the addition of NaCl to the reactor containing the immobilized enzyme and Avicel, the rate of glucose production is much less than that of the control. This phenomenon could be explained by the fact that several components of the cellulase enzyme complex would not be bound to DEAE-Macrosorb at pH 5.0. These could include endoglucanase II and III and cellobiohydrolase II, with pI values of 5.5, 7.4, and 5.9, respectively, which have been shown to be present in the cellulase complex from the *T. reesei* strain L27 [10]. A lower rate of hydrolysis would be observed because these components might act synergistically with those that do bind at pH 5.0 to DEAE-Macrosorb. For example, synergism between the endoglucanase and cellobiohydrolase components, as well as between the two cellobiohydrolase components present in fungal cellulases, has been demonstrated [11]. At pH 5.0, the cellulase components cellobiohydrolase I (pI 4.2) and endoglucanase I (pI 4.7) would be bound to DEAE-Macrosorb and eluted from the support by 0.5 M NaCl. However, although these components, acting alone, could bring about cellulose hydrolysis, the rate and extent of hydrolysis would not be as great as would be attained in the presence of the other components.

3.5 The Effect of pH on the Adsorption of Cellulase Protein to DEAE-Macrosorb

In view of the fact that cellulase adsorbed on DEAE-Macrosorb does not desorb in the presence of Avicel and because of the fact that not all the cellulase components would be bound at pH 5.0, the effect of pH on the adsorption of cellulase protein to DEAE-Macrosorb was studied (Table 3). In these experiments, the molarity of the buffer used was 10 mM (instead of 50 mM, as was used in the previously described adsorption experiments), and thus a greater percentage of protein was bound at pH 5.0 because of the lower ionic strength of the buffer. It can be seen that even at pH 7.0, not all the cellulase protein was adsorbed, even though, according to the manufacturer, the total protein-binding capacity of DEAE-Macrosorb is ~45 mg per g wet weight Macrosorb. This result may be explained by the presence of cellulase components whose pI values exceed 7.0 in the *T. reesei* C 30 cellulase complex. Shoemaker et al. [10] have described an endoglucanase and a β-glucosidase from *T. reesei* whose pI values are 7.4 and 8.7, respectively. In contrast, the fact that some protein adsorbs to DEAE-Macrosorb even at pH 3.5 suggests that there are cellulase components whose isoelectric points are lower than 3.5 in the cellulase complex. In this regard, Sprey and Lambert [12] have subjected *T. reesei* cellulase to preparative isoelectric focusing and found endoglucanase and cellobiohydrolase fractions whose pI values are less than 3.5. Such fractions would remain bound to DEAE-Macrosorb even at pH 3.5.

Table 3. The effect of pH on the adsorption of cellulase protein to DEAE-macrosorb

pH	Bound protein (mg)	Bound protein (% of total)
3.5	1.35	25
5.0	3.55	66
6.0	4.0	74
7.0	4.1	76

10 mM sodium acetate buffer was used for pH 3.5, 5.0, and 6.0; 10 mM hepes buffer was used for pH 7.0; 5.4 mg cellulase protein was initially contacted with the support

3.6 Recovery of Cellulase Protein from Avicel Hydrolysates and Its Reuse

Based upon the above data obtained with DEAE-Macrosorb, a scheme was devised whereby cellulase adsorbed on this support could be used, recovered from the reaction medium, and reused (Fig. 5). Cellulase was adsorbed on DEAE-Macrosorb at pH 7.0 to give 4.1 mg protein bound per g of support. The support was then shaken for 60 min at 23 °C in 20 ml of 10 mM sodium acetate buffer, pH 3.5, after which time the amount of protein in the decanted supernatant was 3.41 mg (≡ 83% of the initial protein bound to the support). Avicel (100 mg) and 1 g of β-glucosidase-Macrosorb were added and glucose formation monitored. In the control, 3.41 mg of soluble cellulase protein was mixed with the substrate and immobilized

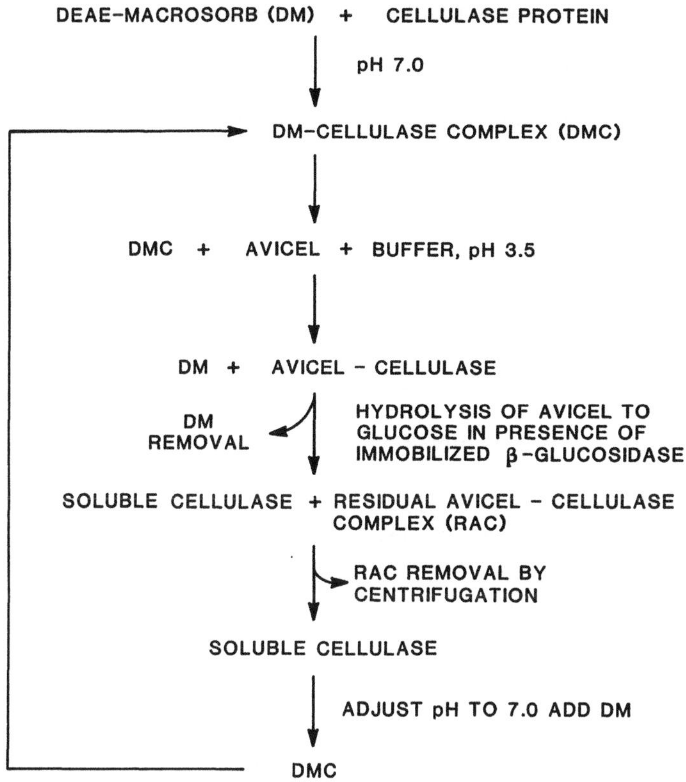

Fig. 5. Scheme for the recovery and reuse of cellulase using DEAE-Macrosorb

Table 4. Avicel hydrolysis by originally immobilized and soluble cellulase protein

Time (h)	Total glucose (mg)	
	Soluble cellulase protein	Originally immobilized cellulase protein
1	7.3	6.3
19	47.2	37.1
43	67.7	54.0
68	73.4	61.6

Avicel (100 mg) was incubated at 30 °C in a shaking water bath in 20 ml of 10 mM, sodium acetate buffer, pH 3.5 containing 1 g β-glucosidase-macrosorb complex and 3.41 mg of cellulase protein

β-glucosidase. It should be noted that cellulose hydrolysis was carried out at pH 3.5 rather than pH 5.0 because the majority of cellulase protein does not bind to DEAE-Macrosorb at pH 3.5. Also, the affinity of *T. reesei* C30 cellulase toward Avicel at pH 3.5 is only 15% lower than that recorded at pH 5.0 [13].

Table 5. Recovery of cellulase protein from cellulosic hydrolysates by readsorption on DEAE-macrosorb

Protein source	Protein in hydrolysate (mg)	Protein readsorbed on 1 g DEAE-macrosorbed (mg)
Soluble cellulase protein (control)	1.6	1.2
Originally immobilized cellulase protein	2.5	1.7

A comparison of the amount of glucose produced by 3.41 mg of cellulase protein that was originally adsorbed on Macrosorb with that produced by the same amount of soluble cellulase protein is shown in Table 4. The lower amount of glucose produced by the originally immobilized enzyme could result from the absence of one or more components present in the soluble cellulase that were not immobilized on DEAE-Macrosorb at pH 7.0. At the end of the hydrolysis, the reaction mixtures were centrifuged, the supernatants adjusted to pH 7.0 with 1 *N* sodium hydroxide, and their protein concentration determined. The DEAE-Macrosorb (1 g), equilibrated with pH 7.0 buffer, was added to each of the supernatants and shaken at 30 °C for 30 min. The protein concentration of the supernatants was redetermined, from which the amount of protein readsorbed to DEAE-Macrosorb was extrapolated. These data, summarized in Table 5, indicated that when the source of cellulase protein was originally soluble or immobilized, 75% and 68% of the hydrolysate protein were recovered, respectively.

Reuse of the recovered cellulase protein was achieved by shaking the support in buffer, pH 3.5, to desorb the protein, decanting the supernatant, and then adding the substrate and 1 g of immobilized β-glucosidase. In control experiments, cellulose hydrolysis was carried out using the same amount of cellulase protein that was desorbed from the support plus the immobilized β-glucosidase. Glucose formation was monitored in both cases. The data in Table 6 show that soluble cellulase protein which has been used to hydrolyze cellulose over a 68-h period can be recovered from solution and reused again.

Of the initial amount of protein used (3.41 mg), approximately 50% was recovered by readsorption to DEAE-Macrosorb (see Table 5). Protein remaining bound on the

Table 6. Reuse of cellulase protein recovered from cellulosic hydrolysates

Protein source	Protein desorbed into solution at pH 3.5 (mg)	Glucose Produced at 68 h (mg)
Soluble cellulase protein (control)	1.0	47.9 (55.4)[a]
Originally immobilized cellulase protein	1.5	45.4 (55.4)[a]

[a] Value in parentheses represents the determined amount of glucose produced by 1.0 and 1.5 mg soluble cellulase protein

cellulosic residue and that which is soluble but not adsorbed account for 26% and 24% of the originally immobilized cellulase protein, respectively. The reason why all of the protein did not readsorb to DEAE-Macrosorb is not known. Since all the protein was originally immobilized on this support, complete readsorption should be expected. It is possible that during the course of hydrolysis some kind of modification occurs to a portion of the protein, preventing its readsorption to the support at pH 7.0.

The data in Table 6 show that the recovered and reused cellulase produced ~45 to 48 mg of glucose from 100 mg of Avicel in 68 h, compared with ~55 mg of glucose produced when the same amount of fresh cellulase protein was used to hydrolyze Avicel. This only represents an approximately 18% reduction in glucose formation by reused cellulase protein and suggests that the unrecovered protein which was bound to the cellulosic residue (0.9 mg, or 26% of originally immobilized cellulose protein) may not be as essential to reclaim as originally thought. It may be more economical to add an amount of fresh enzyme to make up the 18% difference rather than elute cellulase protein from cellulosic residues. However, this remains to be determined.

Although the reasons for the decrease in glucose production by reused enzyme protein compared with that achieved by the same amount of fresh cellulose protein can only be speculated upon at present, two main possibilities may be considered: (1) there may be a lower concentration of a component of cellulase in the recovered enzyme that would result in a lower rate of hydrolysis; and (2) some of the enzyme activity of the reused enzyme may have been lost because of inactivation by shaking [14], especially since a protective agent such as polyethylene glycol was not used during the hydrolysis.

3.7 Partial Recovery of Cellulase Protein Adsorbed on Cellulosic Residue

It has been noted that phosphate buffer at neutral pH is able to effectively desorb cellulase protein which is adsorbed on undigested lignocellulosic residues [7,8]. By combining this step with adsorption to DEAE-Macrosorb, it was shown that cellulase protein adsorbed onto Avicel could be partially recovered. In this experiment, an Avicel residue was recovered from a cellulosic hydrolysate and determined to contain 0.42 mg of protein. After shaking with 10 mM sodium phosphate buffer, pH 7.0, for 30 min at 30 °C, 0.2 mg was solubilized. Recovery of this soluble protein was achieved by its adsorption to DEAE-Macrosorb.

4. Conclusions and Final Discussion

The adsorption of cellulase protein to four different protein adsorbents has been measured. The adsorbents DEAE- and CM-Macrosorb, both ion exchangers, have been shown to have the greatest capacity to bind the protein under the conditions used. However, in this study, only DEAE-Macrosorb was further used to recover cellulase protein from a cellulosic hydrolysate whose pH was adjusted to 7.0. As expected, more protein adsorbs to this support at pH 7.0 compared with results obtained at lower pH values. In this work, reuse of the cellulase protein was achieved by effecting desorption at pH 3.5 and continuing hydrolysis at this pH.

Desorption could also be influenced by increasing the ionic strength of the medium.

It has been shown that the pure microcrystalline cellulose particles (20 μ in diameter) are easily separated from the larger and denser Macrosorb particles. The question as to whether real cellulosic or lignocellulosic substrate particles can also be separated easily from the enzyme support must also be considered. This also seems likely since mechanical pretreatment of cellulosic materials often results in the generation of particles whose sizes are less than those of the Macrosorb particles (160 to 1000 μ). Thus, milling of cotton linters, municipal solid waste, and wood flour can result in particle sizes of 17, 53, and 200 μ, respectively [15]. However, separation of the support from the cellulosic residue may not be necessary if a process is used in which the cellulosic hydrolysate flows through a column of the enzyme adsorbent. In such a system, reuse of the enzyme would be achieved by elution of the protein from the adsorbent together with its return to the reactor for cellulose hydrolysis.

Klyosov [15] has described a process being carried out in the Soviet Union whereby cellulose is enzymatically converted to glucose in a columnar reactor, and released cellulase protein is readsorbed on excess or newly added cellulose. Although the use of cellulose as an adsorbent improved the economy of enzyme utilization, it should be pointed out that the β-glucosidase component of cellulase is poorly adsorbed on cellulose [5] and that the affinity of different endoglucanases from the same [16] and different species [17] differs widely. It is possible, therefore, that key cellulase enzyme components, which poorly adsorb on cellulose but are nevertheless essential for complete hydrolysis of cellulose to glucose, would be lost if cellulose is used as the adsorbent.

In this work, only the adsorption of cellulase protein to protein adsorbents has been measured. Further studies are under way that will provide information on the adsorption of individual cellulase components to supports such as DEAE-Macrosorb and on the stability of the enzyme under the conditions of adsorption. The separation of the cellulase enzyme components based upon differences in their pI values will also lead to an understanding of whether components with pI values less than 3.5 or greater than 7.0 are essential for efficient and complete hydrolysis of cellulose. It may be that such components need not be recovered and reused.

5. Acknowledgements

The secretarial assistance of Ms. Debbie Weaver is gratefully acknowledged. This work was supported by the Office of Basic Energy Sciences, U.S. Department of Energy, under Contract No. DE-ACO5-840R21400 with Martin Marietta Energy Systems, Inc.

6. References

1. Karube, I., Tanaka, S., Shirai, T., and Suzuki, S.: Biotechnol. Bioeng. *19*, 1183 (1977)
2. Woodward, J., and Zachry, G. S.: Enzyme Microb. Technol. *4*, 245 (1982)
3. Fadda, M. B., Dessi, M. R., Maurici, R., Rinaldi, A., and Satta, G.: Appl. Microbial. Biotechnol. *19*, 306 (1984)

4. Rogalski, J., Szczodrak, J., Dawidowicz, A., Ilczuk, Z., and Leonowicz, A.: Enzyme Microb. Technol. *7*, 395 (1985)
5. Woodward, J., Baird, A. R., and Parrish, C.: in Proceedings of World Congress III of Chemical Engineering, Tokyo, Japan *1*, 921 (1986)
6. Lee, J. M., and Woodward, J.: Biotechnol. Bioeng. *25*, 2441 (1983)
7. Sinitsyn, A. P., Bungay, M. L., Clesceri, L. S., and Bungay, H. R.: Appl. Biochem. Biotechnol. *8*, 25 (1983)
8. Deshpande, M. V., and Eriksson, K. E.: Enzyme Microb. Technol. *6*, 338 (1984)
9. Sundstrom, D. W., Klei, H. E., Coughlin, R. W., Biederman, G. J., and Brouwer, C. A.: Biotechnol. Bioeng. *23*, 473 (1985)
10. Shoemaker, S., Watt, K., Tsitovsky, G., and Cox, R.: Biotechnology *1*, 687 (1983)
11. Wood, T. M., and McCrae, S. I.: Biochem. J. *234*, 93 (1986)
12. Sprey, B., and Lambert, C.: FEMS Microbiol. Lett. *23*, 227 (1984)
13. Woodward, J., Whaley, K. S., Zachry, G. S., and Wohlpart, D. L.: Biotechnol. Bioeng. Symp. Ser. *No. 11*, 619 (1981)
14. Reese, E. T.: in Solution Behavior of Surfactants, vol. 2 (Mittal, K. L., and Fendler, E. J., eds.), p. 1487, New York, Plenum (1982)
15. Klyosov, A. A.: Appl. Biochem. Biotechnol. *12*, 249 (1986)
16. Cherroglazov, V. M., Ermolova, O. V., and Klësov, A. A.: Biokhimiya *50*, 1108 (1985)
17. Woodward, J., Marques, H. J., and Picker, C. S.: Prep. Biochem. *16*, 337 (1986)

2.5 The Construction of a Bioreactor for the Production of an Optically Active Compound

Shigeru Kajiwara and Hidekatsu Maeda

Fermentation Research Institute, Agency of Industrial Science & Technology,
Ministry of International Trade & Industry, Higashi 1-1-3, Yatabe, Tsukuba, Ibaraki 305, Japan

Contents

A droplet gel-entrapping method used for enzyme immobilization was improved to simplify the procedure and to increase the enzyme stability. This immobilization technique is suitable for coupled enzyme reactions requiring cofactors. Leucine dehydrogenase (LeuDH) and formate dehydrogenase (FDH) were freeze-dried with bovine serum albumin, dextrin and stabilizers. The freeze-dried enzyme powder was suspended in a methylcellosolve solution containing polyethyleneglycol (# 4000) diacrylate, *N,N'*-methylenebisacrylamide and 2-hydroxyethylacrylate, and the suspension was gelled with initiators. The gel was cut up and the pieces were washed in a buffer to remove the methylcellosolve and the dextrin inside. The maximum conversion ratio for a LeuDH-FDH gel column was determined to be 99.8% by means of the recycling reaction. On long-term operation at 30 °C for leucine production, the initial conversion ratio (7.2%) gradually decreased to 6.6% over the first 10 days. However, the conversion ratio remained almost constant after the 10th day. The effects of flow rate, temperature, pH, and the concentrations of formate, α-ketoisocaproate, ammonium and NAD on the leucine productivity with the gel column were also investigated.

The application of enzymes such as dehydrogenases to asymmetrical reduction processes for the production of chiral compounds has been attempted. In the case of dehydrogenases, co-immobilization with an NADH regenerating enzyme is essential for a more efficient transformation process. The membrane reactor system is well known as a co-immobilization system. However, clogging of the ultrafiltration membrane in the reactor by the enzyme protein occurs and a membrane reactor is well known to be inferior to a packed bed reactor from the viewpoints of the operability and the chemical yield. Therefore, the authors previously tried to develop a simpler and more

Bioproducts and Bioprocesses
Editors: Fiechter, Okada, Tanner

efficient reactor system. The result was a "droplet gel-entrapping method" as a new immobilization technique, in which the coupled enzymes are introduced into small holes formed in the gel and the substrate is diffused into the holes.[1] As the reported method involves a water-oil-water type double emulsion, however, the preparation procedure is very complicated because of the difficulty in maintaining the critical condition of the emulsion. Therefore, we tried to develop a new type of "droplet gel-entrapping method" not involving emulsification, and succeeded in the development of a new method in which a suspension of the enzyme powder in an organic solvent is gelled and the organic solvent is then removed by washing. The immobilization method was also found to make the enzymes more stable. The details are presented in this paper.

1. Materials and Methods

Materials

Yeast formate dehydrogenase, of which the specific activity was 3.1 units mg^{-1}, NAD and NADH were purchased from Boehringer-Mannheim-Yamonouchi Ltd., Tokyo. Leucine dehydrogenase[2,3,4], of which the specific activity was 120 U mg^{-1}, was purchased from Toyobo Co., Ltd., Osaka. Bovine serum albumin (BSA), *N,N'*-methylenebisacrylamide (MBA), 2-hydroxyethylacrylate (HEA), benzoylperoxide, *N,N*-dimethylaniline, methylcellosolve, 3-(cyclohexylamino)propanesulfonic acid (CAPS) and 2-(cyclohexylamino)ethanesulfonic acid (CHES) used were of the highest grade available.

The preparation of polyethyleneglycol(# 4000)diacrylate (PEGDA) was performed as described previously[1].

1.1 Assaying of the Enzyme Activities

The free enzymes were assayed at 30 °C by measuring the absorbance change at 340 nm due to the reduction of NAD by FDH and the oxidation of NADH by LeuDH in thermostated cells[3], respectively. The following substrate solutions were used for the free enzymes: 33.3 mM sodium formate and 1 mM NAD in 0.1 M tris(hydroxymethyl)aminomethane (Tris) — HCl buffer (pH 7.5), in a total volume of 3.0 ml, in the case of FDH; and 10.0 mM sodium α-ketoisocaproate and 0.1 mM NADH in 0.75 M ammonium chloride — NaOH buffer (pH 7.5), in a total volume of 3.0 ml, in the case of LeuDH. The reaction was started by the addition of 20 μl of the enzyme solution. The absorbance change at 340 nm was recorded against time and the initial rate was obtained from the recorded chart.

The immobilized enzymes were assayed by measuring the absorbance change at 340 nm due to the reduction of NAD by FDH and the oxidation of NADH by LeuDH, in packed gel columns at 30 °C, respectively. The following substrate solutions were used for the assays: 1.0 M sodium formate, 10 mM NAD and 10 mM 2-mercaptoethanol in 0.1 M phosphate buffer (pH 7.5) in the case of FDH; and 100 mM sodium α-ketoisocaproate, 10 mM NADH and 10 mM 2-mercaptoethanol in 2.0 M ammonium chloride — NaOH buffer (pH 7.5) in the case of LeuDH. 25 mg of the gel was packed into a column (5 mm i.d. × 70 mm long) and then the substrate solution was passed downward through the column at a flow rate of 15.0 ± 0.5 ml h^{-1}. The temperature of the colum was maintained at 30 °C and the substrate solution was kept at 4 °C. The difference in the amount of NADH between the substrate solution and the effluent was determined with a spectrophotometer 1 h after the start of the reaction. Although the activities of the gel were dependent on the substrate concentration, especially the coenzyme concentration, the maximum activities were observed under the conditions with the substrate concentrations mentioned above. Under the conditions for giving 20% substrate conversion, both activities showed a linear relationship to the gel amount.

1.2 Assays of the Amount of Leucine

The amount of leucine was determined with a Shimadzu LC-6A amino acid analysis system [5]. Samples were applied on a column of Shimadzu ISC-07/S1504, which is a strongly acidic cation-exchange resin, and leucine was eluted with 1.75 M citrate buffer (pH 5.28) at 0.3 ml min^{-1} at 55 °C. The effluent was reacted with *o*-phthalaldehyde at 55 °C and then the resultant fluorescence was quantitated with a Shimadzu RF-530 fluorescence spectrometer. The peak area was integrated with a Shimadzu C-R3A integrator.

1.3 Assays of the Amount of Proteins

Proteins were determined by Lowry's method [6] with bovine serum albumin as a standard.

1.4 Preparation of the LeuDH-FDH Immobilized Enzyme Gel

The enzyme powder was prepared as follows. An enzyme solution containing 80 units of LeuDH, 15 units of FDH, 450 mg of dextrin, 50 mg of BSA, 1.2 mg cysteine, 1.1 mg of dithiothreitol, 0.6 µl 2-mercaptoethanol and 2.6 mg of EDTA in 5 ml of 0.1 M phosphate buffer (pH 7.5) was freeze-dried for 20 h. The enzyme solid was ground into a fine powder with a pestle and mortar, and 510 mg of enzyme powder was obtained. A monomer solution containing 320 mg of PEGDA, 40 mg of MBA and 40 mg of HEA in 1.97 ml of methylcellosolve saturated with nitrogen gas was added to 200 mg of the enzyme powder, which contained 26.6 units of LeuDH activity and 4.95 units of FDH activity, in a sealed bottle. After degassing with an aspirator for 5 min with continuous stirring, 20 µl of methylcellosolve containing 1.0 mg of benzoylperoxide and 10 µl of methylcellosolve containing 1.0 µl of *N*,*N*-dimethylaniline were added to the mixture in that order, and then the mixture was kept at 35 °C under a nitrogen atmosphere. A gel was formed within 35 to 40 min, and the gelation was allowed to further proceed for 5 h at 35 °C. The gel was cut up with a cutter and the pieces were washed in 1000 ml of 0.1 M phosphate buffer (pH 7.5) with continuous stirring for one night to remove methylcellosolve and dextrin from the enzyme powder. After washing, the immobilized LeuDH-FDH gel was collected on a Buchner funnel.

1.5 Effects of Various Conditions on the Leucine Production Rate of the LeuDH-FDH Gel Column

The standard conditions were as follows. 500 mg of the gel was packed into a column (5 mm i.d. × 70 mm long) and then the substrate solution was passed downward through the column at a flow rate of 15.0 ± 0.5 ml h^{-1}. The temperature of the column was maintained at 30 °C. The amount of leucine in the effluent was determined by HPLC. The substrate solution was composed of 10 mM sodium α-ketoisocaproate and 0.3 mM NAD in 0.2 M ammonium-formate buffer (pH 7.5). When the conversion rate was over 20%, the amount of gel in the column was decreased. Other details are given in the legends to Figs. 1 to 3.

1.6 Maximum Conversion of Leucine on the Recycling Reaction in the LeuDH-FDH Gel Column

The substrate solution was composed of 10 mM sodium α-ketoisocaproate, 1 mM NAD, 0.1 mM 2-mercaptoethanol and 0.1 mM EDTA as stabilizers, and 0.01% potassium sorbate as a preservative in 0.2 M ammonium-formate buffer (pH 7.5). 500 mg of the gel was packed into a column (5 mm i.d. × 70 mm long) and then the substrate solution was passed downward through the column at a flow rate of 15.0 ± 0.5 ml h^{-1}. The temperature of the column was kept at 30 °C and the substrate solution was kept at 4 °C. 5 ml of the substrate solution was recycled into the LeuDH-FDH gel column and the amount of leucine in the reaction solution was determined. The recycling rate was 3 times per h.

1.7 Continuous Leucine Production with the LeuDH-FDH Gel Column

The substrate solution was composed of 10 mM sodium α-ketoisocaproate, 100 mM sodium formate, 20 mM ammonium chloride, 0.3 mM NAD, 0.1 mM 2-mercaptoethanol and 0.1 mM EDTA as stabilizers, and 0.01% potassium sorbate as a preservative in 0.05 M Tris-HCl buffer (pH 7.5). 500 mg of the gel was packed into a column (5 mm i.d. × 70 mm long) and then the substrate solution was passed downward through the column at a flow rate of 15.0 ± 0.5 ml h^{-1}. The temperature of the column was kept at 30 °C and other parts of the reactor were maintained at 4 °C. The substrate solution was also kept at 4 °C and replaced with a freshly prepared one every day. The effluent was collected with a fraction collector and the amount of leucine in each fraction was determined by HPLC. Cavitation was prevented by setting a preheating column packed with only glass wool on top of the reaction column and slightly increasing the pressure inside the column by tightening the rubber tube at the flow exit.

2. Results

2.1 Preparation of the LeuDH-FDH Immobilized Enzyme Gel

The authors improved a droplet gel-entrapping method used for enzyme immobilization. At first, leucine and formate dehydrogenases were freeze-dried with BSA, dextrin and stabilizers. The freeze-dried enzyme powder was dispersed in methylcellosolve containing PEGDA, MBA and HEA, and then the mixture was gelled by adding the initiators under continuous stirring. After cutting up the LeuDH-FDH gel, the pieces were washed with buffer to remove the methylcellosolve and the dextrin inside. As a result, many droplets which retained the leucine dehydrogenase, the formate dehydrogenase and BSA were enclosed in the gel, and the diffused substrate was transformed in the droplets. The total wet weight of the gel prepared according to the procedure described under Materials and Methods was 2.33 g. The LeuDH and FDH activities in the washing solution were 4.66 units (17.5% of the total) and 1.35 units (27.2% of the total), respectively. The entrapped LeuDH and FDH activities in the gel were expected to be 9.4 U g^{-1} gel wet and 1.5 U g^{-1} gel, respectively. The LeuDH and FDH activities in the gel were, in fact, 2.5 U g^{-1} wet and 0.79 U g^{-1} wet, respectively. Therefore, the expressed LeuDH and FDH activities in the gel were 26.9% and 51.0% of the expected entrapped activities, respectively. The particle size was determined by the method of Maeda et al. using micrographs [7]. It ranged from about 0.4 mm to about 1.0 mm, and the arithmetic mean was 0.58 mm. The size of enzyme powder particles fixed in the gel was also determined using micrographs. It ranged from about 40 µm to about 90 µm, and the arithmetic mean was 56 µm. The size of the enzyme powder particles is considered to be very similar to that of the droplets in the gel.

2.2 Effects of Various Conditions on the Leucine Production Rate of the LeuDH-FDH Gel Column

The effects of various conditions such as flow rate, temperature, pH, and the concentrations of α-ketoisocaproate, formate, ammonium and NAD on the leucine production rate of the column packed with LeuDH-FDH gel were investigated. The results are shown in Figs. 1 to 3.

The relationship between the flow rate and the leucine production rate is shown in Fig. 1. The results indicate that the leucine production rate of the LeuDH-FDH gel decreased slightly below a flow rate of 7.0 ml h^{-1}. This decrease in the leucine production rate of the gel is considered to be due to film diffusion limitations.

The effect of the temperature of the LeuDH-FDH gel column on the leucine production rate is shown in Fig. 2. The maximum production of leucine was obtained at 51 °C and the leucine production finally stopped at 65 °C. Since one of the coupled enzymes, leucine dehydrogenase from *Bacillus* sp., is stable for 10 min at 60 °C in 50 mM phosphate buffer (pH 6.9), the termination of leucine production at 65 °C is considered to have been due to inactivation of FDH. An Arrhenius plot of the data in Fig. 2 (not shown as a figure) indicated that the inactivation

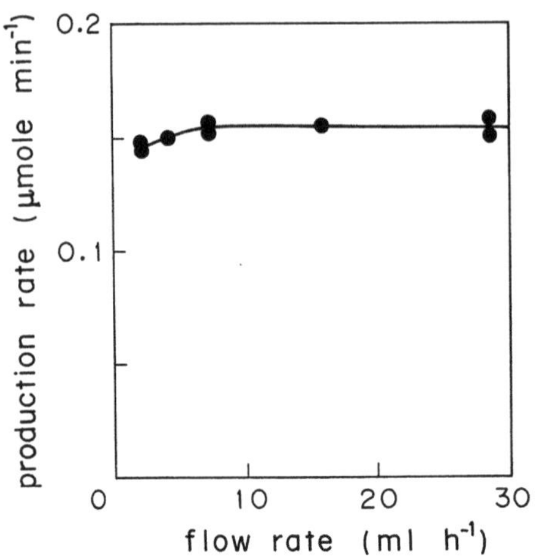

Fig. 1. Effect of the flow rate on the leucine production rate of the LeuDH-FDH gel. The experimental conditions were as given in the text except for the flow rate

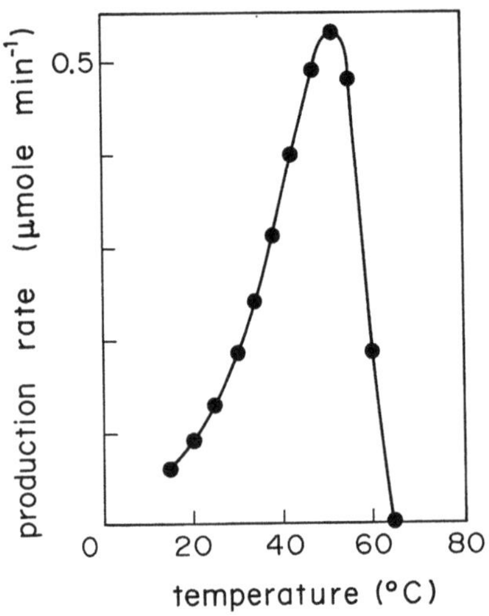

Fig. 2. Effect of temperature on the leucine production rate of the LeuDH-FDH gel. The experimental conditions were as given in the text except for the temperature

appeared over 42 °C. So, all other experiments were performed at 30 °C, because NADH, an important factor for the coupled enzyme reaction, was remained stable.

The effect of the pH of the substrate solution on the leucine production rate of the LeuDH-FDH gel column is shown in Fig. 3. The peak leucine production rate was observed in the pH range of 7.5 to 8.5. The optimum pH and pH profile for the LeuDH-FDH gel were almost identical with those for free FDH.

The relationship between the leucine production rate in the gel column, and the formate, α-ketoisocaproate, ammonium and NAD concentrations were investigated. The leucine production rate of the LeuDH-FDH gel column became constant at over 100 mM formate, 20 mM ammonium and 0.3 mM NAD, respectively.

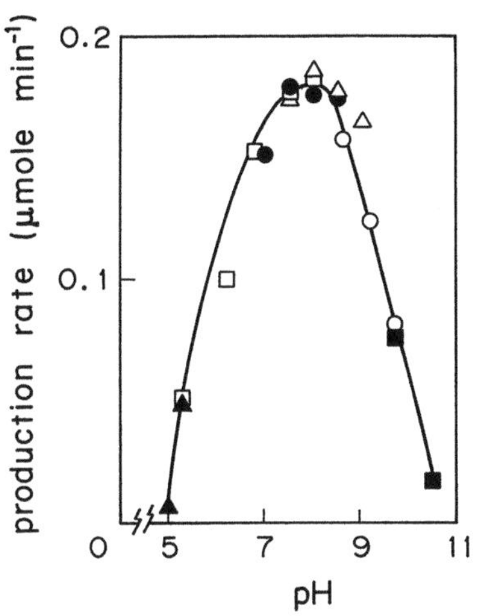

Fig. 3. Effect of pH on the leucine production rate of the LeuDH-FDH gel.
▲ 0.05 M acetate buffer
□ 0.05 M phosphate buffer
△ 0.05 M Tris-HCl buffer
● 0.2 M ammonium-formate buffer
■ 0.05 M CAPS buffer
○ 0.05 M CHES buffer
The experimental conditions were as given in the text except for the composition of the substrate solution and the buffer employed. 0.2 M ammonium chloride and 0.2 M sodium formate were present in all the above kinds of buffers except for the ammonium-formate buffer

2.3 Maximum Conversion of Leucine on the Recycling Reaction in the LeuDH-FDH Gel Column

The recycling reaction for leucine production was carried out using the column packed with LeuDH-FDH gel at 30 °C, as described under Materials and Methods. The conversion ratio increased linearly above 60% conversion and the maximum conversion ratio was observed at 99.8% conversion after 8 h, as shown in Fig. 4. The result was calculated to have been obtained after 24 recyclings.

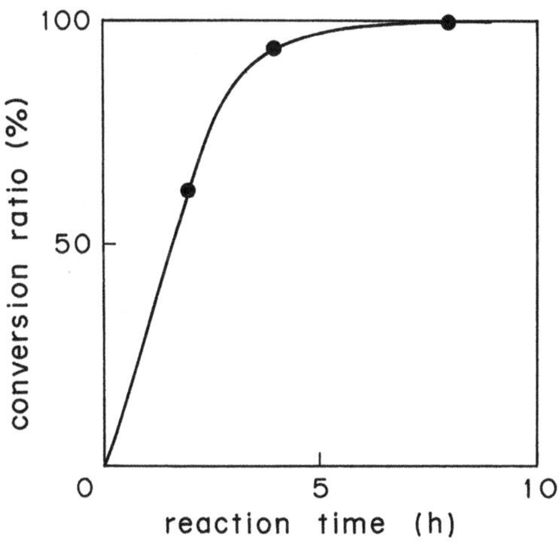

Fig. 4. Maximum conversion of leucine on the recycling reaction in the LeuDH-FDH gel column. The experimental conditions are given in the text

2.4 Continuous Leucine Production with the LeuDH-FDH Gel Column

Continuous leucine production was carried out for 4 weeks using the column packed with LeuDH-FDH gel at 30 °C. Since the decrease in the conversion ratio is generally known to be nearly proportional to the decrease in activity of the gel in the range of

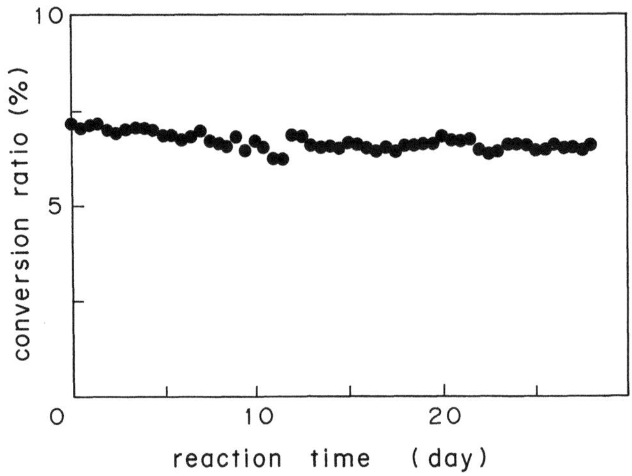

Fig. 5. Continuous leucine production in a column reactor packed with the LeuDH-FDH gel. The experimental conditions are given in the text

0 to 10%, we carefully checked the decrease in conversion ratio. The conversion ratio gradually decreased up to the tenth day at the rate of 0.8% day. As shown in Fig. 5, there was almost no decrease in activity after the tenth day. The LeuDH-FDH gel column maintained about 90% productivity compared with the initial level, even after 28 d. Since the FDH activity is much lower than the LeuDH activity in this coupled enzyme system, the decrease in leucine productivity is considered to reflect the decrease in FDH activity in the LeuDH-FDH gel. The expressed FDH activity calculated from the leucine production rate was 23% of the entrapped FDH activity in the gel at the starting point.

3. Discussion

A droplet gel-entrapping method was found to be suitable for coupled enzyme reactions requiring cofactors, as described above. The following remarkable phenomenon as to the activity expression of the immobilized enzyme gel should be mentioned. 10 mM NAD was required to obtain the maximum FDH activity, when the FDH activity was determined. This means that at least 10 mM NAD is necessary for all the enzyme molecules in the droplets of the gel (3.5 units per 2.33 g wet) to work well. Furthermore, the NAD concentration will increase, if more enzyme is introduced into the droplets. On the other hand, only 0.3 mM NAD was required to obtain the maximum leucine productivity when the coupled enzyme reaction was carried out. This is a much smaller amount than the amount required for the FDH activity assay. Miyawaki et al. have already reported that a very high turnover of NAD was expected theoretically [8], and proved it experimentally with a coupled enzyme system of alcohol dehydrogenase and lactate dehydrogenase [9]. Therefore, the NAD concentration can be greatly decreased in the coupled enzyme reaction system.

Preliminary screening of organic solvents was carried out to find one which did not dissolve the enzyme powder but did dissolve a crosslinkable prepolymer and monomer. Very few of the solvents were suitable for this purpose. Most solvents could not be used due to difficulty in gel formation or considerable leakage of enzymes. Finally, methylcellosolve, *N,N*-dimethylformamide and dioxane were found to be suitable solvents. The gels involving these solvents showed very little leakage and acceptable enzyme activities. The most suitable solvent was methylcellosolve. These three solvents are soluble in both water and organic solvents, and are especially good solvents for dissolving cellulose derivatives, resins and other polymers. On the other hand, the difference between the gel structure formed in an organic solvent and the gel structure in water after exchanging to water will be an interesting subject for future investigation.

Saccharides in the enzyme powder play two roles: firstly, as enzyme stabilizers and, secondly, as a space-maker for droplet formation. When a monosaccharide such as glucose or mannose, and a disaccharide such as sucrose, lactose or maltose were employed in the preparation of the freeze-dried enzyme powder, the enzyme powder could not be ground into fine particles because the powder rapidly absorbed moisture from the air. Mannitol gave a fine enzyme powder but this sugar alcohol slightly dissolved in the methylcellosolve and the gel obtained was soft like slime, and almost all of the enzyme activities leaked out of the gel. Inulin, which is a fructo-oligosaccharide, and dextrin could be applied to this entrapping method.

The physical entrapment of enzymes inside the microcavities of porous fibers has already been reported by Dinelli et al. [10]. Their idea is very similar to ours, however, their immobilization procedure is very different from ours, and the shape of the immobilized enzyme gel is filamentous. On the other hand, Nakamura et al. reported that a soft immobilized enzyme gel was entrapped in a hard gel for immobilization in order to make handling easy, and the kinetic model was proposed [11]. We consider that the kinetic model might be applicable for our kinetic analysis.

An NADH regeneration system involving FDH was used in both the droplet gel-entrapping method previously reported [1] and the method described in this paper. The FDH reaction was considered to be the limiting step of the coupled enzyme reaction. So, the temperature and pH profiles were similar to those for free FDH in both cases. The lowest concentrations of formate and NAD required to obtain the maximum production rate in the case of the previously reported malate dehydrogenase-FDH gel column were 50 mM and 0.1 mM, respectively. However, the lowest concentrations of formate and NAD required to obtain the maximum production rate in the case of the LeuDH-FDH gel column in this study were 100 mM and 0.3 mM, respectively, which are 2–3 times higher. This suggests that the pore size of the gel matrix in the latter gel may be smaller than that of the gel matrix in the former gel. Namely, the increase in the minimum concentration is considered to be due to an increase in the diffusion limitation of the substrate in the gel matrix.

After continuous leucine production for 4 weeks, the productivity of the reactor column packed with the LeuDH-FDH gel was still about 90% of the initial level. The limiting factor for productivity in this column reaction is the FDH activity, and an FDH inactivation rate of 0.83% day in a membrane reactor has already been reported by Wichmann et al. [12] Therefore, the remaining productivity of our LeuDH-FDH column should have been 79%, based on calculation using the results of Wichmann. However, it was 90%. This may be because our new method makes an unstable enzyme such as a dehydrogenase much more stable.

This entrapping method is also expected to be applicable to other kinds of multi-step enzyme reactions.

4. References

1. Kajiwara, S. and Maeda, H.: Biotech. Bioeng. *28*, 1794 (1986)
2. Soda, K., Misono, H., Mori, K. and Sakato, H.: Biochem. Biophys. Res. Commun. *44*, 931 (1971)
3. Ohsima, T., Misono, H. and Soda, K.: J. Biol. Chem. *253*, 5719 (1987)
4. Schütte, H., Hummel, W., Tsai, H. and Kula, M.-R.: App. Microb. Biotech., *22*, 306 (1985)
5. Ishida, Y., Fujita, T. and Asai, K.: J. Chromatogr. *204*, 143 (1981)
6. Lowry, O. H., Rosebrough, N. J., Farr, A. L. and Randall, R. J.: J. Biol. Chem. *193*, 265 (1951)
7. Maeda, H. and Suzuki, H.: Agric. Biol. Chem. *36*, 1581 (1972)
8. Miyawaki, O., Nakamura, K. and Yano, T.: J. Chem. Eng. Japan *15*, 142 (1982)
9. Miyawaki, O., Nakamura, K. and Yano, T.: J. Chem. Eng. Japan *15*, 224 (1982)
10. Dinelli, D., Marconi, W. and Morisi, F.: "Method in Enzymology", Vol. XLIV, (K. Mosbach, ed.) p. 227, Academic Press, New York 1976
11. Nakamura, K., Hibino, K. and Yano, T.: Proc. of PACHEC '83, Vol. IV, p. 197 (The Third Pacific Chemical Engineering Congress) 1983
12. Wichmann, R., Wandrey, C., Bueckmann, A. F. and Kula, M.-R.: Biotech. Bioeng. *23*, 2789 (1981)

2.6 A New Method of Graphical Analysis of Batch Data for Continuous Operation, the "Tangential Method"

Tetsuya Yomo, Itaru Urabe, and Hirosuke Okada

Department of Fermentation Technology, Faculty of Engineering, Osaka University, 2-1 Yamada-oka, Suita-shi, Osaka 565, Japan

Contents

A simple graphical method for the determination of the operational conditions of a continuous reactor based on the batch data of a reaction is described. We term it the "tangential method", as tangential lines play important roles. Reactions are grouped into single- and multi-curve reactions. If the stoichiometric coefficients between the substances concerning the reaction do not change during the course of the reaction, the reaction is defined as a single-curve reaction, and if all the values of the coefficients are not constant during the course, the reaction is defined as a multi-curve reaction.

For a single-curve reaction, the value of the required residence time of a continuous stirred tank reactor (CSTR) is obtained only by drawing a tangential line at the requested point on the batch course of the reaction. For a multi-curve reaction, not only the residence time but also the inlet concentrations are determined graphically. This tangential method is simple, and the precise batch data for the analysis are needed only in the vicinity of the requested point. The graphical method is based on the assumption that the reaction rate is determined by the composition of the reaction mixture. Therefore this method is not applicable to the reaction that contains a substance having age.

The behavior of a continuous reactor with a recycling system is analyzed, and the methods for obtaining the desirable recycling conditions are described.

Bioproducts and Bioprocesses
Editors: Fiechter, Okada, Tanner

The tangential method is applicable for designing multi-stage reactor systems. For a single-curve reaction, a general strategy is described for designing a multi-stage system with the shortest residence time. The same strategy is also applicable to a multi-curve reaction by adding extra feeding lines to the CSTRs that are used in the second or later stages.

The method for predicting the stability of a point on the batch course of a reaction is described. A stable point on the batch course is defined as the point whose state has the possibility to be realized as a steady state in a CSTR. Three kinds of criteria for judging the stability of the point are shown. For a multi-curve reaction, a point on the batch course is judged to be stable if one of the criteria is satisfied for all the batch curves of the substances representing the reaction. A stable point has a restitutive domain which is defined as the set of the fluctuated points from which the original steady state is spontaneously recovered. The fluctuated points are assumed to be on the batch course, otherwise the graphical method cannot be used. For a single-curve reaction, the restitutive domain of a stable point is graphically determined, but for a multi-curve reaction, the restitutive domain cannot be determined.

1. Introduction

Chemical and biochemical reactions are often done in a continuous way for practical purposes. In such a case, we must know the operational conditions for maintaining the optimum steady state of a reaction. In practice, the actual optimum conditions are experimentally obtained by trial and error, but it is important to know the characteristics of the reaction and to know the appropriate conditions to be examined as the first candidate for the experimental operation.

The operational conditions for a continuous reaction can be predicted either by the mathematical model of the reaction or by the graphical analysis of the batch data of the reaction. Each method has its own merits and demerits, and the most important merit of the graphical method is its applicability to a wide variety of reactions regardless of the presence or absence of their mathematical models.

Consider a reaction whose batch data is shown in Fig. 1A. The batch curve is generally written as

$$y = f(t) \tag{1}$$

where t is time from the start of the reaction, and y is the concentration of Y, which is one of the substances participating in the reaction. When this reaction is done in a continuous stirred tank reactor (CSTR) with a residence time τ, the material balance of Y is

$$\tau(dy/dt) = y_i - y + \tau v(y) \tag{2}$$

where y_i is the inlet concentration of Y, dy/dt is the rate of the change in y in the reactor, and v(y) is the rate of the change in y by the reaction (namely, the reaction rate for Y). At steady state, Eq. (2) becomes

$$\tau = (y - y_i)/v(y)\,. \tag{3}$$

Now let us suppose that we want to do the reaction continuously at the same state as that of the batch reaction mixture at $t = t_e$ (and $y = y_e$), i.e., at the

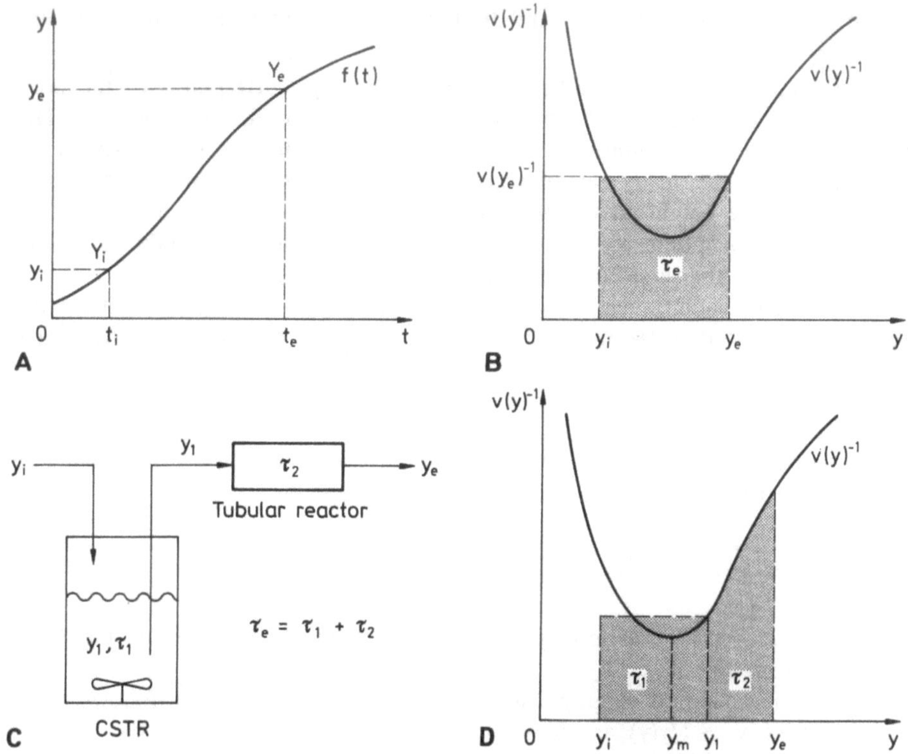

Fig. 1A–D. The classical method of graphical solution. **A** the course of the batch reaction $y = f(t)$; y_i and y_e are the concentrations of a substance Y at time t_i and t_e, respectively. **B** graphical solution of the required residence time τ_e; the τ_e value is obtained as the hatched area. **C** a two-stage continuous reaction system; y_i, y_1, and y_e are the concentrations of Y in the inlet solution, in the CSTR, and in the outlet solution, respectively; τ_1 and τ_2 are the residence times of the CSTR and the tubular reactor, respectively; the overall residence time τ_e is given as $\tau_e = (\tau_1 + \tau_2)$. **D** the graphical solution of τ_e for the two-stage system; the τ_e value is given as the hatched area, and the value reaches a minimum at $y_1 = y_m$ where the value of $v(y)^{-1}$ reaches a minimum

point $Y_e(t_e, y_e)$ on the bach curve shown in Fig. 1A. From Eq. (3), the value of the required residence time τ_e is given by

$$\tau_e = (y_e - y_i)/v(y_e) \tag{4}$$

where $v(y_e)$ is the rate for Y in the reaction mixture whose composition is represented by y_e; here the "composition" means both the kinds and the concentrations of all the substances participating in the reaction. We assume that the reaction rate is determined by the composition of the reaction mixture; this means that the reaction rate in a continuous reactor is the same as that in a batch reactor if the two reactors have the same reaction mixture. On the basis of the above assumption, the following equation is formed at $y = y_e$:

$$v(y_e) = (df/dt)_{y=y_e} \equiv (df/dt)_{t=t_e} \tag{5}$$

where $(df/dt)_{y=y_e}$ and $(df/dt)_{t=t_e}$ are the batch reaction rate for Y at $y = y_e$ and at $t = t_e$, respectively. By Eqs. (4) and (5), the τ_e value is related to the batch data at point Y_e, and thus the steady state at $y = y_e$ corresponds to the point $Y_e(t_e, y_e)$ by the identities of both the composition and the reaction rate. This is the principle of the graphical solution. In this report, we often express the characteristics of the steady state of a continuous reaction as those of the corresponding point on the batch curve; for instance, "τ_e is the τ value of point Y_e" and "the stability of point Y_e", the latter example means "the stability of the steady state with the same composition as point Y_e".

On the other hand, when the reaction shown in Fig. 1A is done in a tubular reactor with a steady-state outlet concentration of y_e, the τ_e value is directly related to the reaction time of the batch course:

$$\tau_e = t_e - t_i \tag{6}$$

where t_i is the time when the value of y reaches y_i on the course. The τ_e value in Eq. (6) is also related to the reaction rate in the reactor by

$$\tau_e = \int_{y_i}^{y_e} [1/v(y)]\, dy\,. \tag{7}$$

The values of v(y) at various y values are related to the rates of the batch reaction by Eq. (5).

The classical method of graphical solution of the τ_e value is as follows: draw the curve of y vs 1/v(y) as shown in Fig. 1B by using batch data, and the value of τ_e for a CSTR is obtained as the hatched area. In this case, the composition of the inlet solution must be adjusted to that of the batch reaction mixture at time t_i shown in Fig. 1A. It is obvious that these values satisfy Eq. (4).

This method is also used for obtaining the conditions that make the τ_e value to be minimum by connecting a tubular reactor and a CSTR as shown in Fig. 1C. From Eqs. (4) and (7), the overall residence time, τ_e, of this system corresponds to the hatched area in Fig. 1D, and the value changes depending on the value of y_1. Fig. 1D indicates that the τ_e value reaches a minimum when y_1 is adjusted to y_m.

Now let the substances participating in a reaction be Y_j(j = 1, 2, ... , n), and the batch course of the reaction be generally expressed by

$$y_j = f_j(t)\ (j = 1, 2, \dots, n) \tag{8}$$

$$dy_j/dy_k = r_{jk}(t)\ (j, k = 1, 2, \dots, n) \tag{9}$$

where y_j is the concentration of Y_j, and $r_{jk}(t)$ is the stoichiometric coefficient between Y_j and Y_k. For a simple reaction, all $r_{jk}(t)$ (j, k = 1, 2, ... , n) are constant during the batch course. In this case, the courses for all the substances in the reaction are related to one another by the constant stoichiometric coefficients r_{jk} (j, k = 1, 2, ... , n); therefore the course of the reaction can be represented by the course (or curve) for anyone of the substances. We call such simple reactions "single-curve reactions". The application of the classical method has been limited to these single-curve reactions.

There are more complex reactions in which all the values of $r_{jk}(t)$ (j, k = 1, 2, ... , n) are not constant. When a graphical solution is applied to such a reaction, at most "n" curves must be drawn as the batch data; each curve shows the course of the reaction for each substance. The number of the curves can be reduced if some pairs of Y_j and Y_k have constant r_{jk}. We call such complex reactions "multi-curve reactions", and if at least n curves are needed to express a reaction, we call it an "n-curve reaction".

As described in Sect. 2.2, we can graphically discriminate between a single- and a multi-curve reaction. A multi-curve reaction may partly be a single-curve reaction. In this report, we discriminate between them only at the point or the region on the course of a reaction when the point or the region is considered as a state or states in a CSTR.

For a multi-curve reaction, some of the batch curves may not be monotone, but for a single-curve reaction, all the curves are monotone. This is because the composition of a reaction mixture unequally determines the set of the reaction rates for all the substances, and therefore the course of a batch reaction does not pass the two or more states (or points) which have the same composition but have different sets of the tangential slopes. It should be noted that a reaction stops at the state (or point) where all the tangential slopes are zero; we call it the end point.

We describe here a new graphical method. As tangential lines are important in this method, we call it the "tangential method". This method is simpler than the classical method as we do not need to calculate the values of v(y) for the whole course of a reaction, and the required values are directly shown by the lengths along the axes. This report also describes the application of this method to the design and analysis of multi-stage reaction systems. Finally the methods for predicting the stability of steady states in a CSTR are described.

2. Determination of the Operational Conditions of a Continuous Reactor

2.1 Single-Curve Reaction

2.1.1 Basic Procedure

Question. Consider a single-curve reaction, $y = f(t)$, shown in Fig. 2. It is requested that this reaction is done in a CSTR at a steady-state of y_e with an inlet concentration of y_i. Find the value of the residence time τ_e that satisfies the above requirements.

Solution. The following procedure is shown in Fig. 2. Locate the point $Y_e(y_e, t_e)$ on $y = f(t)$ by using the given value of y_e; the point shows the value of t_e. Draw the tangential line $y = g(t)$ at point Y_e and locate the point $G(y_i, t_g)$ on the tangential line by using the given value of y_i. Point G shows the value of t_g, and the value of τ_e is obtained directly from the figure as

$$\tau_e = t_e - t_g . \tag{10}$$

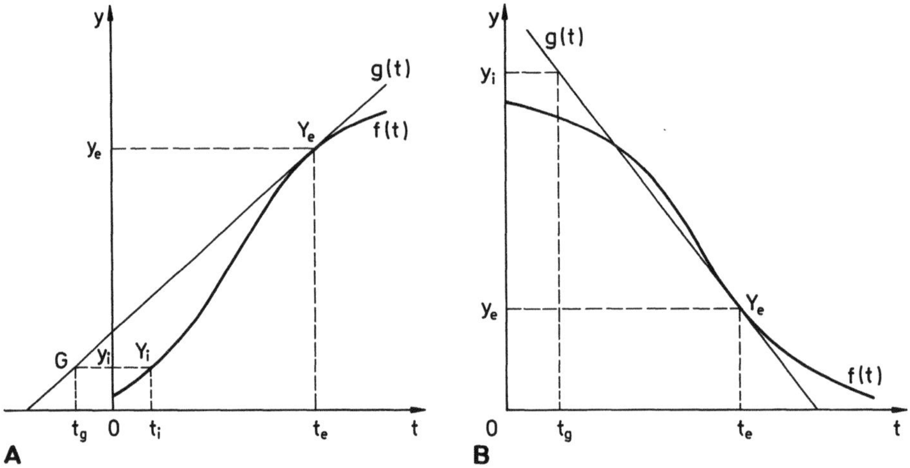

Fig. 2A, B. The tangential method of graphical solution for a single-curve reaction. The curve of $y = f(t)$ and the values of y_e and y_i are given. The coordinates of point Y_i are (t_i, y_i). $y = g(t)$ is the tangential line at the point $Y_e(t_e, y_e)$ on $y = f(t)$. The value of t_g is obtained as the corrdinates of the point $G(t_g, y_i)$ on $y = g(t)$, and the required τ_e value is obtained as $\tau_e = (t_e - t_g)$. **A** the case of $df/dt > 0$; **B** the case of $df/dt < 0$

Proof. The tangential line is written as

$$g(t) = (df/dt)_{t=t_e} (t - t_e) + y_e \,. \tag{11}$$

At $t = t_g$, Eq. (11) becomes

$$g(t_g) = y_i = (df/dt)_{t=t_e} (t_g - t_e) + y_e \,. \tag{12}$$

From Eqs. (10) and (12), the graphically obtained τ_e value is expressed by

$$\tau_e = (y_e - y_i)/(df/dt)_{t=t_e} \,. \tag{13}$$

By using Eq. (5), Eq. (4) is derived from Eq. (13).

In the case of Fig. 2A, the substance Y is regarded as a product of the reaction. If Y is a substrate, the course will be such a curve as shown in Fig. 2B. In both cases, the above question can be solved by the same procedure.

We can choose arbitrary values for y_i and y_e; this means that we can put the points Y_i and Y_e at any position on the batch curve (of course, $t_i < t_e$). But after fixing the y_i value, the composition of the inlet solution must be adjusted to that of the batch reaction mixture at point Y_i. As shown in Fig. 1B, the value of y_i does not necessarily have the corresponding point on the curve. In such cases, the concentrations of the other substances must be calculated by using the constant stoichiometric coefficients, and such conditions can be realized if none of these calculated values is negative.

2.1.2 Isotau Points and Isotau Curves

Let us consider the same question as in the previous section. If the same residence time, τ_e, is obtained at other steady states, i.e., at $y = y_j$ ($j = 1, 2, \ldots ; y_j \neq y_e$), we must be careful not to operate the reactor at these steady states. In such a case, the composition of the reaction mixture must be adjusted to that of the intended steady state of y_e at the time of starting the continuous operation with the residence time of τ_e.

The points $Y_j(t_j, y_j)$ ($j = 1, 2, \ldots$) on the batch curve of $y = f(t)$ have the same τ value as point Y_e, and so we call them the "isotau points" of Y_e. We can find the isotau points of Y_e by using the "isotau curves" of Y_e, $y = h(t)$, which are defined as

$$h(t) = \exp\left[(t + C)/\tau_e\right] + y_i \tag{14}$$

where C is a variable constant. The isotau curves of Y_e with different C values form a set of curves parallel to the t axis. One of these curves, $y = h_e(t)$, passing point Y_e is expressed by

$$h_e(t) = (y_e - y_i)\exp\left[(t - t_e)/\tau_e\right] + y_i\,. \tag{15}$$

The tangential line of the curve $y = h_e(t)$ at point Y_e is expressed as

$$y = \left[(y_e - y_i)/\tau_e\right](t - t_e) + y_e\,. \tag{16}$$

Since τ_e satisfies Eq. (13), Eq. (16) is the same as Eq. (11). This means that the two curves $y = h_e(t)$ and $y = f(t)$ touch each other at point Y_e. Similarly, the curve $y = h_j(t)$ touches the curve $y = f(t)$ at point Y_j. This indicates that the isotau curves $y = h(t)$ touch the curve $y = f(t)$ at the isotau points of Y_e. Thus we can search the curve $y = f(t)$ for these isotau points Y_j ($j = 1, 2, \ldots$) by moving the curve $y = h_e(t)$ parallely to the t axis. As described in Sect. 4.3, the isotau curves and the isotau points of Y_e play important roles in examining the stability of point Y_e, and another method of finding the isotau points are shown in Fig. 14.

2.2 Multi-Curve Reaction

The tangential method is also applicable to multi-curve reactions. In this case, not only the residence time but also the inlet concentrations of the substances are graphically determined.

Question. Consider a two-curve reaction shown in Fig. 3; the reaction converts a substrate S into a product P, and the course is expressed by the following equations:

$$s = f_s(t) \quad \text{and} \quad p = f_p(t) \tag{17}$$

$$dp/ds = r_{ps}(t) \tag{18}$$

where s is the concentration of S, and p is the concentration of P. It is requested that this reaction is done in a CSTR at a steady-state of p_e with

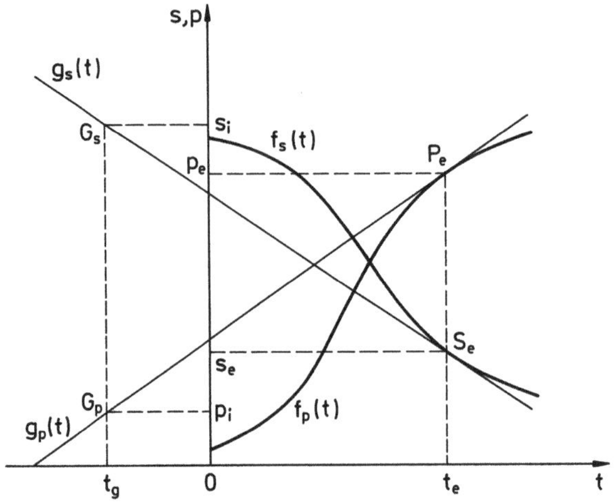

Fig. 3. Graphical solution for a two-curve reaction. s and p, the concentrations of substances, S and P, respectively. The curves of $s = f_s(t)$ and $p = f_p(t)$, and the values of p_e and s_i are given. $s = g_s(t)$, the tangential line at $S_e(t_e, s_e)$ on $s = f_s(t)$; $p = g_p(t)$, the tangential line at $P_e(t_e, p_e)$ on $p = f_p(t)$. The value of t_g is obtained as the coordinates of the point $G_s(t_g, s_i)$ on $y = g_s(t)$, the required value of p_i is obtained as the coordinates of the point $G_p(t_g, p_i)$ on $y = g_p(t)$, and the required value of τ_e is obtained as $\tau_e = (t_e - t_g)$

an inlet concentration of s_i. Find the value of the residence time τ_e and the inlet concentration p_i.

Solution. The following procedure is shown in Fig. 3. Locate the point $P_e(t_e, p_e)$ on $p = f_p(t)$; the point shows the value of t_e. Locate the point $S_e(t_e, s_e)$ on $s = f_s(t)$ by using the t_e value. Draw the two tangential lines $p = g_p(t)$ and $s = g_s(t)$ at the points P_e and S_e, respectively. Locate the point $G_s(t_g, s_i)$ on $s = g_s(t)$ by using the given value of s_i. Point G_s shows the value of t_g, and the τ_e value is obtained as $(t_e - t_g)$. Then locate the point $G_p(t_g, p_i)$ on $p = g_p(t)$ by using the value of t_g; the point shows the values of p_i.

Proof. Fig. 3 shows that the value of τ_e and p_i thus obtained satisfy the following equations:

$$\tau_e = t_e - t_g = (s_e - s_i)\,(df_s/dt)^{-1}_{t=t_e} \tag{19}$$

$$= (p_e - p_i)\,(df_p/dt)^{-1}_{t=t_e}\,. \tag{20}$$

Similarly to Eq. (5),

$$v_s(s_e, p_e) = (df_s/dt)_{t=t_e} \tag{21}$$

$$v_p(s_e, p_e) = (df_p/dt)_{t=t_e} \tag{22}$$

where $v_s(s_e, p_e)$ and $v_p(s_e, p_e)$ are the reaction rates for S and P, respectively, in the reaction mixture with the composition of $s = s_e$ and $p = p_e$. From Eqs. (19)–(22)

$$\tau_e = (s_e - s_i)/v_s(s_e, p_e) = (p_e - p_i)/v_p(s_e, p_e)\,. \tag{23}$$

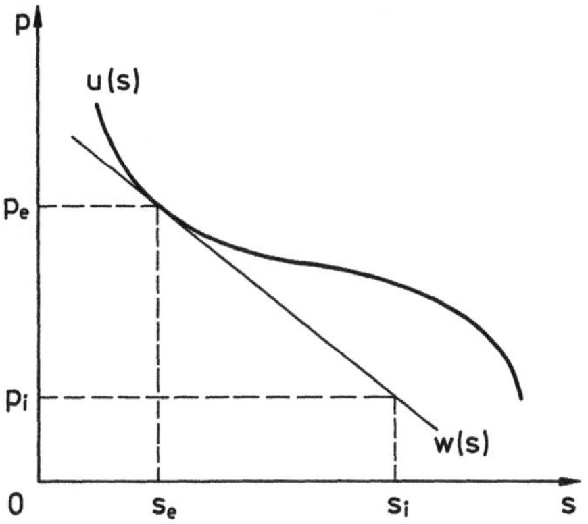

Fig. 4. The course of a two-curve reaction expressed on the concentration diagram. The curve of $p = u(s)$ is obtained from the batch data of a reaction. $p = w(s)$ is the tangential line at the point (s_e, p_e) on $p = u(s)$. The required composition of the inlet solution for a CSTR operating at the point (s_e, p_e) is expressed as the coordinates of a point on the tangential line, for example, the point (s_i, p_i)

Equation (23) indicates the same meaning as Eq. (4). Thus the obtained values satisfy the requirements in the question.

In the above procedure, the p_i value is determined so that point P_e has the same τ value as point S_e. The p_i value can initially be given instead of the s_i value.

Figure 4 shows the relationship between s and p obtained from the batch data of a two-curve reaction. Now let the curve in the figure be expressed by

$$p = u(s)\,. \tag{24}$$

The tangential line $p = w(s)$ at the point (s_e, p_e) on $p = u(s)$ is expressed by

$$w(s) = (du/ds)_{s=s_e}\,(s - s_e) + p_e\,. \tag{25}$$

From Eq. (5)

$$\begin{aligned}(du/ds)_{s=s_e} &= [(df_p/dt)/(df_s/dt)]_{t=t_e} \\ &= v_p(s_e, p_e)/v_s(s_e, p_e)\,.\end{aligned} \tag{26}$$

Therefore Eq. (25) becomes

$$w(s) = [v_p(s_e, p_e)/v_s(s_e, p_e)]\,(s - s_e) + p_e\,. \tag{27}$$

On the other hand, the following equation is formed from Eq. (23):

$$v_p(e_e, p_e)/v_s(s_e, p_e) = (p_e - p_i)/(s_e - s_i)\,. \tag{28}$$

Transforming Eq. (28)

$$p_i = [v_p(s_e, p_e)/v_s(s_e, p_e)]\,(s_i - s_e) + p_e\,. \tag{29}$$

Equations (27) and (29) indicate that the tangential line $p = w(s)$ passes the point (s_i, p_i); this means that any composition of the inlet solution required for maintaining the steady state at the point (s_e, p_e) is shown as the coordinate values of a point on the tangential line.

For a single-curve reaction, Eq. (24) shows a straight line, because r_{ps} in Eq. (18) is constant. When a multi-curve reaction contains a single-curve region, the region corresponds to the segment of a straight line on $p = u(s)$. Therefore we can discriminate between a single- and multi-curve reaction by drawing the curve of $p = u(s)$. We can also discriminate between them by using the batch curves of all the substances in the reaction. For a single-curve reaction, for example, the concentration of all the substances in the feeding solution are shown by the points on their own curves at the same value of t. This means that the required composition of the inlet solution are always expressed by the coordinates of the points on $p = u(s)$; in short, the point (s_i, p_i) is on $p = u(s)$. It should also be pointed out that in a tubular reactor, the reaction proceeds along the curve of $p = u(s)$, and therefore the point (s_i, p_i) is on $p = u(s)$.

For a general multi-curve reaction that is expressed by Eqs. (8) and (9), the number of batch curves in Fig. 3 and the dimension of Fig. 4 increase at most to n. However, the fundamental procedure for the determination of the required values of the residence time and the inlet concentrations is the same as shown in Figs. 3 and 4.

2.3 Reactors with a Recycling System

In the continuous reaction systems described so far, we can choose arbitrary values for the concentration of only one of the substances in the inlet solution, and the values for those of the other substances are determined depending on the fixed value of the selected substance. This is because the set of these values must correspond to a point on the tangential line $p = w(t)$ for a CSTR, or a point on the curve $p = u(s)$ for a tubular reactor as described in the previous section. However, if a recycling system is added to a continuous reactor as shown in Fig. 5, we can choose arbitrary values for the concentrations of all the substances in the feeding solution (i.e., p_i and s_i in Fig. 5); in this case, either a CSTR or a tubular reactor can be used as the reactor.

The material balances for the reactor system shown in Fig. 5 are

$$F_b = F_i + F_a \tag{30}$$

$$F_c = F_b \quad \text{and} \quad F_e = F_i \tag{31}$$

$$s_b F_b = s_i F_i + s_a F_a \tag{32}$$

$$p_b F_b = p_i F_i + p_a F_a \tag{33}$$

$$s_c F_c = s_e F_e + s_a F_a \tag{34}$$

$$p_c F_c = p_e F_e + p_a F_a \tag{35}$$

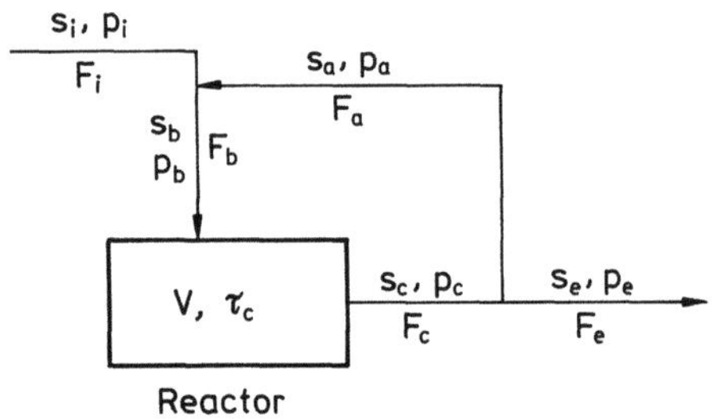

Fig. 5. Continuous reactor with a recycling system. The reactor is either a CSTR or a tubular reactor. F is a flow rate. The subscripts i, a, b, c and e show the names of the lines. τ_c is the residence time of the reactor, and V is the working volume of the reactor

where F is a flow rate, and the subscripts i, a, b, c and e show the inlet line to the system, the recycling line, the inlet line to the reactor, the outlet line from the reactor and the outlet line from the system, respectively. From these equations, the following relationships are obtained:

$$F_i/F_a = (s_a - s_b)/(s_b - s_i) = (p_a - p_b)/(p_b - p_i) \tag{36}$$

$$F_i/F_a = (s_a - s_c)/(s_c - s_e) = (p_a - p_c)/(p_c - p_e)\,. \tag{37}$$

The residence time of the reactor, τ_c, and the overall residence time of the system, τ_e, are defined as

$$\tau_c = V/F_b \quad \text{and} \quad \tau_e = V/F_i \tag{38}$$

where V is the working volume of the reactor; the volume of the lines in the system is assumed to be negligible. From Eqs. (38) and (30), the relationship between τ_c and τ_e is expressed by

$$\tau_e = \tau_c(F_b/F_i) = \tau_c(1 + F_a/F_i)\,. \tag{39}$$

2.3.1 CSTR

Question. Consider a two-curve reaction which is expressed by Eqs. (17) and (18). It is requested that the reaction is done in a CSTR with a recycling system as shown in Fig. 5, where the values of V, s_i, p_i and p_c are given. Determine the values of the other parameters in order to satisfy the above requirements.

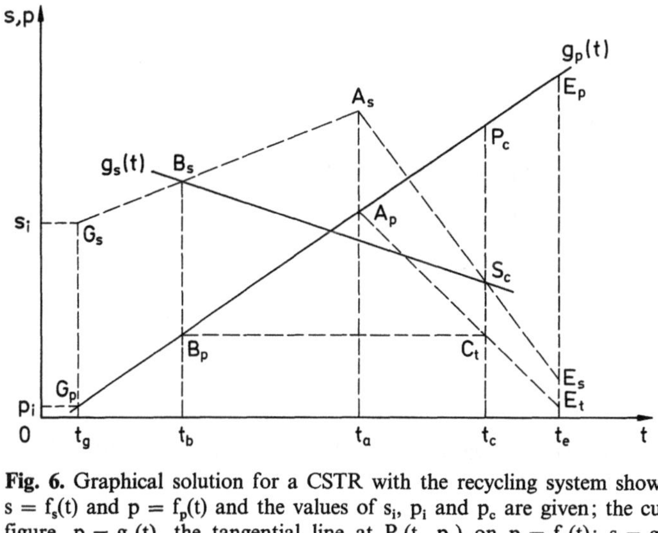

Fig. 6. Graphical solution for a CSTR with the recycling system shown in Fig. 5. The curves of $s = f_s(t)$ and $p = f_p(t)$ and the values of s_i, p_i and p_c are given; the curves are not shown in this figure. $p = g_p(t)$, the tangential line at $P_c(t_c, p_c)$ on $p = f_p(t)$; $s = g_s(t)$, the tangential line at $S_c(t_c, s_c)$ on $s = f_s(t)$. The coordinates of the other points are as follows: $G_p(t_g, p_i)$, $G_s(t_g, s_i)$, $A_p(t_a, p_a)$, $A_s(t_a, s_a)$, $B_p(t_b, p_b)$, $B_s(t_b, s_b)$, $C_t(t_c, p_b)$, $E_t(t_e, p_i)$, $E_p(t_e, p_e)$ and $E_s(t_e, s_e)$. The points G_p, A_p, B_p and E_p are on the line of $p = g_p(t)$; the points B_s is on $s = g_s(t)$ and also on the line segment A_sG_s; the points A_p, C_t and E_t are on a line; the points A_s, S_c and E_s are on a line. For p_a and s_a, arbitrary values are chosen in the range of $s_a > g_s(t_a)$. The values of τ_c and τ_e are obtained as $\tau_c = (t_c - t_b)$ and $\tau_e = (t_e - t_g)$. The values of s_e and p_e are obtained as the coordinates of the points E_s and E_p, respectively

Solution. The following procedure is shown in Fig. 6; in this figure, the curves $s = f_s(t)$ and $p = f_p(t)$ are not shown. Locate the point $P_c(t_c, p_c)$ on $p = f_p(t)$; this point shows the value of t_c. Locate the point $S_c(t_c, s_c)$ on $s = f_s(t)$; the s_c value is given from the point. Draw the two tangential lines $p = g_p(t)$ and $s = g_s(t)$ at the points P_c and S_c, respectively. Locate the point $G_p(t_g, p_i)$ on $p = g_p(t)$ by using the given value of p_i; this point shows the value of t_g. Then locate the point $G_s(t_g, s_i)$ by using the given value of s_i and the t_g value. Next locate the point $A_p(t_a, p_a)$ on $p = g_p(t)$, here an arbitrary value for p_a can be chosen; the value of t_a is thus determined. Locate the point $A_s(t_a, s_a)$, here an arbitrary value can be chosen for s_a in the range of $s_a > g_s(t_a)$ for the case of $s_i < g_s(t_g)$ (in the case of Fig. 6), or in the range of $s_a < g_s(t_a)$ for the case of $s_i > g_s(t_g)$. Connect the two points G_s and A_s with a straight line, locate the point $B_s(t_b, s_b)$ where the line A_sG_s intersects the line $s = g_s(t)$; this point shows the values of s_b and t_b. Then locate the point $B_p(t_b, p_b)$ on $p = g_p(t)$ by using the t_b value; this point shows the value of p_b. The value of τ_c is obtained from the figure as

$$\tau_c = t_c - t_b . \tag{40}$$

The values of F_b and F_c are obtained from Eqs. (38) and (31).

The values of τ_e, s_e and p_e are graphically obtained as follows. Locate the point $C_t(t\ , p_b)$ by using the values of t_c and p_b, and connect the two points A_p and C_t. Locate the point $E_t(t_e, p_i)$ on the line A_pC_t by using the p_i value. Point E_t shows the value of t_e, and the value of τ_e is obtained from the figure as

$$\tau_e = t_e - t_g . \tag{41}$$

Locate the points $E_s(t_e, s_e)$ and $E_p(t_e, p_e)$ on the lines A_sS_c and $y = g_p(t)$, respectively, by using the t_e value. These points show the values of s_e and p_e. The values of F_i, F_e and F_a are easily obtained from Eqs. (38), (31) and (30). Thus, the values of all the parameters in Fig. 5 are determined.

Proof. From Fig. 6, the following equations are formed:

$$(s_a - s_b)/(s_b - s_i) = A_sB_s/B_sG_s \tag{42}$$

$$(p_a - p_b)/(p_b - p_i) = A_pB_p/B_pG_p \tag{43}$$

$$A_sB_s/B_sG_s = A_pB_p/B_pG_p . \tag{44}$$

From these equations

$$(s_a - s_b)/(s_b - s_i) = (p_a - p_b)/(p_b - p_i) \tag{45}$$

Therefore, the graphically determined values of s_a, s_b, p_a and p_b satisfy Eq. (36). It should be noted that from Eqs. (36), (42) and (44), the ratio of F_i/F_a is shown in the figure as

$$F_i/F_a = A_sB_s/B_sG_s = A_pB_p/B_pG_p \tag{46}$$

By the same procedure as shown by Eqs. (19)–(23), the following equation is obtained:

$$\tau_c = (s_c - s_b)/v_s(s_c, p_c) = (p_c - p_b)/v_p(s_c, p_c) \tag{47}$$

Eq. (47) indicates the same meaning as Eq. (23).

On the triangle $\Delta A_pG_pE_t$ in Fig. 6, the following equation is formed:

$$(t_e - t_g)/(t_c - t_b) = G_pE_t/B_pC_t = A_pG_p/A_pB_p \tag{48}$$

From Eq. (46)

$$A_pG_p/A_pB_p = (A_pB_p + B_pG_p)/A_pB_p = 1 + F_a/F_i\,. \tag{49}$$

From Eqs. (40), (41), (48) and (49), Eq. (39) is derived. Therefore the graphically obtained τ_e value satisfies Eq. (39).

From Fig. 6, the following equations are also formed:

$$\begin{aligned}(s_a - s_c)/(s_c - s_e) &= A_sS_c/S_cE_s \\ &= A_pC_t/C_tE_t = A_pB_p/B_pG_p\end{aligned} \tag{50}$$

$$\begin{aligned}(p_a - p_c)/(p_c - p_e) &= A_pP_c/P_cE_p \\ &= A_pC_t/C_tE_t = A_pB_p/B_pG_p\,.\end{aligned} \tag{51}$$

From Eqs. (50), (51) and (46), Eq. (37) is derived. Therefore the graphically obtained values of s_e and p_e satisfy Eq. (37).

In the above question, the value of V is given, but the value of F_i or F_a can be given instead. In the above solution, the points G_p and A_p are put on $p = g_p(t)$ prior to the points G_s and A_s, respectively, but the order can be reversed, and the points G_s and A_s can be put on $s = g_s(t)$ initially, because "s" and "p" are exchangeable.

Figure 7 shows the relationships between the concentrations in the lines named with the subscripts i, a, b, c and e, here the values of the coordinates of the points I, A, B, C and E show the concentrations of the substances S and P in these lines named with the corresponding subscripts. The geometric characteristics of this plot are as follows: the points A, B and I are on a straight line, and the points A, C and E are on another line; point B is between A and I, and point C is between A and E; line BC is parallel to line IE. Point C must be in a stable domain (see Chap. 4) on the curve of $p = u(s)$; the curve is drawn by using

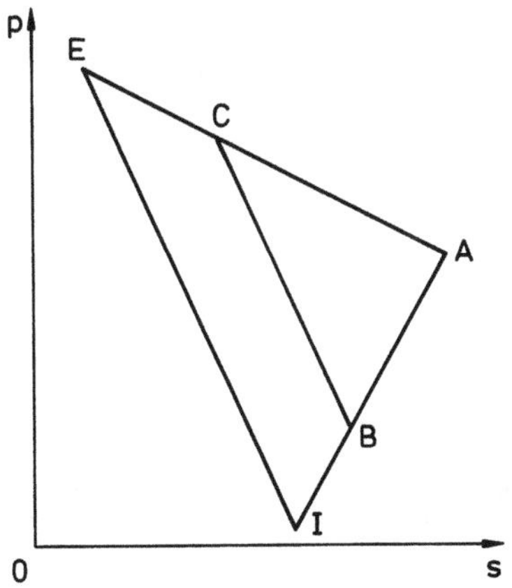

Fig. 7. Concentration diagram. The coordinates of the points $I(s_i, p_i)$, $A(s_a, p_a)$, $B(s_b, p_b)$, $C(s_c, p_c)$ and $E(s_e, p_e)$ show the concentrations of the substances S and P in the lines named with the subscripts i, a, b, c and e, respectively (see Fig. 5). The points B and C are on the line segments AI and AE, respectively, and line BC is parallel to line IE. For a CSTR, point C is on the experimental curve of $p = u(s)$, and point B is on the tangential line $p = w(s)$ drawn at point C on $p = u(s)$ (see Fig. 4); for a tubular reactor, both point B and point C are on $p = u(s)$

the batch data (see Fig. 4). Point B must be on the tangential line p = w(s) drawn at point C on p = u(s).

This diagram is useful not only for the analysis of the system but also for the determination of desirable recycling conditions, because the order of locating the points is arbitrary unless the above requrements are satisfied. For example, if the points C and I are given, the directions of the lines BC and IE are also given, and we can locate the other points in anyone of the following orders: E, A and B; E, B and A; A (and B) and E; B, A and E; B, E and A. It should be noted that the τ_c value is determined when the points B and C are fixed, and the τ_e value is determined when the points I and E are fixed.

Figure 7 indicates that the composition of the recycled solution must not be the same as that of the outlet solution, otherwise the points A and E coincide with point C; if a portion of the outlet solution is recycled without changing the composition, point B can not be on the tangential line p = w(s), and consequently the required steady state is not obtained. On the other hand, if point I is on p = w(s), we do not need to add the recycling system.

The same procedure described above is needed also for single-curve reactions to modify the inlet composition. The number of curves to be analyzed and the number of the dimension of Fig. 7 are determined by the number of substances whose inlet concentrations must be modified. The difference between single- and multi-curve reactions is only that the ratio of the slopes of any pair of tangential lines is constant or not; this means in Fig. 7 that for a single-curve reaction, the points B and C are on the straight line of p = u(s) as described in Sect. 2.2.

When the inlet concentrations of more than two substances must be modified, at first apply the above analysis to the most important pair (for example, S_1 and P_1), and determine the optimal conditions for them. Then, take another pair containing S_1 or P_1 (for example, S_2 and P_1). In this case, the positions of the points $S_{2c}(t_c, s_{2c})$ and $B_{s2}(t_b, s_{2b})$ and the values of t_g, t_a and t_e are already fixed. Therefore we can choose an arbitrary value only for s_{2i} (point G_{s2}), s_{2a} (point A_{s2}) or s_{2e} (point E_{s2}) in the range not to make the other value negative. The values of the parameters for the other substances can be determined by repeating the same procedure as used for the pair of S_2 and P_1.

2.3.2 Tubular Reactor

Question. Consider a two-curve reaction shown in Fig. 8; the reaction is expressed by Eqs. (17) and (18). It is requested that the reaction is done in a tubular reactor with a recycling system as shown in Fig. 5, where the values of V, s_i, p_i and p_c are given. Determine the values of the other parameters in order to satisfy the above requirements.

Solution. The following procedure is shown in Fig. 8. Locate the point $P_c(t_c, p_c)$ on $p = f_p(t)$, and then the point $S_c(t_c, s_c)$ on $s = f_s(t)$. Choose an arbitrary value for t_b, and locate the points $B_s(t_b, s_b)$ and $B_p(t_b, p_b)$ on $s = f_s(t)$ and $p = f_p(t)$, respectively, by using the t_b value. The value of τ_c is determined as $\tau_c = (t_b - t_c)$. Draw the line P_cB_p, and locate the point $G_p(t_g, p_i)$ on the line by using the given value of p_i. Locate the point $G_s(t_g, s_i)$ by using the given value of s_i and the t_g value. Locate the point $A_p(t_a, p_a)$ on the line P_cB_p so that point B_p is between

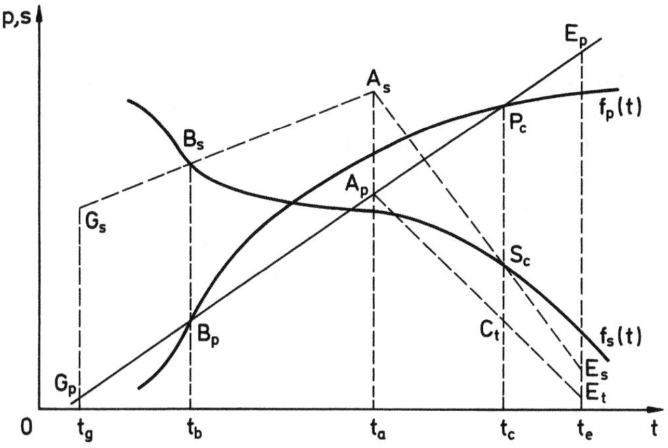

Fig. 8. Graphical solution for a tubular reactor with a recycling system shown in Fig. 5. The curves of $s = f_s(t)$ and $p = f_p(t)$ and the values of p_e, s_i and p_i are given. The coordinates of the points are as follows: $P_c(t_c, p_c)$, $S_c(t_c, s_c)$, $B_p(t_b, p_b)$, $B_s(t_b, s_b)$, $G_p(t_g, p_i)$, $G_s(t_g, s_i)$, $A_p(t_a, p_a)$, $A_s(t_a, s_a)$, $C_t(t_c, p_b)$, $E_t(t_e, p_i)$, $E_p(t_e, p_e)$ and $E_s(t_e, s_e)$. The points P_c and B_p are on $p = f_p(t)$; the points S_c and B_s are on $s = f_s(t)$; the points P_c, B_p, G_p, A_p and E_p are on a line; the point B_s is on the line segment G_sA_s; the points A_p, C_t and E_t are on a line; the points A_s, S_c and E_s are on a line. For t_b and t_a, arbitrary values are chosen in the ranges of $t_b < t_c$ and $(t_b - t_a)(t_b - t_c) < 0$. The values of τ_c and τ_e are obtained as $\tau_c = (t_c - t_b)$ and $\tau_e = (t_e - t_g)$. The values of s_e and p_e are obtained as the coordinates of the points E_s and E_p, respectively

A_p and G_p, here an arbitrary value can be chosen for p_a in the range that satisfies the above condition. Draw the line G_sB_s, and locate the point $A_s(t_a, s_a)$ on the line by using the t_a value. Thus the values of τ_c, s_c, s_b, p_b, s_a and p_a are determined. The values of F_b and F_c are obtained from Eqs. (38) and (31), and the ratio of F_i/F_a is obtained as the ratio of $(t_a - t_b)/t_b$. The values of τ_e $(= t_e - t_g)$, s_e and p_e are obtained by the same procedures described in Fig. 6, and the points used for the determination of these values are shown in Fig. 8. The values of F_i, F_e and F_a are obtained from Eqs. (38), (31) and (30). Thus the values of all the parameters in Fig. 5 are determined.

The proof for the above solution is the same as described for a CSTR. The relationships between the concentrations in the lines are also the same as shown in Fig. 7, and the degree of freedom for the determination of these parameters is almost the same; the only difference is that point B in Fig. 7 is not on the tangential line $p = w(s)$, but on the curve of $p = u(s)$ for a tubular reactor. This causes the difference in the procedures shown in Figs. 6 and 8 that the t_b value is fixed prior to the t_g and t_a values for a tubular reactor; however, by drawing the curve $p = u(s)$ in the concentration diagram shown in Fig. 7, we can easily locate either point A or point E prior to point B. It should be pointed out that for a tubular reactor, the direction of line IE is not determined by locating point C.

2.4 Reproducibility and Age

These graphical solutions are based on batch data, and it is tacitly assumed that the data is reproducible. But all the reactions are not reproducible, and all the course of a reaction is not required to be reproducible for graphical solution. The lack of the reproducibility of a reaction is the lack of constant relationships between the rates and the composition of a reaction mixture. This may be due to the intrinsic nature of the reaction, or due to unknown factors. The tangential method, however, is applicable to such reactions if they have the partial reproducibility described below.

For the determination of the operational conditions of a CSTR working at a requested steady state, the minimum requirement for the reproducibility of the batch course of the reaction is that the batch course just passes the state where the composition and the reaction rates for all the substances are the same as those at the requested steady state. Even the reproducibility of the appearance of the state is not required. In short, the minimum requirement are the following two points: a constant reaction rate is obtained at the requested composition of a reaction mixture and such data are obtained at least once in repeated batch experiments.

For example, consider a single-curve reaction $y = f(t)$, and let the requested steady-state value of y be y_e. The above requirement means that even if the reaction draws different curves on different batch experiments, each curve must have the same tangential slope at $y = y_e$; this requirement can be examined graphically. It should be noted that there is no requirement for the reproducibility of the relationship between y and t, the time from the start of the reaction.

The results of the above analysis have practical meanings that the precise batch data are needed only in the vicinity of the requested point, and that the lack of the reproducibility in the other ranges, for example, lag phase and so on, does not affect the results of the graphical solution, i.e., the requested steady state of the continuous reaction.

For a tubular reactor, the minimum requirement for the reproducibility of the batch course is that the course always passes at least two states having constant compositions with a constant time interval. If there are such two states, the earier state gives the inlet composition, the latter state gives the outlet composition, and the time span between the states gives the residence time. We can search these states by moving the batch curves parallely to the t axis, and find them as the points that all the curves pass. It should be noted that the reproducibility of the relationship between the reaction rate and the composition is not required.

Theoretically, these graphical solutions are applicable only to a reaction which does not contain a substance having age; in most cases, substances with age are living organisms. This limitation comes from the essential difference between the reactions in a batch reactor and in a CSTR, and the assumption that the reaction rate is determined by the composition of the reaction mixture is no longer satisfied for the mixture containing a substance having age. However, if the physiological state or the activities of a living organism is determined by the environmental conditions, the operational conditions of the continuous culture of the organism can be predicted from its batch data by these graphical methods.

When the organism is fed continuously to the reactor, the time needed for its adaptation to a new environment must be short enough comparing with the mean residence time of the continuous culture; otherwise, the distribution in the physiological state of the organism at the steady state of the continuous culture will considerably be different from that of the organism at the corresponding point of the batch course.

3. Design of a Continuous Reactor System with the Shortest Residence Time

3.1 Single-Curve Reaction

The residence time of a single stage continuous reaction can be shortened by using a multi-stage system of tubular and stirred tank reactors, and the system with the shortest residence time can be designed by the tangential method as well as the classical method (see Fig. 1C).

Question. Consider a single-curve reaction, $y = f(t)$. It is requested that the reaction is done continuously in a multi-stage system of CSTRs and tubular reactors with inlet and outlet concentrations of y_i and y_e, respectively. Design a reactor system which makes the value of the overall residence time τ_e to be minimum.

Solution. An example of the following procedure is shown in Fig. 9A. Locate the points $Y_e(t_e, y_e)$ and $Y_i(t_i, y_i)$ on $y = f(t)$ by using the values of y_e and y_i, respectively. In the region between the points Y_e and Y_i, look for the inflection points on $y = f(t)$, and name them $M_j(t_{mj}, y_{mj})$ ($j = 1, 2, \dots, n$). They are the points that satisfy the following conditions:

$$(d^2f/dt^2)_{t=t_{mj}} = 0 \quad \text{and} \quad (y_e - y_i)(d^3f/dt^3)_{t=t_{mj}} < 0 \tag{52}$$

where the order of the points is $t_e > t_{m1} > t_{m2} > \dots > t_{mn} > t_i$; in the case of Fig. 9A, $n = 2$. It is assumed that the inflection points are in stable domains (see Chap. 4). Draw the tangential lines $y = g_{mj}(t)$ at the points M_j ($j = 1, 2, \dots, n$) and compare the slopes. Let the values of $|df/dt|_{t=t_{mk}}$ at the point $M_k(t_{mk}, y_{mk})$ be the largest of all the values of $|df/dt|_{t=t_{mj}}$ ($j = 1, 2, \dots, n$); in the case of Fig. 9A, $k = 2$. Devide the curve of $y = f(t)$ into the following three domains: the outlet domain ($t_e \geqq t \geqq t_{m1}$), the middle domain ($t_{m1} > t \geqq t_{mk}$) and the inlet domain ($t_{mk} > t \geqq t_i$). The above question is first solved in each domain, and local conclusions are obtained; then the final conclusion is obtained by summing up these local conclusions.

If none of the points M_j ($j = 1, 2, \dots, n$) is found in the region between Y_e and Y_i, i.e., $n = 0$, see "Final conclusion".

Outlet Domain. Draw the tangential line $y = g_e(t)$ at the point $Y_e(t_e, y_e)$. On the curve of $y = f(t)$, look for the point $Y_q(t_q, y_q)$ where the slope of the tangential line $y = g_q(t)$ is the same as that of $y = g_e(t)$. If point Y_q is found, locate the point (t_{gq}, y_q) on $y = g_e(t)$ by using the y_q value; in this case, the local conclusion is the following two-stage system: the first stage is a tubular reactor with

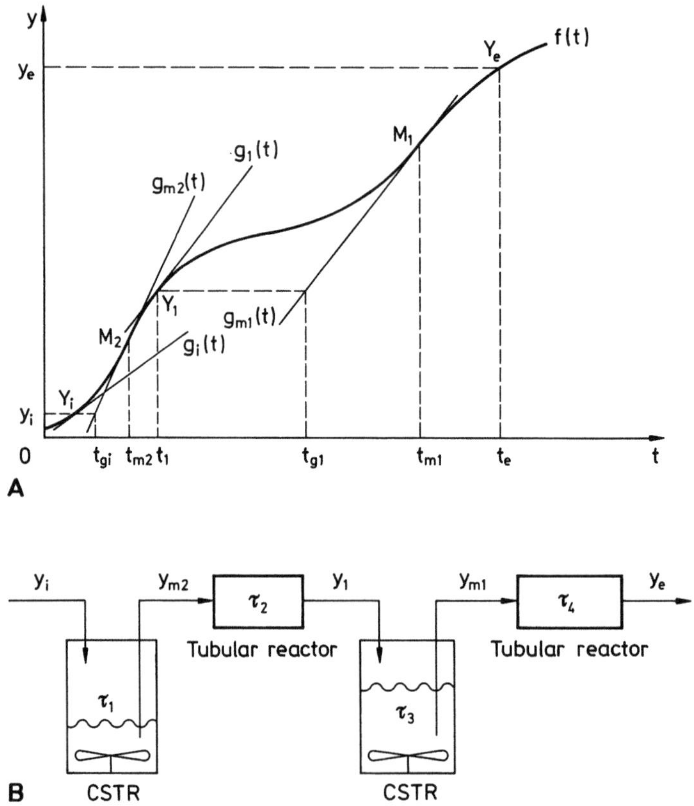

Fig. 9 A, B. Graphical solution of a reactor system with the shortest residence time for a single-curve reaction. **A** graphical solution; the curve of $y = f(t)$ and the values of y_e and y_i are given; $M_1(t_{m1}, y_{m1})$ and $M_2(t_{m2}, t_{m2})$ are inflection points that satisfy Eq. (52); $y = g_{m1}(t)$, $y = g_{m2}(t)$, $y = g_1(t)$ and $y = g_i(t)$ are the tangential lines at the points, M_1, M_2, $Y_1(t_1, y_1)$ and $Y_i(t_i, y_i)$, respectively; $y = g_1(t)$ has the same slope as $y = g_{m1}(t)$; the tangential line $y = g_e(t)$ at the point $Y_e(t_e, y_e)$ is not shown; the values of t_{g1} and t_{gi} are obtained from the coordinates of the points (t_{g1}, y_1) on $y = g_{m1}(t)$ and (t_{gi}, y_i) on $y = g_{m2}(t)$, respectively. **B** the required multi-stage system; τ_1, τ_2, τ_3 and τ_4 are the residence times for the first, second, third and forth reactors, respectively, and the values are obtained from Fig. 9 A as follows: $\tau_1 = (t_{m2} - t_{gi})$, $\tau_2 = (t_1 - t_{m2})$, $\tau_3 = (t_{m1} - t_{g1})$ and $\tau_4 = (t_e - t_{m1})$; the final conclusion is: do the reaction in the four-stage continuous reactor system operating under the above conditions; the shortest overall residence time τ_e is given by $\tau_e = (\tau_1 + \tau_2 + \tau_3 + \tau_4)$

the τ value of $(t_q - t_{m1})$, and the second stage is a CSTR with the τ value of $(t_e - t_{gq})$. If point Y_q is not found, and if $|df/dt|_{t=t_e} < |df/dt|_{t=t_{m1}}$, the local conclusion is a single tubular reactor with the τ value of $(t_e - t_{m1})$; this is the case of Fig. 9 A.

Exceptional Case. If $|df/dt|_{t=t_e} > |df/dt|_{t=t_{m1}}$, look for point Y_q in the other domains; if there is more than one such point, take the nearest point from Y_e. If point Y_q is found in the middle domain, the local conclusion is described under "Middle

Domain". If point Y_q is found in the inlet domain, locate the point (t_{gq}, y_q) on $y = g_e(t)$; in this case, the final conclusion is obtained to be the two-stage system of a tubular reactor with the τ value of $(t_q - t_i)$ and a CSTR with the τ value of $(t_e - t_{gq})$. If point Y_q is not found in the region between Y_e and Y_i, locate the point (t_{gi}, y_i) on $y = g_e(t)$ by using the y_i value; the final conclusion is a single CSTR with the τ value of $(t_e - t_{gi})$.

Middle Domain. Devide this domain at the inflection points into subdomains, and name them subdomain j $(j = 1, 2, \dots, k - 1)$ whose range is $t_{mj} > t \geqq t_{m(j+1)}$. In each domain j $(j = 1, 2, \dots, k - 1)$, look for the point $Y_j(t_j, y_j)$ where the slope of the tangential line $y = g_j(t)$ is the same as that of $y = g_{mj}(t)$. If point Y_j is found, locate the point (t_{gj}, y_j) on $y = g_j(t)$ by using the y_i value; in this case, the sublocal conclusion for subdomain j is the two-stage system of a tubular reactor with the τ value of $(t_j - t_{m(j+1)})$ and a CSTR with the τ value of (t_{mj}, t_{gj}). If point Y_j is not found in subdomain j, the sublocal conclusion for the subdomain is a single CSTR.

If point Y_q is in subdomain j, locate the point (t_{gq}, y_q) on $y = g_e(t)$; in this case, the sublocal conclusion for the sum of the outlet domain and subdomains 1, 2, ..., and j is obtained to be the two-stage system of a tubular reactor with the τ value of $(t_q - t_{m(j+1)})$ and a CSTR with the τ value of $(t_e - t_{gq})$.

Summation of the Sublocal Conclusions. Fuse a subdomain (for example, subdomain j) for which the sublocal conclusion is a single CSTR with the next subdomain (for example, subdomain j + 1), and make the sublocal conclusion for the fused subdomain by the same method as described above. Continue this fusion process until the sublocal conclusion for the fused subdomain becomes the two-stage system of a tubular reactor and a CSTR. After the fusion process, all the sublocal conclusions are two-stage systems of a pair of tubular reactor and a CSTR. The local conclusion for the middle domain which consists of "h" subdomains is the 2h-stage system of h tubular reactors and h CSTRs connected alternately with a tubular reactor first.

Inlet Domain. Draw the tangential line $y = g_i(t)$ at the point $Y_i(t_i, y_i)$. On $y = f(t)$, look for the point $Y_n(t_n, y_n)$ where the slope of the tangential line $y = g_n(t)$ is the same as that of $y = g_{mk}(t)$. If point Y_n is found, locate the point (t_{gn}, y_n) on $y = g_{mk}(t)$ by using the y_n value; in this case, the local conclusion is the two-stage system of a tubular reactor with the τ value of $(t_n - t_i)$ and a CSTR with the τ value of $(t_{mk} - t_{gn})$. If point Y_n is not found, locate the point (t_{gi}, y_i) on $y = g_{mk}(t)$ by using the y_i value; in this case, the local conclusion is a single CSTR with the τ value of $(t_{mk} - t_{gi})$; this is the case of Fig. 9A.

Final Conclusion. Connect the reactor systems obtained as the local conclusions for the three domains in the order of the inlet, middle and outlet domains. The resulted multi-stage reactor system is the final conclusion for a general single curve reaction. The overall residence time is the sum of the residence times of all the reactors in the system.

The final conclusions obtained for the case of $|df/dt|_{t=t_e} > |df/dt|_{t=t_{m1}}$ are described under "Exceptional Case".

If there is none of the inflection points, M_j $(j = 1, 2, \dots, n)$, in the region between Y_e and Y_i, look for the point $Y_q(t_q, y_q)$ in the region. If point Y_q is found, locate the point (t_{gq}, y_q) on the tangential line, $y = g_e(t)$, by using the y_q value; in this case, the final conclusion is the two-stage system of a tubular reactor with the τ value of $(t_q - t_i)$ and a CSTR with the τ value of $(t_e - t_{gq})$. If point Y_q is not found, locate the point (t_{gi}, y_i) on $y = g_e(t)$; in this case, the final conclusion is as follows: a single CSTR with the τ value of $(t_e - t_{gi})$ if $|df/dt|_{t=t_e} > |df/dt|_{t=t_i}$, or a single tubular reactor with the τ value of $(t_e - t_i)$ if $|df/dt|_{t=t_e} < |df/dt|_{t=t_i}$; if $(df/dt)_{t=t_e} = (df/dt)_{t=t_i}$, either reactor can by used because $y = f(t)$ shows a straight line in this region.

Figure 9B shows the conclusion for the case of Fig. 9A, where $n = 2$, $k = 2$, and the point Y_2 is not present.

In the above solution, it is assumed that the inflection points M_j ($j = 1, 2, \ldots, n$) are in stable domains. However, this requirement for stability is only for the inflection points whose states are used as the steady states of CSTRs. If this assumption is not valid, we cannot solve the questions in this chapter.

3.2 Multi-Curve Reaction

3.2.1 Two-Stage System with a CSTR and a Tubular Reactor

Question. Consider the two-curve reaction shown in Fig. 10 and expressed by Eqs. (17) and (18). It is requested that the reaction is done continuously in a multistage system with the inlet and outlet concentrations of p_i and p_e, respectively. Design a reactor system with the shortest overall residence time.

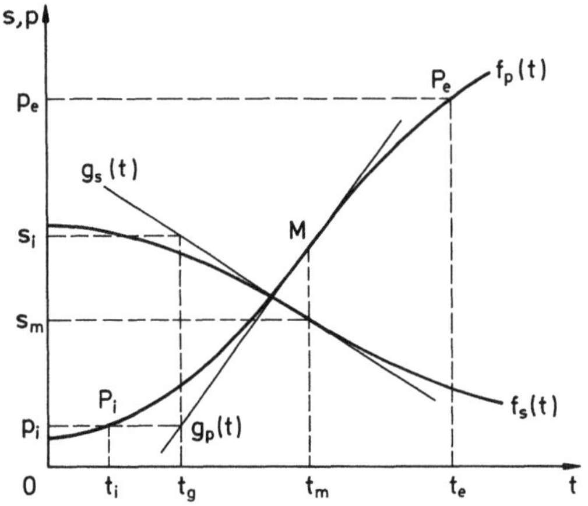

Fig. 10. Graphical solution of a reactor system with the shortest residence time for a two-curve reaction (a special case). The curves of $s = f_s(t)$ and $p = f_p(t)$ and the values of p_i and p_e are given. $M(t_m, p_m)$ is an inflection point of $p = f_p(t)$ that satisfies Eq. (52). $p = g_p(t)$, the tangential line at point M on $p = f_p(t)$; $s = g_s(t)$, the tangential line at the point (t_m, s_m) on $s = f_s(t)$. The values of t_e, t_i and t_g are obtained as the coordinates of the points $P_e(t_e, p_e)$ on $p = f_p(t)$, $P_i(t_i, p_i)$ on $y = f(t)$ and (t_g, p_i) on $y = g_p(t)$. The required values of τ_1 and τ_2 are obtained as $\tau_1 = (t_m - t_g)$ and $\tau_2 = (t_2 - t_m)$, and the required value of s_i is obtained as the coordinate of the point (t_g, s_i) on $y = g_s(t)$.

Solution. In the region between the points $P_e(t_e, p_e)$ and $P_i(t_i, p_i)$ on $p = f_p(t)$, there is only one inflection point, $M(t_m, p_m)$, that satisfies Eq. (52), and there are not the points $P_q(t_q, p_q)$ and $P_1(t_1, p_1)$ whose tangential slopes are the same as those of the points P_e and M, respectively. The curve of $p = f_p(t)$ also shows that $|df_p/dt|_{t=t_e} < |df_p/dt|_{t=t_m}$. Therefore the question can be solved by the following procedure. Locate the point (t_m, s_m) on $s = f_s(t)$ by using the value of t_m. If the point (t_m, s_m) is a stable point, draw the two tangential lines $p = g_p(t)$ and $s = g_s(t)$ at point M and the point (t_m, s_m), respectively. Locate the point (t_g, p_i) on $p = g_p(t)$ by using the p_i value, and the point (t_g, s_i) on $s = g_s(t)$ by using the t_g value. The final conclusion is that set the concentration of S in the inlet solution to be s_i, and do the reaction in the two-stage system of a CSTR with the τ_1 value of $(t_m - t_g)$ and a tubular reactor with the τ_2 value of $(t_e - t_m)$. The value of the shortest overall residence time, τ_e, is given by

$$\tau_e = \tau_1 + \tau_2 = t_e - t_g \,. \tag{53}$$

If the point (t_m, s_m) is not a stable point, this question can not be solved by this procedure.

3.2.2 Multi-Stage System

For the batch curve of $p = f_p(t)$ shown in Fig. 11, the continuous reactor system with the shortest residence time is determined by the method described in Sect. 3.1 to be a three-stage system of a tubular reactor with the τ_1 value of $(t_1 - t_i)$,

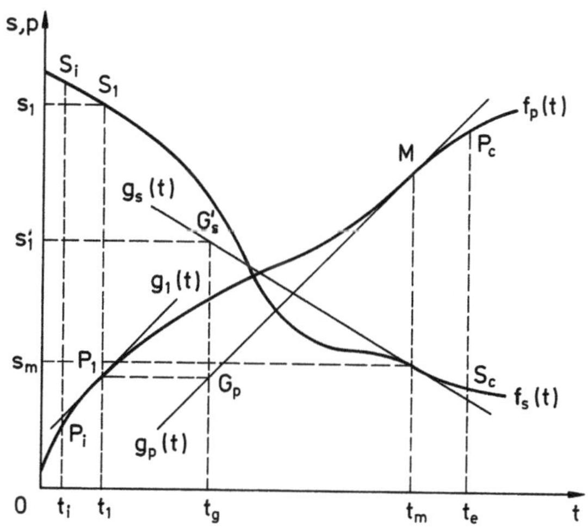

Fig. 11. Graphical analysis of a general two-curve reaction for designing a reaction system with the shortest residence time. The curves of $s = f_s(t)$ and $p = f_p(t)$ and the values of p_i and p_e are given. $M(t_m, p_m)$ is an inflection point of $p = f_p(t)$ that satisfies Eq. (52). $p = g_p(t)$, $s = g_s(t)$ and $p = g_1(t)$ are tangential lines at the points $M(t_m, p_m)$, (t_m, s_m) and $P_1(t_1, p_1)$, respectively; $p = g_1(t)$ has the same slope as $p = g_p(t)$. The coordinates of the other points are as follows: $P_c(t_c, p_c)$, $S_c(t_c, s_c)$, $P_i(t_i, p_i)$, $S_i(t_i, s_i)$, $S_1(t_1, s_1)$, $G_p(t_g, p_1)$ and $G_s'(t_g, s_1')$. The points P_c and P_i are on $p = f_p(t)$, the points S_c, S_i and S_i are on $s = f_s(t)$, the point G_p is on $p = g_p(t)$ and the point G_s' is on $s = g_s(t)$

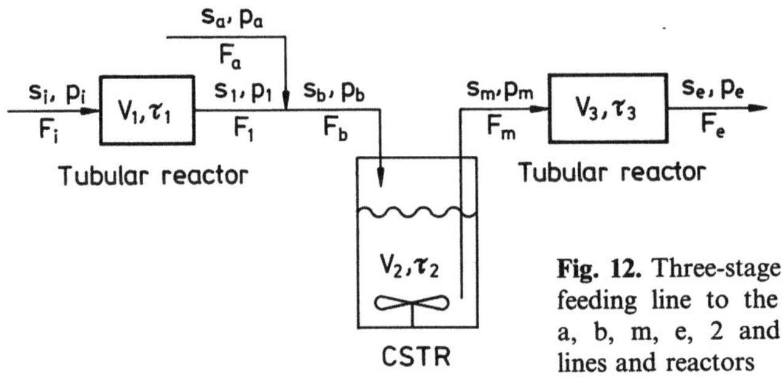

Fig. 12. Three-stage reactor system with an extra feeding line to the CSTR. The subscripts i, 1, a, b, m, e, 2 and 3, show the names of the lines and reactors

a CSTR with the τ_2 value of $(t_m - t_g)$ and a tubular reactor with the τ_3 value of $(t_e - t_m)$ (see Fig. 11). In this case, however, the first tubular reactor and the second CSTR cannot directly be connected, because the concentration of S in the outlet solution from the first reactor (s_1 in Fig. 11) is different from the required value of the inlet concentration of S (s_1' in Fig. 11). It should be noted that if a CSTR is placed at the first stage of a system, the required conditions can be obtained as shown in Fig. 10. Thus, for multi-curve reactions, the required operational conditions can be determined only for the reactor systems in which a CSTR is not used in the second or a later stage.

To overcome this difficulty, we add an extra feeding line to a CSTR when it is used in the second stage as shown in Fig. 12. By this feeding, the composition of the inlet solution can be changed to the required one; this is the same idea used in Sect. 2.3.

The material balances for the reaction system are written as follows:

$$F_b = F_1 + F_a \tag{54}$$

$$F_1 = F_i \quad \text{and} \quad F_e = F_m = F_b \tag{55}$$

$$p_b F_b = p_1 F_1 + p_a F_a \tag{56}$$

$$s_b F_b = s_1 F_1 + s_a F_a \tag{57}$$

where the subscripts i, 1, a, b, m and e, show the inlet line to the first reactor, the outlet line from the first reactor, the extra feeding line, the inlet line to the CSTR, the outlet line from the CSTR and the outlet line from the third reactor, respectively. From these equations, the following relationships are obtained:

$$F_1/F_a = (s_a - s_b)/(s_b - s_1) = (p_a - p_b)/(p_b - p_1) \tag{58}$$

The residence times for the first, second, and third reactors are defined as

$$\tau_1 = V_1/F_i, \quad \tau_2 = V_2/F_b \quad \text{and} \quad \tau_3 = V_3/F_m \tag{59}$$

$$\tau_e = \tau_1 + \tau_2 + \tau_3 \tag{60}$$

where V_1, V_2 and V_3 are the working volumes of the reactors (see Fig. 12). These equations are the same as Eqs. (30), (32), (33) and (36), if the subscript 1 is replaced by i. Therefore, the operational conditions of this system are determined by the same procedure as described in Sect. 2.3.1

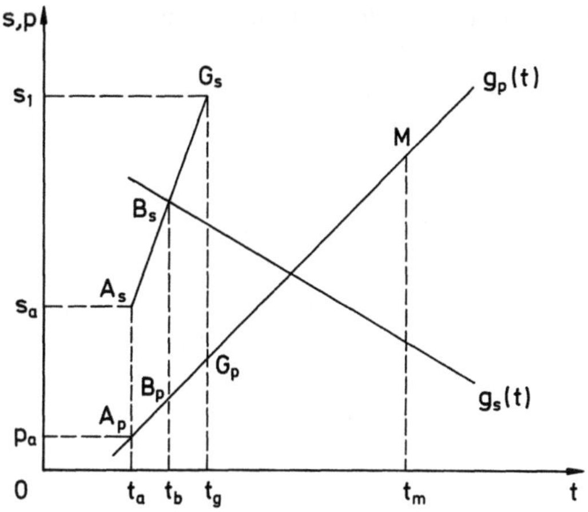

Fig. 13. Graphical solution for a CSTR with an extra feeding system shown in Fig. 12. The curves of $s = g_s(t)$ and $p = g_p(t)$ and the values of s_1, p_1, s_m and p_m are given or determined as shown in Fig. 11. The coordinates of the points newly located in this figure are as follows: $A_p(t_a, p_a)$, $A_s(t_a, s_a)$, $G_s(t_g, s_1)$, $B_p(t_b, p_b)$ and $B_s(t_b, s_b)$. The points A_p and B_p are on $p = g_p(t)$; the point B_s is on $y = g_s(t)$ and also on the line segment A_sG_s. For p_a and s_a, arbitrary values are chosen in the ranges of $p_a < p_m$ and $s_a < g_s(t_a)$. The τ_2 value is obtained as $\tau_2 = (t_m - t_b)$

Question. Consider the two-curve reaction as shown in Fig. 11. As the results of the analysis shown in Fig. 11, it is requested that the reaction is done in a three-stage continuous system shown in Fig. 12, where the values of F_i, V_1, τ_1, τ_3, s_i, p_i, s_1, p_1, s_m, p_m, s_e and p_e are given or determined as shown in Fig. 11. Determine the values of F_a, τ_2, s_a, p_a and the other parameters which satisfy the above requirements.

Solution. The following procedure is shown in Figs. 11 and 13. Draw the two tangential lines $p = g_p(t)$ and $s = g_s(t)$ at the point $M(t_m, p_m)$ on $p = f_p(t)$ and at the point (t_m, s_m) on $s = f_s(t)$, respectively. Locate the points $G_p(t_g, p_1)$ and $A_p(t_a, p_a)$ on $p = g_{mp}(t)$, here an arbitrary value for p_a can be chosen in the range of $p_a < p_m$; then locate the points $G_s(t_g, s_1)$ and $A_s(t_a, s_a)$, here an arbitrary value for s_a can be chosen in the range of $s_a > g_s(t_a)$ for the case of $s_1 < g_s(t_g)$, or in the range of $s_a < g_s(t_a)$ for the case of $s_1 > g_s(t_g)$ (the case of Fig. 13). Connect the two points A_s and G_s with a straight line, locate the point $B_s(t_b, s_b)$ where the line A_sG_s crosses with the line $s = g_s(t)$, and locate point $B_p(t_b, p_b)$ on $p = g_p(t)$ by using the t_b value. The value of τ_2 is obtained from the figure as

$$\tau_2 = t_m - t_b \,. \tag{61}$$

The values of F_1, F_a, F_b, F_m, F_e, V_2 and V_3 are obtained from Eqs. (46), (54), (55) and (59). Thus the values of all the parameters in Fig. 11 are determined.

The validity of the above solution is confirmed by the same procedure described under Proof in Sect. 2.3.1. The above solution is easily extended to the cases general multi-curve reactions by the similar procedure described in Sect. 2.3.1.

In the reactor system shown in Fig. 12, the residence time of the CSTR, τ_2, depends on the values of p_a and s_a. In Fig. 13, for example, the τ_2 value increases with the increase in the s_a value. On the other hand, the increase in the p_a value always decreases the τ_2 value; but practically the p_a value should be zero.

In conclusion, the strategy of designing the continuous reactor system with the shortest residence time described in Sect. 3.1 is also applicable to multi-curve reactions by adding an extra feeding line to the CSTR. The conditions of the extra feeding alter the τ_2 value and also the amount of the substances that go in and go out the CSTR (and the reactor system). Therefore we cannot generally compare the efficiency of the reactor systems with different extra feeding conditions. In practical meaning, however, the optimum system will be designed on the basis of the above analysis.

4. Criteria of the Stability of a Steady State in a CSTR

When we choose a point on the batch course of a reaction as a requested state of a continuous reaction in a CSTR, the state must be stable; otherwise we cannot obtain the steady state intended. Therefore, it is desirable to examine the stability of the point on the batch course before starting the continuous operation. This requirement is only for CSTRs, and for tubular reactors we can choose any point on the course as a steady state of the outlet solution.

Generally not all the states on the batch course of a reaction are realized as the steady states of continuous reactions. In this report, a stable point on a batch course is defined as the point whose state has the possibility to be realized as a steady state. The set of the stable points forms domains on the course, and we call them "stable domains". We used the word "possibility" because we judge the stability on the basis of the batch data alone. A point in the stable domains, however, has competence to be a candidate for an experimental operation.

Consider a single-curve reaction, $y = f(t)$. If a state at $y = y_e$ is realized as a steady state, the following expression must be satisfied in a CSTR:

$$[(d/dy)\,(dy/dt)]_{y=y_e} < 0\,. \tag{62}$$

If the above expression is not satisfied at $y = y_e$, the state cannot be realized. When a reactor is operating at $\tau = \tau_e$, Eq. (2) becomes

$$dy/dt = v(y) - (y - y_i)/\tau_e\,. \tag{63}$$

Differentiating the equation with y

$$(d/dy)\,(dy/dt) = dv/dy - 1/\tau_e\,. \tag{64}$$

At $y = y_e$

$$[(d/dy)\,(dy/dt)]_{y=y_e} = (dv/dy)_{y=y_e} - 1/\tau_e\,. \tag{65}$$

From Eqs. (65) and (5)

$$[(d/dy)\,(dy/dt)]_{y=y_e} = [(d/dy)\,(df/dt)]_{y=y_e} - 1/\tau_e \tag{66}$$

Therefore, Eq. (62) becomes

$$[(d/dy)\,(df/dt)]_{y=y_e} - 1/\tau_e < 0\,. \quad (67)$$

This expression is used as the definition that the point at $y = y_e$ on the batch course is stable, but actually Eq. (67) or (62) is a necessary condition. Conversely, if Eq. (67) is not satisfied, the point is unstable. For a multi-curve reaction, a point on the batch course is defined to be stable if Eq. (62) or (67) is satisfied for all the curves representing the reaction.

Suppose that a reaction is at a steady state in a CSTR. When the state is shifted to a new state by fluctuation, there are two possibilities whether or not the original steady state is spontaneously recovered. Here we assume that all the new states made by fluctuation are on the batch course of the reaction, because we cannot use the graphical method for the states that are not on the course. If we want to predict whether or not the steady state is recovered from a new state, the whole course from the new state to the original steady state or to the other states must be on the batch course. As described in Sect. 4.3, single curve reactions satisfy the above conditions. If the original steady state is recovered from a new state, we call the new state a "restitutive state (or point)". The set of the restitutive points of a steady state forms a domain on the batch course, and we call it the "restitutive domain" of a steady state. The size of the restitutive domain of a steady state shows a kind of capacity of the stability of the steady state. If a new state made by fluctuation or a state on the return course from the new state is not on the batch course, we can not predict whether the new state is a restitutive state or not, but this does not mean that the new state is not a restitutive state.

Stability and reproducibility are different categories. When we choose a point on a batch course as a requested steady state, the point must have both stability and reproducibility.

4.1 Stability Criterion I

Question. Consider a single-curve reaction, $y = f(t)$. It is requested that the reaction is done in a CSTR at a steady state of y_e and an inlet concentration of y_i. Judge the stability of the point $Y_e(t_e, y_e)$ on the batch course.

Solution. Put the point $Y_x(t_x, y_x)$ on $y = f(t)$ in the sufficiently small neighbourhood of the point $Y_e(t_e, y_e)$. Find out the residence times τ_e and τ_x of the points Y_e and Y_x, respectively, by the tangential method. Now define the discriminant, D, as

$$D = (y_x - y_e)\,(\tau_x - \tau_e)\,(df/dt)_{y=y_x} \quad (68)$$

where $(df/dt)_{y=y_x}$ is the tangential slope at point Y_x. The conclusion is as follows: if $D > 0$, the point Y_e is stable; if $D < 0$, the point Y_e is unstable.

Proof. When $y = y_x$ and $\tau = \tau_e$, Eq. (62) becomes

$$(dy/dt)_{y=y_x} = v(y_x) - (y_x - y_i)/\tau_e \quad (69)$$

where $(dy/dt)_{y=y_x}$ is the rate of the change in y in the reactor at $y = y_x$, and $v(y_x)$ is the reaction rate for Y at $y = y_x$. By using Eq. (5) at $y = y_x$, Eq. (69) becomes

$$(dy/dt)_{y=y_x} = (df/dt)_{y=y_x} - (y_x - y_i)/\tau_e \,. \tag{70}$$

On the other hand, the value of τ_x is determined so as to satisfy Eq. (13) at $y = y_x$; thus

$$y_x - y_i = \tau_x (df/dt)_{y=y_x} \,. \tag{71}$$

From Eqs. (70) and (71)

$$(dy/dt)_{y=y_x} = (\tau_e - \tau_x)(1/\tau_e)(df/dt)_{y=y_x} \,. \tag{72}$$

From Eqs. (72) and (68)

$$(y_x - y_e)(dy/dt)_{y=y_x} = (-D)/\tau_e \,. \tag{73}$$

If the following expression is satisfied, Eq. (62) is also satisfied:

$$(y_x - y_e)(dy/dt)_{y=y_x} < 0 \,. \tag{74}$$

Therefore, from Eqs. (73) and (74) and from the fact of $\tau_e > 0$, the stability of the point $Y_e(t_e, y_e)$ is predicted by the sign of the discriminant D.

It should be pointed out that for a single-curve reaction, the same τ_x value must be obtained when the procedures described under "Solution" is used for the batch curves of the other substances concerned. By this method, we can confirm that the reaction is a single-curve reaction at least between the points Y_e and Y_x.

4.2 Stable Domains and Stability Criterion II

Question. In the same situation described under "Question" in the previous section, find out the stable domains on the batch curve.

Solution. Draw the following logarithmic curve as shown in Fig. 14:

$$y = F(t) = y_0 \ln |f(t) - y_i| \tag{75}$$

where y_0 is a positive constant having a dimension of concentration. Look for the inflection points of $y = F(t)$, and name them $M_j(t_{mj}, y_{mj})$ and $W_j(t_{wj}, y_{wj})$, where $j = 1, 2, \ldots$ They are the points that satisfy the following conditions:

$$(d^2F/dt^2)_{t=t_{mj}} = 0 \quad \text{and} \quad (d^3F/dt^3)_{t=t_{mj}} < 0 \tag{76}$$

$$(d^2F/dt^2)_{t=t_{wj}} = 0 \quad \text{and} \quad (d^3F/dt^3)_{t=t_{wj}} > 0 \tag{77}$$

where the order of the points is $t_{mj} > t_{wj} > t_{m(j+1)} > t_{w(j+1)}$. The curve shown in Fig. 14 has the points W_1, M_2, W_2 and M_3.

The stable domains on $y = F(t)$ are between W_j and M_{j+1} ($j = 1, 2, \ldots$) except the both ends, and the other domains are unstable; in the case of Fig. 14, the stable domains are between W_1 and M_2 and between W_2 and M_3. This means that the stable domains or points have negative second derivatives on $y = F(t)$; this is another expression of stability (criterion II). If the curve of $y = f(t)$ is drawn in the same figure, the stable domains on $y = F(t)$ can easily

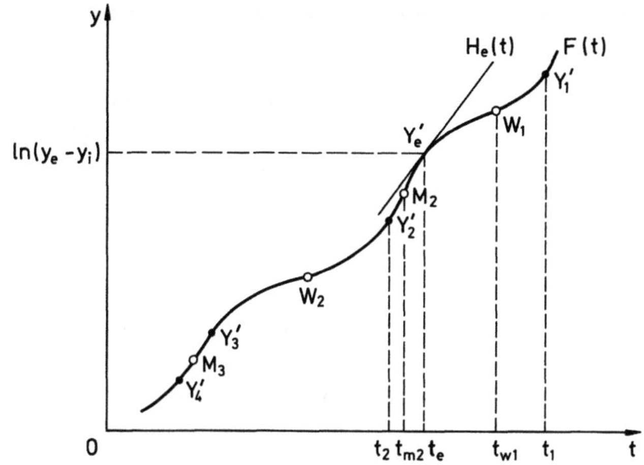

Fig. 14. Graphical solution of the stable domains of a reaction and the restitutive domain of a steady state. The curve of $y = f(t)$ (not shown) and the values of y_e and y_i are given ($y_e > y_i$). The curve of $y = F(t)$ is expressed by $F(t) = \ln[f(t) - y_i]$ ($y_0 = 1$ in Eq. (75)). $y = H_e(t)$ is the tangential line at $Y'_e(t_e, \ln(y_e - y_i))$ on $y = F(t)$. M_2, M_3, W_1 and W_2 are inflection points defined by Eqs. (76) and (77). Y'_1, Y'_2, Y'_3 and Y'_4 are the points having the same tangential slope as Y'_e. The stable domains are between W_1 and M_2 and between W_2 and M_3 except the ends, i.e., $t_{w1} > t > t_{m2}$ and $t_{w2} > t > t_{m3}$. The restitutive domain of Y'_e is between Y'_1 and Y'_2 except the ends, i.e., $t_1 > t > t_2$

be transferred to those on $y = f(t)$, and are in the range of $t_{wj} > t > t_{m(j+1)}$ ($j = 1, 2, \ldots$).

Proof. Differentiating Eq. (75), the following equation is obtained if $y_e > y_i$:

$$(d^2F/dt^2)_{t=t_e} = y_0[(d^2f/dt^2)_{t=t_e}/(y_e - y_i) - (df/dt)^2_{t=t_e}/(y_e - y_i)^2] \,. \tag{78}$$

The second derivative of $y = f(t)$ is transformed as follows:

$$\begin{aligned} d^2f/dt^2 &= (dy/dt)\,[(d/dy)\,(df/dt)] \\ &= (df/dt)\,[(d/dy)\,(df/dt)] \end{aligned} \tag{79}$$

at $t = t_e$ and $y = y_e$

$$(d^2f/dt^2)_{t=t_e} = (df/dt)_{y=y_e}[(d/dy)\,(df/dt)]_{y=y_e} \,. \tag{80}$$

From Eqs. (78), (80), (13) and (5)

$$(d^2F/dt^2)_{t=t_e} = (y_0/\tau_e)\,([(d/dy)\,(df/dt)]_{y=y_e} - 1/\tau_e) \,. \tag{81}$$

The same equation is obtained in the case of $y_e < y_i$, too. From the definition of stability shown by Eq. (67), the following conclusion is obtained: if $(d^2F/dt^2)_{t=t_e} < 0$ the point Y_e is stable, and if $(d^2F/dt^2)_{t=t_e} \geqq 0$ the point Y_e is unstable; the equality of the latter expression is satisfied at $t = t_{mj}$ and $t = t_{wj}$ ($j = 1, 2, \ldots$). Thus the stable domains on $y = F(t)$ are between W_j and M_{j+1} ($j = 1, 2, \ldots$)

except the both ends, and those on $y = f(t)$ are in the ranges of $t_{wj} > t > t_{m(j+1)}$ $(j = 1, 2, \ldots)$.

For a multi-curve reaction, examine the stability of a requested point on the batch course for all the curves. If the point is judged to be stable for all the curves, the point is stable. If the point is judged to be unstable for any one of the curves, the point is unstable. The stable domains are obtained as those common between all the curves.

4.3 Restitutive Domain of a Steady State

For a single-curve reaction, the whole course from a fluctuated state to the original steady state or to the other states is on the batch course, but for a multi-curve reaction, the states on the course from the fluctuated state are not on the batch course (see Proposition below). Therefore, we can determine the restitutive domain of a steady state only for a single-curve reaction.

Question. Consider a single-curve reaction, $y = f(t)$. It is requested that the reaction is done in a CSTR at a steady state of y_e with an inlet concentration of y_i and a residence time of τ_e. Determine the restitutive domain of the steady state.

Solution. The following procedure is shown in Fig. 14. Draw the curve of $y = F(t)$ which is expressed by Eq. (75), and draw the tangential line $y = H_e(t)$ at the point $Y'_e(t_e, y_0 \ln(y_e - y_i))$ on $y = F(t)$. Look for the points where the slopes of the tangential lines are the same as that of $y = H_e(t)$, and name them $Y'_j(t_j, y_0 \ln(y_j - y_i))$, where $j = 1, 2, \ldots$ and $t_j > t_{j+1}$. If more than one point is found in the both sides of point Y'_e, choose the two neighbouring points Y'_j and Y'_{j+1} so that the domain between them contains the point Y'_e. The restitutive domain on $y = F(t)$ of the steady state at $y = y_e$ is between Y'_j and Y'_{j+1} except the both ends, that is, $t_j > t_e > t_{j+1}$ on $y = f(t)$. If no point is found in the region of $t > t_e$, the region on $y = F(t)$ [and on $y = f(t)$] is the restitutive domain, and the theoretical boundary point is the end point where $(df/dt)_{t=t_x} = 0$. If no point is found in the region of $t < t_e$, the region on $y = F(t)$ is the restitutive domain; in this case, the region on $y = f(t)$ is $t_i < t < t_e$, and we can not examine the region of $t \leqq t_i$ by this method (for this case, see Sect. 4.4), here the t_i value is obtained as the coordinates of the point $Y_i(t_i, y_i)$ on $y = f(t)$. When the fluctuated point is on the boundary point, we can nor predict whether the original state is recovered or not. When the fluctuated point is out of the restitutive domain, the original state is not recovered.

Proof. Let us suppose that the composition of the reaction mixture is shifted from y_e to y_x, i.e., from the point $Y_e(t_e, y_e)$ to the point $Y_x(t_x, y_x)$ on $y = f(t)$, by fluctuation. For a single-curve reaction, we can judge whether point Y_x is a restitutive point or not by Eq. (74): if Eq. (74) is satisfied in the range between the points Y_e and Y_x, the point Y_x is a restitutive point.

The following proof is for the case of $y_e > y_i$; for the case of $y_e < y_i$, the procedure is the same. In this case, the following expression is formed because the batch curve of a single-curve reaction is monotone:

$$df/dt > 0 \,. \tag{82}$$

As point Y_e is a stable point, the following expression is formed from stability criterion II

$$(d^2F/dt^2)_{t=t_e} < 0 . \tag{83}$$

Differentiating Eq. (75)

$$(dF/dt)_{t=t_e} = y_0/\tau_e . \tag{84}$$

The tangential line $y = H_e(t)$ is expressed as

$$H_e(t) = (y_0/\tau_e)(t - t_e) + y_0 \ln (y_e - y_i) . \tag{85}$$

The points Y'_j and Y'_{j+1} are located so as to satisfy the following equation:

$$(dF/dt)_{t=t_j} = (dF/dt)_{t=t_{j+1}} = y_0/\tau_e . \tag{86}$$

From Eqs. (83) and (86) and the relative positions of the points Y'_e, Y'_j and Y'_{j+1}, the following expressions are formed for the point Y'_x if the point is in the region between Y'_j and Y'_{j+1} on $y = F(t)$:

$$(dF/dt)_{t=t_x} = y_0/\tau_x < y_0/\tau_e (t_j > t_x > t_e) \tag{87}$$

$$= y_0/\tau_x > y_0/\tau_e (t_e > t_x > t_{j+1}) . \tag{88}$$

From Eqs. (73) and (68)

$$(y_x - y_e)(dy/dt)_{y=y_x} = -(y_x - y_e)(\tau_x - \tau_e)(df/dt)_{y=y_x} . \tag{89}$$

From Eqs. (89), (82), (87) and (88), Eq. (74) is satisfied. This means that point Y'_x is a restitutive point if it is in the region between the points Y'_j and Y'_{j+1} on $y = F(t)$. Therefore, the region between Y'_j and Y'_{j+1} except the both ends (i.e., $t_j > t_e > t_{j+1}$) is the restitutive domain of the point Y'_e (or Y_e). If point Y'_x is on one of the boundary points, the right hand side of Eq. (89) becomes zero because $\tau_x = \tau_e$; in this case, we cannot predict which direction point Y'_x will move. If point Y'_x goes beyond the boundary points, the point enter the unrestitutive domain, and the point can not return to Y'_e; for example, if point Y'_x is in the region between Y'_{j+1} and Y'_{j+2}, the inequality signs in Eqs. (87) and (88) are reversed, and therefore Eq. (74) is not satisfied; if point Y'_x is in the region between Y'_{j+2} and Y'_{j+3}, Eq. (74) is satisfied, but point Y'_x can not return to Y'_e beyond the point Y'_{j+2}, and a new steady state at Y'_{j+2} is obtained.

Eq. (84) shows that the slope of the tangential line at Y'_e gives the value of τ_e that is the τ value of the point Y_e on $y = f(t)$. The points Y'_j ($j = 1, 2, \ldots$) are made to have the same tangential slope as Y'_e, therefore these points also give the same τ_e value to the points $Y_j(t_j, y_j)$ ($j = 1, 2, \ldots$) on $y = f(t)$. In short, the points Y_j ($j = 1, 2, \ldots$) are the isotau points of Y_e.

From Eqs. (15) and (78), $y = H_e(t)$ is related to one of the isotau curves of Y_e, $y = h_e(t)$, by the following equation:

$$H_e(t) = y_0 \ln [h_e(t) - y_i] . \tag{90}$$

Thus isotau points of Y_e can be found by using $y = H_e(t)$ and $y = F(t)$ as well as $y = h_e(t)$ and $y = f(t)$. On the other hand, the inflection points on $y = F(t)$ correspond to the points on $y = f(t)$ where the curve of $y = f(t)$ touches and intersects the isotau curves drawn at these points. Thus the curve of $y = h_e(t)$ can be used to judge the stability of point Y_e and to find the restitutive domain of the point.

As shown in the above results, the curve of $y = F(t)$ is very useful. It visualizes the characteristic points on $y = f(t)$ such as the boundary points of stable domains or restitutive domains and isotau points; the τ value of a point is also obtained from the tangential slope at the corresponding point.

It should be noted that the positions of isotau points and the positions of the border points of stable domains on $y = f(t)$ change depending on the y_i value since the shape of the curve of $y = F(t)$ changes. This means that the ranges of a stable domain and a restitutive domain are controllable by the y_i value.

Proposition. If the whole course from the fluctuated state to the original syeady state or to the other states is on the batch course of the reaction, the reaction is a single-curve reaction at least during the course from the fluctuated state. The reverse is also true.

Proof. Consider a two-curve reaction expressed by Eqs. (17) and (18). The reaction is at a steady state of $s = s_e$ and $p = p_e$, with the inlet concentrations of s_i and p_i and the residence time of τ_e. Now let us suppose that the composition of the reaction mixture is shifted to s_x and p_x by fluctuation. From Eq. (70)

$$(ds/dt)_{s=s_x} = (df_s/dt)_{s=s_x} - (s_x - s_i)/\tau_e \tag{91}$$

$$(dp/dt)_{p=p_x} = (df_p/dt)_{p=p_x} - (p_x - p_i)/\tau_e \,. \tag{92}$$

From Eq. (18)

$$(df_p/dt)_{p=p_e} = r_{ps}(t_e)\,(df_s/dt)_{s=s_e} \tag{93}$$

$$(df_p/dt)_{p=p_x} = r_{ps}(t_x)\,(df_s/dt)_{s=s_x} \,. \tag{94}$$

If the states on the course from the point (s_x, p_x) are on the batch course, the following equation must be satisfied at the fluctuated point:

$$(dp/dt)_{p=p_x} = r_{ps}(t_x)\,(ds/dt)_{s=s_x} \,. \tag{95}$$

From Eqs. (91), (92) and (95)

$$(df_p/dt)_{p=p_x} - (p_x - p_i)/\tau_e = r_{ps}(t_x)\,[(df_s/dt)_{s=s_x} - (s_x - s_i)/\tau_e] \,. \tag{96}$$

From Eqs. (96) and (94)

$$p_x - p_i = r_{ps}(t_x)\,(s_x - s_i) \,. \tag{97}$$

On the other hand, the tangential line $p = w_e(s)$ at the point (p_e, s_e) on $p = u(s)$ is written from Eqs. (25) and (18) as

$$w_e(s) = r_{ps}(t_e)\,(s - s_e) + p_e \,. \tag{98}$$

Similarly, the tangential line $p = w_x(s)$ at the point (s_x, p_x) on $p = u(s)$ is written as

$$w_x(s) = r_{ps}(t_x)\,(s - s_x) + p_x \,. \tag{99}$$

As described in Sect. 2.2, the point (s_i, p_i) is on $y = w_e(s)$. This is expressed by

$$w_e(s_i) = r_{ps}(t_e)\,(s_i - s_e) + p_e = p_i \,. \tag{100}$$

Equation (97) shows that at $s = s_i$ Eq. (99) becomes

$$w_x(s_i) = r_{ps}(t_x)(s_i - s_x) + p_x = p_i \,. \tag{101}$$

Equations (100) and (101) mean that the two tangential lines $y = w_e(s)$ and $y = w_x(s)$ intersect at the point (s_i, p_i). Since Eq. (101) is formed for all the points on the course from the point (s_x, p_x) to the point (s_e, p_e) or to the other points, the course segment from the point (s_x, p_x) to the point (s_e, p_e) or to the other points is on the line of $p = w_x(s)$. As described in the proposition, the course segment from the point (s_x, p_x) is also on $p = u(s)$, and therefore the reaction is a single curve reaction with a constant stoichiometric coefficient of $r_{ps}(t_x)$ at least on the the course segment. When the point (s_x, p_x) is in the restitutive domain of point Y_e, the line $p = w_x(s)$ coincides with $p = w_e(s)$.

It is obvious that the reverse is also true.

4.4 Extension of the Classical Method and Stability Criterion III

The characteristic points on $y = f(t)$, which are visualized on $y = F(t)$ as described in the previous section, are also visualized on the curve of $z = v(y)$ as shown in Fig. 15. The stable domains are between W'_j and M'_{j+1} except the both ends. Here M'_j and W'_j $(j = 1, 2, \ldots)$ are the points where the lines from the point $(y_i, 0)$ touch the curve of $z = v(y)$, and M'_j and W'_j are in the region of negative and positive second derivatives, respectively. This is the third expression of stability criterion (criterion III; see Proposition described below). When the line from the point $(y_i, 0)$ touches the curve of $z = v(y)$ at an inflection point, the corresponding two neighbouring points overlap each other at the inflection point, and the domain between them becomes the point.

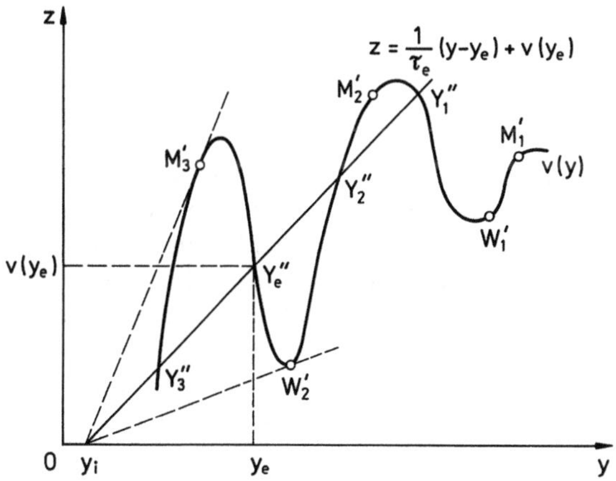

Fig. 15. Graphical analysis of the stability of steady state in a CSTR. The curve of $y = f(t)$ (not shown) and the values of y_e and y_i are given. The curve of $z = v(y)$ is drawn by calculating the values of $v(y)$ from the curve of $y = f(t)$. M'_1, M'_2, M'_3, W'_1 and W'_2 are the points where the lines from the point $(y_i, 0)$ touch the curve of $z = v(y)$. The line that passes the points $(y_i, 0)$ and $Y''_e(y_e, v(y_e))$ is expressed by $z = [(1/\tau_e)(y - y_e) + v(y_e)]$; Y''_1, Y''_2 and Y''_3 are the points where this line intersects $z = v(y)$

Suppose that we want to do the reaction in a CSTR at the point $Y_e(t_e, y_e)$ which corresponds to the point $Y_e''(y_e, v(y_e))$ in a stable domain on $z = v(y)$. Connect the two points Y_e'' and $(y_i, 0)$; the line is expressed from Eq. (4) by

$$z = (1/\tau_e)(y - y_e) + v(y_e)\,. \tag{102}$$

Therefore, the value of the required residence time, e, is obtained from the slope of this line. The points Y_j'' $(j = 1, 2, \ldots)$ where the line intersects the curve of $z = v(y)$ correspond to the isotau points Y_j $(j = 1, 2, \ldots)$ of Y_e, and thus the restitutive domain of point Y_e'' (and Y_e) is obtained (in Fig. 15, between Y_2'' and Y_3'' except the both ends). It should be pointed out that by this method we can examine the region of $t \leqq t_i$ whether the region is restitutive or not. The merit of this method is that the effects of the y_i value are easily examined.

In the classical method, the curve of $1/v(y)$ vs y is used as shown in Fig. 1, but as shown in Fig. 15, the curve of $z = v(y)$ expresses the information on the stability and operational conditions more clearly. However, to draw the curve of $z = v(y)$ is not easy. In conclusion, the tangential method is the best method for graphical solution.

Proposition. If the tangential slope (α_e) at Y_e'' is smaller than the slope (β_e) of the line expressed by Eq. (102), the point Y_e'' (and consequently point Y_e) is a stable point.

Proof. Consider a single-curve reaction, $y = f(t)$. The difference between the two slopes α_e and β_e is

$$\alpha_e - \beta_e = [(d/dy)(df/dt)]_{y=y_e} - 1/\tau_e\,. \tag{103}$$

From the definition of stability of Eq. (67), the following conclusion is obtained: $\alpha_e < \beta_e$ the point Y_e'' (namely Y_e on $y = f(t)$) is stable, and if $\alpha_e \geqq \beta_e$ the point is unstable.

5. Conclusions

The tangential method is a simple graphical method for determination of the operational conditions of a continuous reactor on the basis of the batch data of a reaction. The course of the batch reaction is directly analyzed by this method, and therefore we do not need to calculate the reaction rates from the batch data. The main work in this method is to draw the tangential line at the requested point on the batch curve, and the values of the conditions for a CSTR operating at the requested state are directly shown by the lengths along the axes. The precise batch data for drawing the tangential line are needed only in the vicinity of the requested point.

The tangential method is used for a multi-curve reaction as well as a single-curve reaction. The methods for discrimination between the two reaction types are described; the course of a reaction drawn as a concentration diagram (see Fig. 4) clearly shows the characteristics of the two types.

The tangential method is applicable for designing complex reaction systems. For a continuous reactor with a recycling system, the desirable recycling conditions can be obtained. The recycling system enables us to choose arbitrary values for the concentrations of the substances in the inlet and outlet solutions. A concentration diagram (see Fig. 7) clearly shows the relationships between the concentrations in the inlet, recycled, and outlet solutions. A multi-stage reactor system with the shortest residence time is designed. For a single-curve reaction, a general strategy is obtained. For a multi-curve reaction, the same strategy is applicable by adding extra feeding lines to CSTRs. The extra feeding line enables us to adjust the composition of the inlet solution to the required one for the operation of the CSTR.

The tangential method is applicable for predicting the stability of a point on the batch course of a reaction. A stable point on the batch course is defined, and a necessary condition for a point to be a stable point is shown. On the basis of the necessary condition, three kinds of criteria for judging the stability of the points are described. Criterion I uses the tangential method directly to the batch curves, criterion II uses a logarithmic plot of the batch curves, and criterion III uses the plot of velocity vs. concentration. The plots used for the criteria II and III show stable domains as well. A stable point has a restitutive domain. If the fluctuated states are on the batch course of the reaction, and if the reaction is a single-curve reaction, the restitutive domain of a steady state can be determined by using the plot for criterion II or III. The boundary points of the domain are the isotau points of the steady state. For a multi-curve reaction, the restitutive domain cannot be determined.

6. Acknowledgements

The authors wish to thank Dr. Suteaki Shioya for helpful discussions.

7. Nomenclature

C	constant [Eq. (14)]
CSTR	continuous stirred tank reactor
D	discriminant [Eq. (68)]
F	volumetric flow rate ($l\ h^{-1}$)
p	concentration of a product, P ($mol\ l^{-1}$)
$r_{jk}(t)$	stoichiometric coefficient between two substances, Y_j and Y_k, at time t ($mol\ mol^{-1}$)
s	concentration of a substrate, S ($mol\ l^{-1}$)
V	working volume of a reactor (l)
v(y)	reaction rate in the reaction mixture whose composition is represented by y ($mol\ l^{-1}\ h^{-1}$)
y	concentration of a substance, Y ($mol\ l^{-1}$)
y_0	positive constant [Eq. (75)] ($mol\ l^{-1}$)
τ	residence time (h)

2.7 Enzymes in Organic Synthesis: Aldol Reactions as a Route to Unusual Sugars

C.-H. Wong

Department of Chemistry, Texas A & M University, College Station, Texas 77843, USA

Contents

Enzymes have been increasingly used as catalysts in practical organic synthesis. This talk illustrates several new practical syntheses of biologically relevant carbohydrates using aldolases as catalysts and summarizes the status of several generic problems in this area.

The potential of enzymes being used as synthetic catalysts stems from the fact that they generally catalyze selective transformations under mild condition without the need for multiple protection and deprotection steps. This potential has been recognized for many years, and today a number of biologically relevant molecules are indeed industrially made by enzymatic methods [1]. In addition, many new synthetic reactions using existing or newly isolated enzymes as catalysts are being discovered and will provide a potentially useful addition to the group of catalysts used in practical synthesis. The enzymatic approach to certain synthetic processes will thus complement conventional synthetic technology on the one hand, and other biologically-based syntheses (bioproduction, recombinant DNA technology, and cell culture) on the other.

1. Asymmetric Carbon-Carbon Bond Formation Catalyzed by Aldolases

Asymmetric carbon-carbon bond formation in aldol addition reactions is one of the most useful and efficient methods in synthetic organic chemistry. Most of the organic aldol reactions developed recently are stoichiometric and require the use of a metal or metal-like enolate complex to achieve stereoselectivity. Since the enolate complex is unstable and tends to dissociate in aqueous solution, the aldol reactions must be carried out in organic solvents at low temperature. This requirement limits the application of organic aldol reactions to synthesis of molecules

Bioproducts and Bioprocesses
Editors: Fiechter, Okada, Tanner

soluble in organic media. For those polyfunctional compounds, such as carbohydrates, the synthesis still relies upon other methods.

The aldol addition catalyzed by the enzymes aldolases, however, are performed in aqueous solution at neutral pH using polyfunctional substrates without protection of the functional groups. Both chemical and enzymatic aldol reactions thus could complement each other and extend the synthetic utility of aldol reactions.

There have been about 30 different aldolases isolated so far; each of them catalyzes an aldol reaction using different substrates with different stereospecificity in carbon-carbon bond formation [1,2]. Most of the reactions are involved in the biosyntheses of carbohydrates and amino acids.

Although there seems no way of changing the stereochemistry of the centers formed in aldolase-catalyzed condensations, different aldols can be prepared by using analogous substrates. The enzyme fructose-1,6-diphosphate (FDP) aldolase, for example, catalyzes the aldol condensation using dihydroxy acetone phosphate as one substrate and a variety of aldehydes as the second. This allows preparation of a number of uncommon sugars and isotopically labelled sugars [3-5]. The isotopically labelled sugars are useful for study of biosynthesis and for determination of structures of

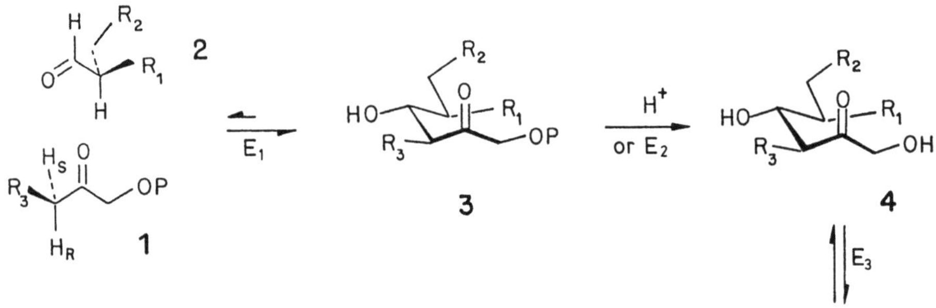

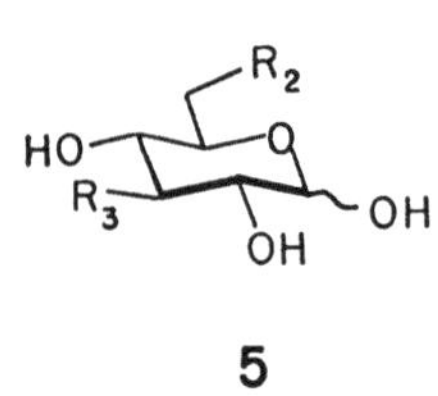

cmpd	R_1	R_2	R_3
1a	--	--	OH
1b	--	--	H
2a	H	OH	--
2b	OH	H	--
2c	OH	OCH_3	--
2d	OH	F	--
3a - 4a	H	OH	OH
3b - 5b	OH	H	OH
3c - 5c	OH	OCH_3	OH
3d - 5d	OH	F	OH
3e - 5e	OH	OH	H

E_1, FDP Aldolase
E_2, Acid Phosphatase
E_3, Glucose Isomerase

Fig. 1. Synthesis of unusual sugars using FDP-aldolase and glucose isomerase as catalysts sugars using NMR techniques. The same strategy can be employed for the synthesis of sugars with different labelling patterns and for rapid synthesis of radioactively labelled sugars for use in clinical application. The 6-deoxysugar and derivatives prepared can be converted to a caramel flavor component Furaneol, which has been a useful additive in the food technology.

We have recently developed the synthesis of unusual sugars using FDP aldolase and glucose isomerase as catalysts (Fig. 1). This combined enzymatic process allows preparation of hexoses with modification at 3, 5, or 6 position via stereospecific C—C bond formation and isomerization [6]. The compounds prepared are of interest as probes to elucidate complex biochemical processes, as inter-

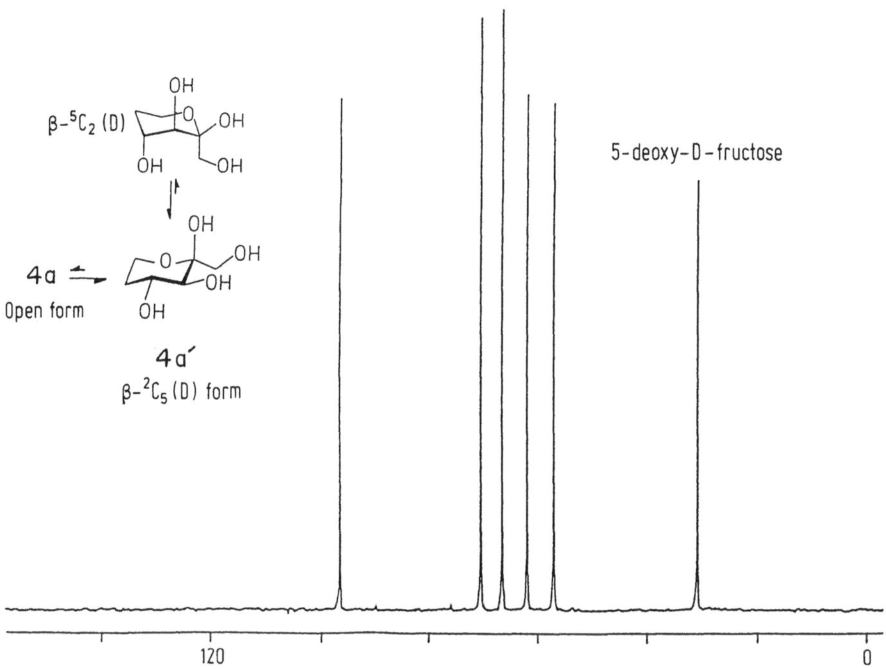

Fig. 2. ^{13}C NMR spectrum of 5-deoxy-D-fructose indicates that it exists in water as a $^{2}C_5$-β-pyranose form (see Ref. 6)

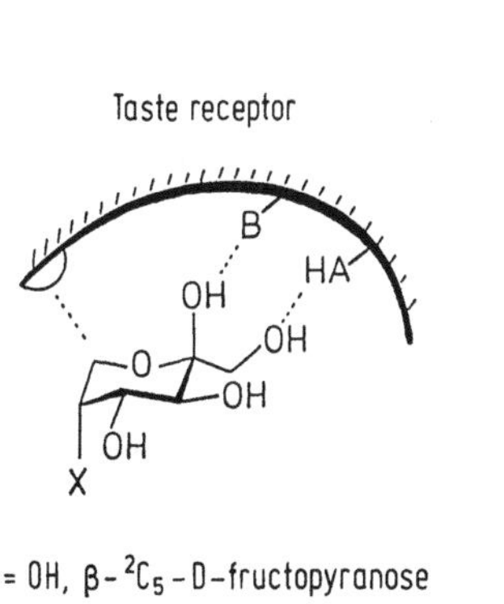

Fig. 3. H-Bonding and lipophilic interactions between substrate and the taste receptor. The receptor model was proposed by Shallenberger (Nature *216*, 480 (1977))

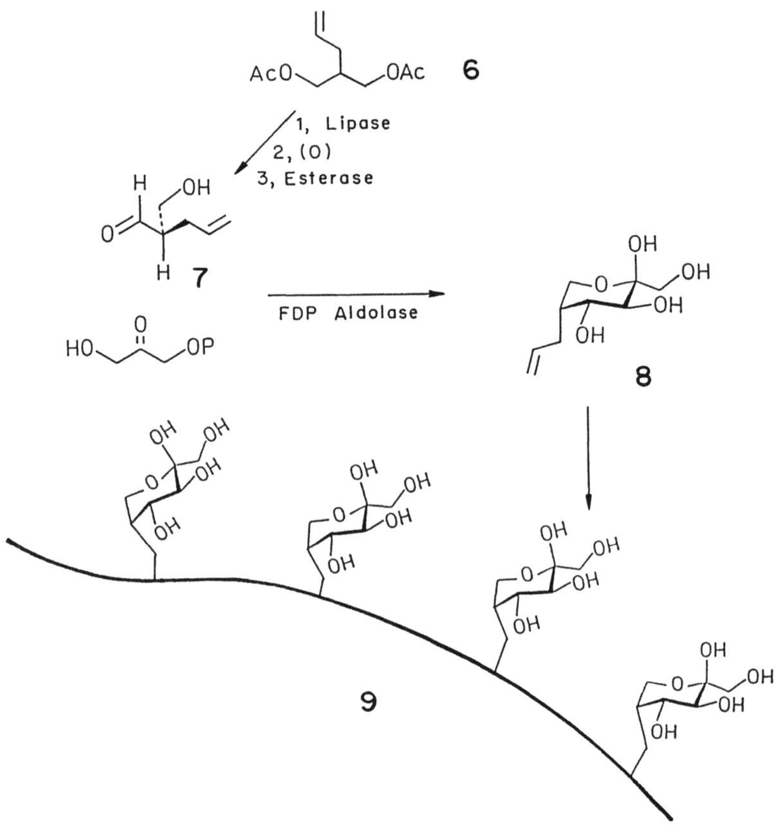

Fig. 4. Synthesis of 5-deoxy-5-allyl-D-fructose (**8**) and a possible radical polymerization to a poly-fructose derivative (**9**). The aldehyde (**7**) was prepared from (**6**) via pancreatic lipase-catalyzed enantioselective hydrolysis, Swern oxidation of the alcohol, and pig liver esterase-catalyzed hydrolysis of the remaining ester

mediates in synthesis, as potential new sweetners, and as new pharmaceuticals. Of the compounds prepared as shown in Fig. 1, 5-deoxy-D-fructose (4a) is particularly of interest. It exists in solution exclusively as β-$^2C_5$(D) form (Fig. 2), a conformation similar to that of the sweet component of D-fructose (Fig. 3), and thus tastes nearly as sweet as D-fructose [7]. Since 4a is not a substrate of glycolytic enzymes, it may be useful as a low-calorie sweetner. Further modification of the 5 position of 4a to 8 allows preparation of a new type of sugar, which may be polymerized to a new type of polysugar (Fig. 4).

Aldolase-catalyzed reactions are also useful for the preparation of selectively modified sugars containing an O-acyl group at a specific position. We have developed a new procedure for selective acetylation of free sugars at the primary position in organic media using pancreatic lipase as catalyst and isopropenyl acetate as acylating reagent [8]. The use of this acylating agent has the advantage that the byproduct isopropenol further converts to acetone spontaneously and thus makes the reaction irreversible. This irreversible transesterification minimizes the undesired

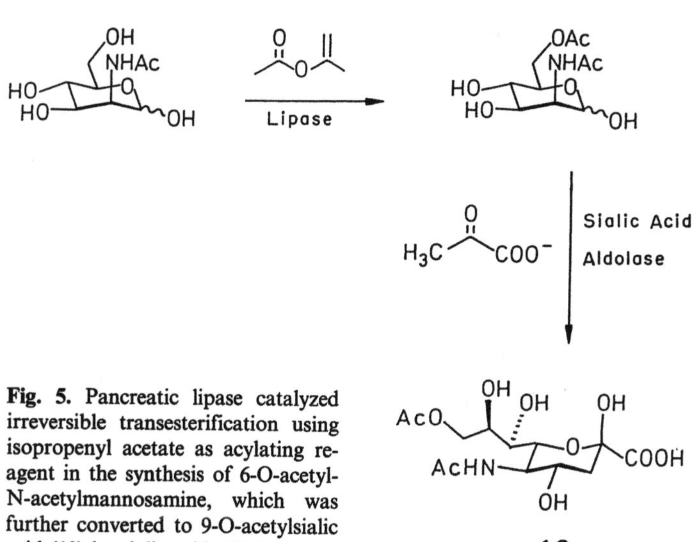

10

Fig. 5. Pancreatic lipase catalyzed irreversible transesterification using isopropenyl acetate as acylating reagent in the synthesis of 6-O-acetyl-N-acetylmannosamine, which was further converted to 9-O-acetylsialic acid (**10**) by sialic acid aldolase

reverse transesterification (which will reduce the optical purity of both the product and unreacted substrate in kinetic resolutions) and offers a better process for controlling the chemical and optical yields. Figure 5 illustrates the selective acetylation of N-acetylmannosamine for use in the sialic acid aldolase-catalyzed synthesis of 9-O-acetyl sialic acid (10), an unusual sialic acid found in tumor cells.

One of the reagents required for the FDP-aldolase catalyzed reactions is dihydroxyacetone phosphate, which must be prepared from dihydroxyacetone, either by chemical or by enzymatic phosphorylation. It also can be generated in situ enzymatically from FDP catalyzed by FDP aldolase and triose phosphate isomerase.

In a typical reaction, the concentration of substrates can be in the range of 0.1 to 0.5 M, and the reaction pH can be in the range of 6 to 7. The productivity of the reactor depends on the amount of catalyst used. In general, 1000 U (20 mg) of the FDP aldolase (IU = 1 μmol of product formed per min) will produce

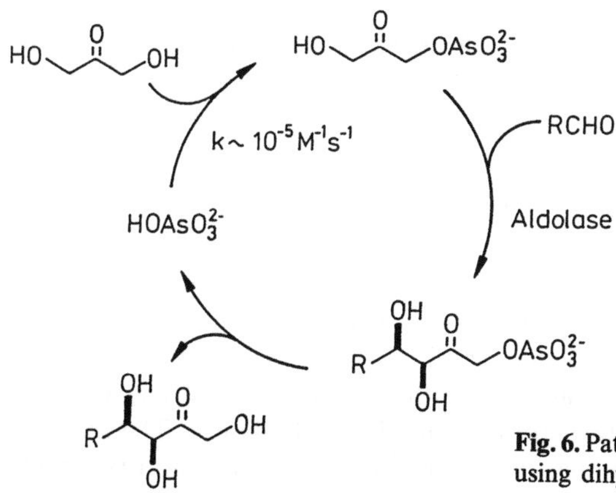

Fig. 6. Pathway for FDP aldolase-catalyzed reactions using dihydroxyacetone and arsenate as substrates

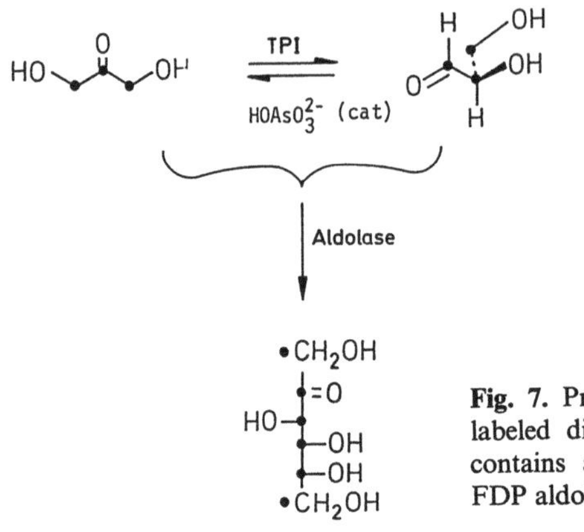

Fig. 7. Preparation of 13C-labeled D-fructose from 13C-labeled dihydroxyacetone. The reaction mixture (pH 7) contains arsenate, triosephosphate isomerase (TPI) and FDP aldolase

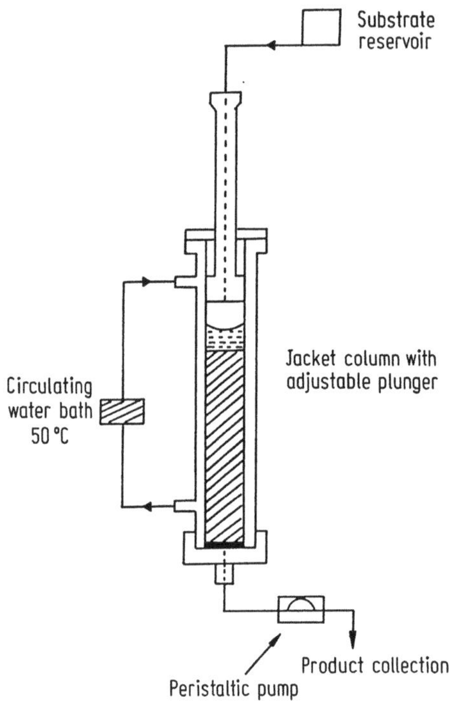

Fig. 8. Schematic illustration of a column reactor for aldolase and glucose isomerase reactions

approximately 1 mol of product a day. The enzymes are quite stable in the immobilized forms, which can be recovered at the conclusion of each reaction and used repeatedly.

Since DHAP is quite expensive, development of inexpensive route to DHAP or new substrate analogs should improve the economic feasibility of aldolase reactions. We recently have found that dihydroxyacetone in the presence of catalytic amounts of inorganic arsenate is a very effective substrate for FDP-aldolase [6)]. The reaction proceed through the mechanisms shown in Fig. 6. Dihydroxy-acetone first reacts spontaneously with inorganic arsenate to dihydroxyacetone arsenate which is then accepted as a substrate for FDP aldolase. After aldol reactions, the arsenate moiety is spontaneously dissociated and used for the next cycle of reaction. This eliminates the need for the preparation of DHAP which is less stable and more expensive than dihydroxyacetone. This strategy was used for the synthesis of ^{13}C-labeled D-fructose from ^{13}C-labeled dihydroxyacetone (Fig. 7). In this case, triosephosphate isomerase (TPI) was added to isomerize dihydroxyacetone arsenate to D-glyceraldehyde 3-arsenate, both of them were then condensed to D-fructose arsenate catalyzed by FDP-aldolase and the fructose arsenate was further converted to D-fructose spontaneously. Figure 8 illustrates the column reactor used for synthesis of sugars using immobilized aldolases or glucose isomerase as catalysts.

2. Current Activities and Future Opportunities

In summary; procedures are now available for enzymatic synthesis of sugars and derivatives using aldolases as catalysts. In the future, more and more synthetic reactions will involve enzymatic or other biological methods, particularly for the preparation of chiral synthons, sensitive intermediates, specialty and fine chemicals, pharmaceuticals, and a number of biologically derived substances for use in biotechnology. We note that enzymatic catalysis in organic synthesis is a process chemistry; its principal objective is to develop practical and efficient routes to specific products. To achieve this goal, a number of technical problems require solution. Current research in this area is to establish the real practicality of enzymes as synthetic catalysts and to explore new enzyme reactions suitable for practical organic synthesis. The immediate future will see the application of site-directed mutagenesis and recombinant DNA technology to the large-scale production of synthetically useful enzymes (for example, the FDP-aldolase gene has been cloned) [9] and to the improvement or modification of enzyme properties for synthetic purpose. The distant future may see the application of advanced protein chemistry to design new enzymes and to produce such enzymes to catalyze new reactions.

3. Acknowledgement

This research was supported by NSF (CHE8318217). The names of those who helped carry out the work are listed in the references.

4. References

1. Whitesides, G. M., Wong, C.-H.: Angew Chem. Int. Ed. Engl. *24*, 617 (1985)
2. Wong, C.-H. in: Enzymes as Catalysts in Organic Synthesis (Schneider, M. P. ed.), p. 199. D. Reidel Publishing Co. 1986
3. Wong, C.-H., Whitesides, G. M.: J. Org. Chem. *48*, 3199 (1983)
4. Wong, C.-H., Mazenod, F. P., Whitesides, G. M.: J. Org. Chem. *48*, 3493 (1983)

5. Wong, C.-H.: Industrial Chemical News *6*, 24, (1985); Durrwachter, J. R., Sweers, H. M., Nozaki, K., Wong, C.-H.: Tetrahedron Lett. *27*, 1261 (1986)
6. Durrwachter, J. R., Drueckhammer, D. G., Nozaki, K., Sweers, H. M., Wong, C.-H.: J. Am. Chem. Soc. *108*, 7812 (1986)
7. Martin, O. R., Korppi-Tommola, S. L., Szarek, W. A.: Can. J. Chem. *60*, 857 (1982)
8. Sweers, H. M., Wong, C.-H.: J. Am. Chem. Soc. *108*, 6421 (1986)
9. Tolan, D. R., Amsden, A. B., Putney, S. D., Urdea, M. S., Penhoet, E. E.: J. Biol. Chem. *259*, 1127 (1984)

2.8 Production of L-Phenylalanine Using Enzymes

Tetsuya Tosa

Research Laboratory of Applied Biochemistry, Tanabe Seiyaku Co. Ltd., 16-89, Kashima 3-chome, Yodogawa-ku, Osaka, 532 Japan

Contents

At present, both biological and chemical synthesis comprised stereospecific enzyme reactions are employed for the industrial production of L-phenylalanine. If effective combination of inexpensive raw materials and enzymes suitable for construction of bioreactor can be accomplished, the chemical and enzymatic method using bioreactor is considered to be preferable. In Tanabe Seiyaku Co., Ltd. since the beginning of 1960's the studies on the commercial production of L-phenylalanine have been carried out. In this paper, some examples on the enzymatic conversion of inexpensive raw materials into L-phenylalanine are presented.

1. Introduction

L-Phenylalanine is an important amino acid as an essential amino acid for human nutrition, and used as a component of amino acid infusions for medical purpose. Further now it is used as an important intermediate of the diet sweetener, aspartame, which is a peptide of L-phenylalanine methyl ester and L-aspartic acid. Thus the market of L-phenylalanine is rapidly increasing as well as that of L-aspartic acid. Eldib estimates [1] that the worldwide annual demand for L-phenylalanine to make aspartame was only 50 tons in 1981, but it would be about 3,000 t in 1985. By the end of 1990's it is expected to be almost 2.5 times. Therefore,

Bioproducts and Bioprocesses
Editors: Fiechter, Okada, Tanner

Table 1. Studies on enzymatic production of L-phenylalanine in Tanabe Seiyaku Co. Ltd.

Raw material	Enzyme
N-Acetyl-DL-phenylalanine	Aminoacylase
DL-Phenylalanine	D-Amino acid oxidase & aminotransferase
DL-Phenylserine	Phenylserine dehydratase & aminotransferase
trans-Cinnamic acid	Phenylalanine ammonia-lyase
Phenylpyruvic acid	Aminotransferase
Acetamidocinnamic acid	Acetamidocinnamate amidohydrolase & aminotransferase
DL-5-Benzylhydantoin	Hydantoinase & carbamoylase

many chemical, pharmaceutical and food industries have strongly payed the attention for the economical production method of L-phenylalanine.

Table 1 shows the studies on enzymatic production of L-phenylalanine in Tanabe Seiyaku Co., Ltd. Namely, first is the aminoacylase reaction which is still operating in commercial base. This is the first industrial application of bioreactor using immobilized enzyme in the world [2]. In the second example chemically synthesized DL-phenylalanine is used as a raw material. In this case D-phenylalanine is converted to phenylpyruvic acid by the action of D-amino acid oxidase, and L-phenylalanine is obtained by aminotransferase reaction. In the third example, chemically systhesized DL-phenylserine is converted to phenylpyruvic acid by the action of phenylserine dehydratase, and L-phenylalanine is produced by aminotransferase reaction. However, these methods using DL-phenylalanine and DL-phenylserine as raw materials were not considered to be advantageous methods compared with the optical resolution of N-acetyl-DL-phenylalanine by the action of aminoacylase, because their productivities were relatively low. On the L-phenylalanine production from trans-cinnamic acid, phenylpyruvic acid, acetamidocinnamic acid and DL-5-benzylhydantoin, we have earnestly studied for long time and found the promising method. So, we would like to explain these methods in this paper.

2. Phenylalanine Ammonia-lyase

First example is L-phenylalanine production from trans-cinnamic acid by phenylalanine ammonia-lyase (PAL) as follows.

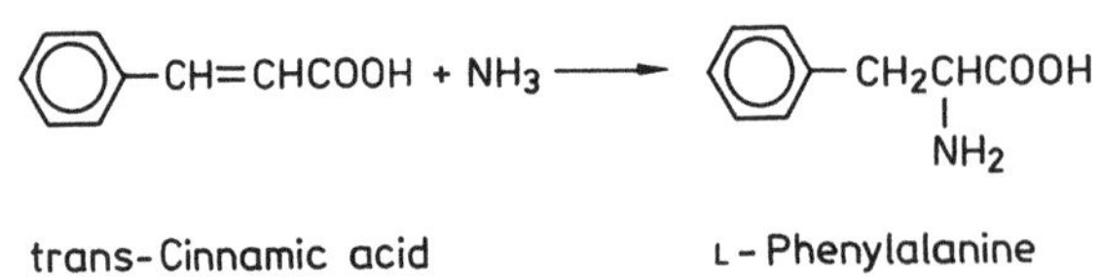

In Japan, in 1960's Ogata and Yamada's group at Kyoto University earnestly studied the phenylalanine ammonia-lyase from microorganisms [3,4]. Following their studies, we have been studying microbial PAL since 1970 [5-9].

2.1 Induction and Stabilization of Enzyme

Rhodosporidium toruloides IFO 0559 could grow abundantly in a simple chemically defined medium consisting of sources of carbon, inorganic nitrogen, and mineral salts, but PAL activity in the cells was not formed. The induction of PAL activity was observed by the addition of L-phenylalanine in such a medium.

The rate of growth and the formation of PAL in a medium consisting 1.0% yeast extract, 1.0% peptone, 0.05% L-phenylalanine, and 0.5% NaCl (pH 6.0) are shown in Fig. 1A. PAL activity was detectable after 10 h of growth, reached a maximum around 16 to 18 h, and then rapidly reduced when growth reached stationary phase. When a little amount of L-isoleucine was added to this medium, however, PAL activity increased somewhat and was sustained for a relatively long time as shown in Fig. 1B.

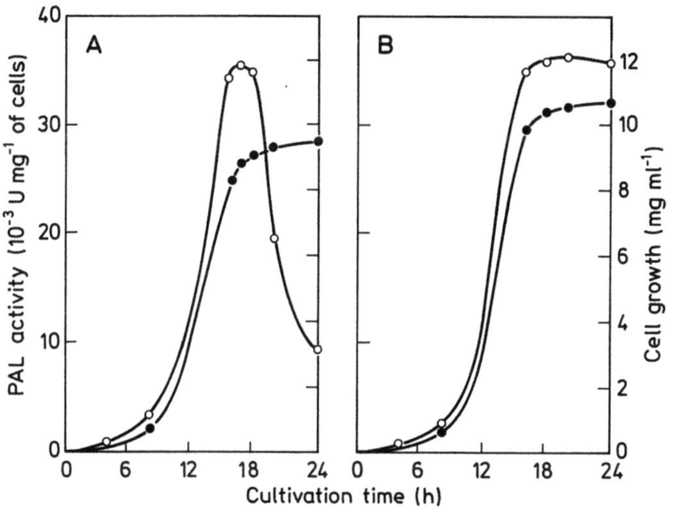

Fig. 1. Typical time course of the formation of PAL **(A)** and effect of addition of L-isoleucine on stability of PAL during cultivation **(B)**. Cultivation was carried out in the medium with or without 0.5% L-isoleucine at 30 °C. Symbols: ○, PAL activity; ●, cell growth

2.2 Conditions for Conversion

Initial velocity of the conversion of trans-cinnamic acid to L-phenylalanine by PAL activity of intact cells was varied with the concentration of trans-cinnamic acid, ammonia, and pH. As the results, optimum trans-cinnamic acid concentration was 50 to 60 mM, optimum ammonia concentration was 7 M, and optimum pH value was 10. Compared with other usual enzyme reaction, these conditions are considered to be very severe.

Figure 2 shows a typical time course of the reaction occurred under the optimum conditions. 110 mM of L-phenylalanine is accumulated from 150 mM of cinnamic

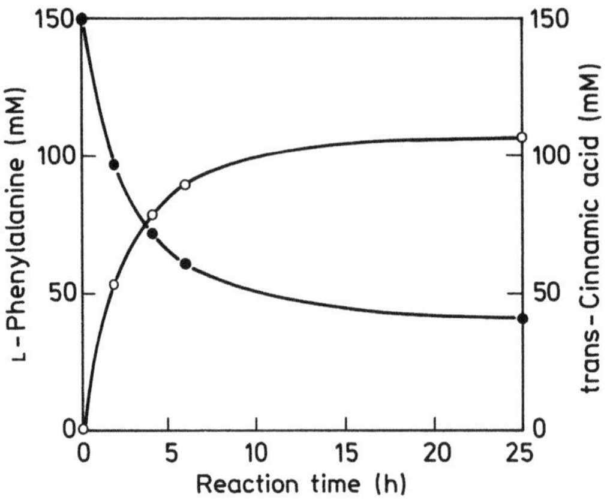

Fig. 2. Conversion of trans-cinnamic acid to L-phenylalanine by PAL in *R. toruloides* IFO 0559. The reaction was carried out at 30 °C. Symbols: ○, L-Phenylalanine; ●, trans-cinnamic acid

acid in the presence of 7.5 M ammonia at pH 10. We investigated the scale up and carried out the cost estimation. As the results we decided that this method was not advantageous as the industrial method, because the enzyme was very unstable in such severe conditions and the cost of trans-cinnamic acid was high.

However, since late 1983, Genex is said to produce L-phenylalanine from trans-cinnamic acid and ammonia by this enzyme reaction [10,11].

3. Aminotransferase

Next example is L-phenylalanine production from phenylpyruvic acid by aminotransferase. When phenylpyruvic acid is used as a raw material, two kinds of enzyme reaction systems are investigated as follows.

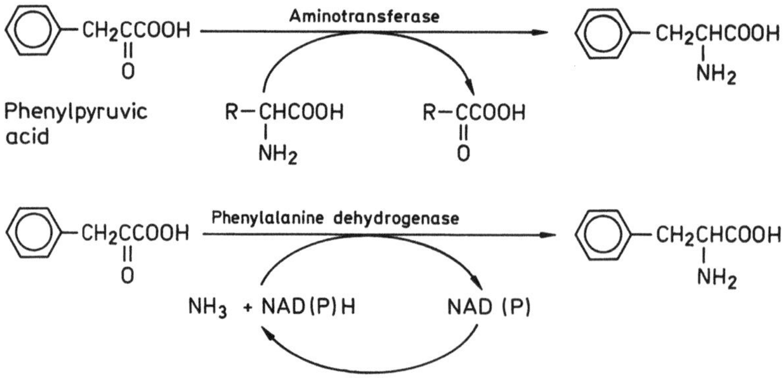

Some aminotransferases can transfer the amino group of L-amino acid to phenylpyruvic acid and give L-phenylalanine [12-15]. Aspartate aminotransferase and aromatic amino acid aminotransferase are well known [16-18].

The other is phenylalanine dehydrogenase reaction accompaning with reduction systems. Recently Nishimura and co-workers at Tokyo University of Agriculture and Technology presented the non-enzymatic reduction process under the high pressure of hydrogen gas [19]. Therefore, this phenylalanine dehydrogenase reaction has advantage that an other amino acid as an amino donor is not necessary.

3.1 Screening of Microorganisms

In order to find out microorganisms having higher aminotransferase activity, we screened many microorganisms and found *Paracoccus denitrificans* in our collections [20]. Further, we constructed an aminotransferase-hyperproducing mutant pFPr-1 by NTG treatment [21], and used it for the subsequent experiments.

3.2 Effect of Amino Donor

As shown in Fig. 3, when L-glutamic acid was used as an amino donor, the initial velocity was higher than those of L-aspartic acid and L-alanine, but the final conversion ratio did not reach over 60%. On the other hand, when L-aspartic acid was used, the conversion ratio reached 92%. L-Alanine and others were not suitable amino donors for this enzyme reaction. From these results, it is considered that L-phenylalanine is produced efficiently from phenylpyruvic acid and L-aspartic acid by the action of aminotransferase in *P. denitrificans* pFPr-1.

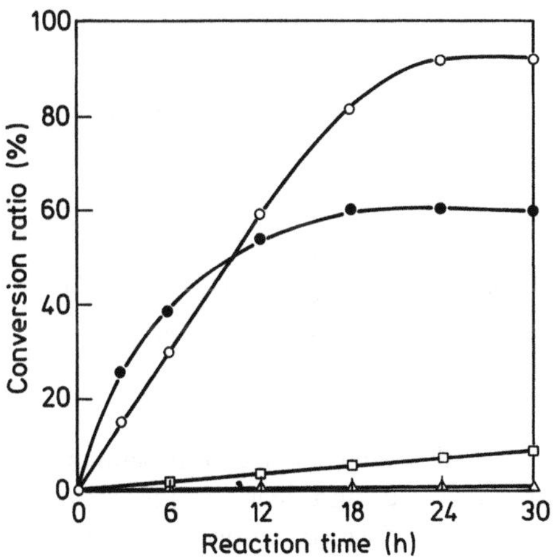

Fig. 3. Effect of amino donor on initial velocity of aminotransferase activity of *P. denitrificans* and conversion ratio. Reaction mixture contained 0.2 M phenylpyruvic acid, 0.2 M respective L-amino acid, 0.1 mM PALP, and *P. denitrificans* cells (pH 8.0). Symbols: ○, L-Aspartic acid; ●, L-glutamic acid; □, L-alanine; △, others

3.3 Conditions for Conversion

The best conditions for the conversion of phenylpyruvic acid and L-aspartic acid to L-phenylalanine by aminotransferase of *P. denitrificans* were pH 8.5 and 30 °C. However, 1.0 M of phenylpyruvic acid and 1.5 M of L-aspartic acid were added periodically to the medium, because this aminotransferase activity is inhibited by high concentration of phenylpyruvic acid. The reaction mixture was agitated, but the velocity of the enzymatic conversion appeared to be almost the same as that of the stationary incubation.

Finally the reaction mixture solidified with the increase of crystals of L-phenylalanine, and the accumulated L-phenylalanine did not decompose, even with prolonged incubation. After the crystals were dissolved by the addition of water and warming, the reaction mixture was filtered to remove the cells. The filtrate was treated with active charcoal and filtered. L-Phenylalanine crystallized out from the filtrate by concentration under reduced pressure. Production yield of pure L-phenylalanine from phenylpyruvic acid was about 80%.

4. Acetamidocinnamate Amidohydrolase

The third example is the enzymatic conversion of L-phenylalanine from acetamidocinnamic acid (ACA). Generally, ACA is converted to N-acetyl-DL-phenylalanine by chemical reduction, and N-acetyl-DL-phenylalanine is hydrolyzed by aminoacylase to L-phenylalanine. Thus, if ACA is completely converted to L-phenylalanine by enzymatic reaction of one step, the production cost of L-phenylalanine is considered to be lower than the aminoacylase route, because the reduction step of ACA and the racemization step of N-acetyl-D-phenylalanine are expected to be omitted. Therefore, the application of enzymes to convert ACA to L-phenylalanine is considered to be advantageous for the commercial production of L-phenylalanine [22].

4.1 Isolation of Hyperactivity Mutant

Thus, several years ago we started the study on the enzymatic conversion of ACA to L-phenylalanine by microbial cells. As the results, it was found that this reaction proceeds by two kinds of enzymes as follows.

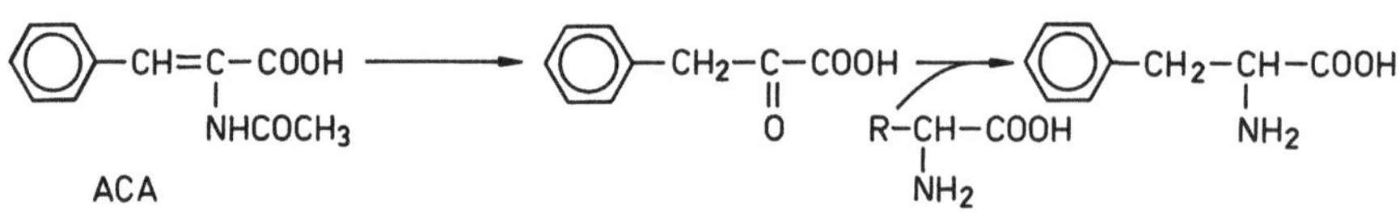

Namely, first is that acetyl group of ACA is hydrolyzed by acetamidocinnamate amidohydrolase, a kind of peptidase or aminoacylase, and then, amino group is splitted non-enzymatically and phenylpyruvic acid is formed. Following this reaction, phenylpyruvic acid is converted to L-phenylalanine by aminotransferase.

At first, we tried to establish the production of L-phenylalanine from ACA by one kind of microorganism, but it was found to be difficult to increase both enzyme activities at the same time. So, we considered the possibility of L-phenylalanine production using two microorganisms. Namely a first microorganism having acetamidocinnamate amidohydrolase activity and a second microorganism having aminotransferase activity were screened again [20].

As the results of screening for microorganisms having acetamidocinnamate amidohydrolase, we found a strain belonging to the genus *Corynebacterium* from soil and

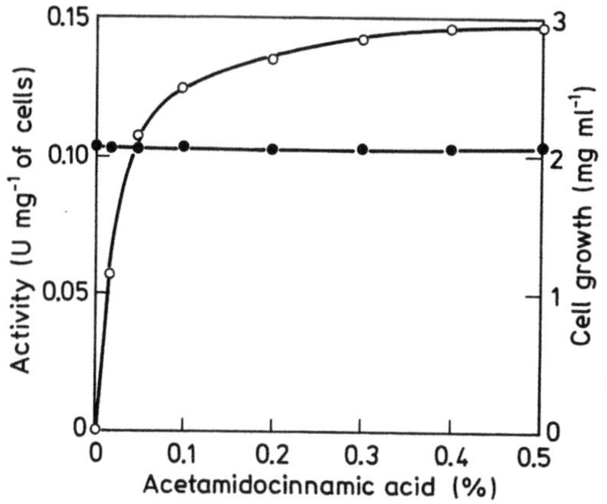

Fig. 4. Induction of acetamidocinnamate amidohydrolase in *C.* sp. strain S-5. ACA was added to the medium, and cultivation was carried out at 30 °C for 24 h. Symbols: ○, Acetamidocinnamate amidohydrolase activity; ●, cell growth

named it as *C.* sp. S-5 [23]. Strain S-5 could grow abundantly in a simple chemically defined medium consisting of a source of carbon, inorganic nitrogen and minerals. However, acetamidocinnamate amidohydrolase activity was not observed in such a medium. Induction of acetamidocinnamate amidohydrolase activity was observed by the additions of ACA in a medium as shown in Fig. 4. In addition, the enzyme formation of the strain is controlled by catabolite repression. To increase the productivity of L-phenylalanine, we isolated *C.* sp. strain S-5 mutant with high enzyme activity by selecting cells that grew faster in a medium containing ACA as a nitrogen source in the presence of glucose. Because when the strain was cultured in a medium containing glucose and ACA as the sole nitrogen source, it showed low in enzyme activity and elongated doubling time.

Isolation method of acetamidocinnamate amidohydrolase-hyperproducing mutant was carried out as follows. The cells of wild-type strain S-5 at an exponential phase were suspended in Tris-malate buffer and treated with NTG. Survival ratio was about 20% and the mutagenized culture was then inoculated into a chemostat vesel containing the enrichment medium. After incubation at 30 °C for 5 d, cells grew fast, were enriched by continuous culture. The dilution rate used was 0.1 h^{-1}. Cells obtained from the chemostat vesel were spread on a solid medium prepared by adding agar to the enrichment medium. After 2 to 3 d of incubation at 30 °C, any large colonies formed were purified by single-colony isolation. As a result, after 5 d of continuous culture, 208 large colonies were isolated. Among them, strain C-23 had highest activity.

4.2 Characteristics of Mutant

In order to clarify the control systems of acetamidocinnamate amidohydrolase formation in mutant strain C-23, wild-type strain S-5 and mutant strain C-23 were cultured in medium containing glucose with or without ACA. As shown in Table 2, strain C-23 showed 10 ~ 12-fold increase in both specific and total activities than those of the wild-type strain. Further, acetamidocinnamate amidohydrolase formation in strain C-23 is not required ACA as an enzyme inducer

Table 2. Effect of ACA on acetamidocinnamate amidohydrolase formation in two strains of *C.* sp.[a]

Conc. of ACA (%)	Strain S-5 (parent)			Strain C-23 (mutant)		
	Cells ($mg\ ml^{-1}$)	Activity		Cells ($mg\ ml^{-1}$)	Activity	
		SA[b]	TA[c]		SA	TA
0	2.2	0.000	0.00	1.8	0.72	1.3
0.5	2.4	0.058	0.14	2.1	0.67	1.4

[a] ACA was added to the medium containing 1% glucose, 0.2% $(NH_4)_2SO_4$, 0.1% KH_2PO_4, and $MgSO_4 \cdot 7\ H_2O$ (pH 7.0). Cultivation was carried out at 30 °C for 24 h
[b] Specific activity ($U^d\ mg^{-1}$ of cells)
[c] Total activity ($U\ ml^{-1}$ of culture medium)
[d] Unit; one unit of activity was defined as that activity which converts one micromole of substrate to product per min

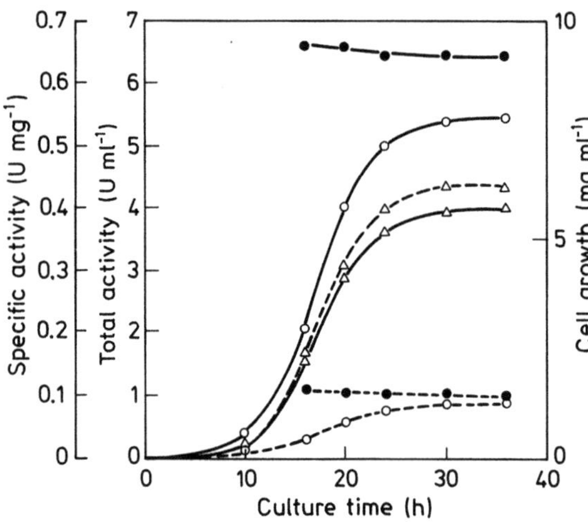

Fig. 5. Typical time course of the formation of acetamidocinnamate amidohydrolase in two strains of *C.* sp. Both strain S-5 (– – –) and strain C-23 (———) were cultured under optimum conditions, respectively. Symbols: ○, Total activity; ●, specific activity; △, cell growth

and is resistant to catabolite repression. These characteristics such as high enzyme activity and resistance to catabolite repression are very advantageous for industrial purpose.

Figure 5 shows a profile of acetamidocinnamate amidohydrolase formation in both strain C-23 and strain S-5. Growth rate of strain C-23 was not changed compared with that of the wild-type strain. Maximal total activity of the enzyme was attained in the early stationary phase of growth. This mutant maintained a high specific activity during culture. Although data was not shown, an increase of the enzyme activity in strain C-23 was considered to be due to increasing of the enzyme protein from results that enzymatic properties in strain C-23 were the same as those of wild-type strain S-5.

On the basis of these results, it seemed that strain C-23 is more advantageous for L-phenylalanine production than wild-type strain S-5.

4.3 Conditions for Conversion

Figure 6 shows a typical time course of L-phenylalanine production from ACA using two kinds of microorganism under the optimum conditions. *C.* sp. strain C-23 was used as an acetamidocinnamate amidohydrolase source, and *P. denitrificans* strain pFPr-1 previously stated was used as an aminotransferase source.

As shown in Fig. 6, ACA was consumed rapidly, and L-phenylalanine was accumulated gradually compared to ACA consumption. After 48 h, reaction was finished, and 760 µmol ml^{-1} of L-phenylalanine was produced in a molar yield of 95% from 0.8 M ACA. The difference between the amount of consumed ACA and that of produced L-phenylalanine in the process of this reaction was considered to be that phenylpyruvic acid was accumulated in the reaction mixture.

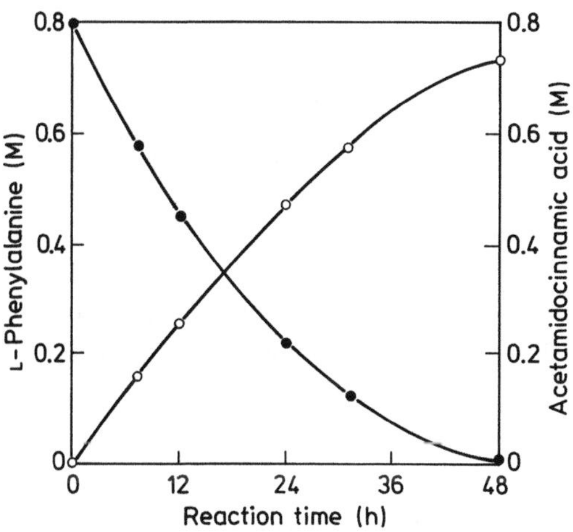

Fig. 6. Production of L-phenylalanine from acetamidocinnamic acid by a two-step enzyme reaction using *C.* sp. strain C-23 and *P. denitrificans* pFPr-1 cells. Symbols: ○, L-Phenylalanine; ●, acetamidocinnamic acid

4.4 Continuous Production Using Immobilized Cells

Succeeding to the batchwise conversion of ACA by two kinds of microorganisms, we attemped the continuous conversion of ACA by immobilized cells. Immobilized *C.* sp. cells, immobilized *P. denitrificans* cells and co-immobilized two kinds of these cells were prepared by ϰ-carrageenan method [21].

Figure 7 shows various types of bioreactors considered for L-phenylalanine production from ACA using these immobilized cells. In A method reactor, acetamidocinnamate amidohydrolase reaction proceeds in the first reactor, and aminotransferase reaction occurs in the second reactor. Namely this is sequential reactor. B and C methods are a mono bed system packed with mixture of two immobilized cells and co-immobilized cells, respectively. The productivity of co-immobilized cells was found

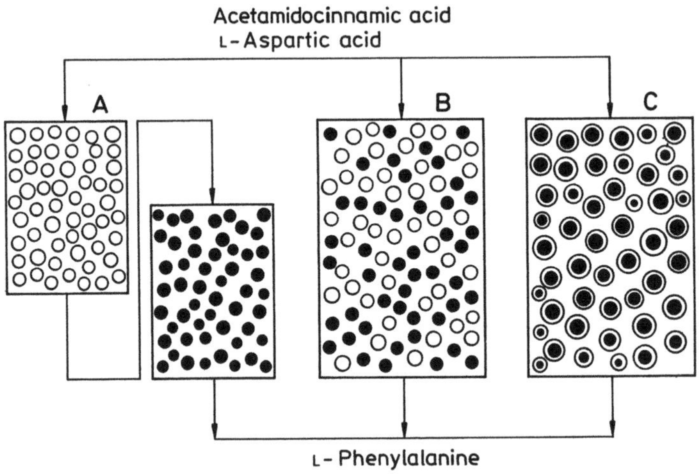

Fig. 7. Various bioreactors for continuous production of L-phenylalanine from acetamidocinnamic acid by a two-step enzyme reaction. Symbols: ○; Immobilized *C.* sp. cells; ●, immobilized *P. denitrificans* cells; ⊙, co-immobilized *C.* sp.-*P. denitrificans* cells

to be higher than that of the mixture of the two immobilized cells as shown in Table 3. It is reasonable that phenylpyruvic acid produced by the acetamidocinnamate amidohydrolase reaction is immediately converted to L-phenylalanine by the aminotransferase reaction without diffusion resistance of phenylpyruvic acid within gel. The operational stability of L-phenalalanine productivity of co-immobilized cells was studied and the half-life of the activity was 3 weeks at 30 °C. When 150 mM of ACA and 300 mM of L-aspartic acid were charged into column packed with co-immobilized cells, 147 mM of L-phenylalanine were produced with 98% of conversion ratio.

Table 3. Comparison of L-phenylalanine productivity in three bioreactors

Bioreactor		L-Phenylalanine productivity[b]
Immobilized cells[a]	Column	(μmol min^{-1} g^{-1} gel)
Two immobilized cells	Two beds	0.70
Mixture of two immobilized cells	Mono bed	0.74
Co-immobilized cells	Mono bed	1.28

[a] Activity yield of $\varkappa$-carrageenan method was about 55% in each case

[b] A solution containing 0.15 M ACA, 0.3 M L-aspartic acid and 0.1 mM PLP (pH 8.0) was passed through the columns packed each with immobilized cells at flow of SV = 2.0 h^{-1} at 30 °C, and an amount of L-phenylalanine in the effluent was measured

5. Combination of Hydantoinase and Carbamoylase

As the last example of this series, we present the production of phenylalanine from DL-5-benzylhydantoin.

Among the studies on the production of optically active amino acids from the respective hydantoins[24-27], p-hydroxy-D-phenylglycine which is an important raw material for synthetic penicillins has been industrially produced by Kanegafuchi Chemical Industry in Singapore using Yamada's method[27]. Since a few years ago, we are also studying the production of L-phenylalanine from the respective hydantoin, namely, DL-5-benzylhydantoin as follows.

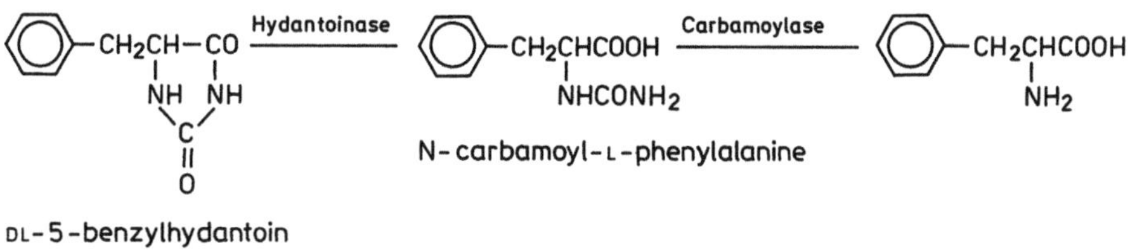

First reaction is a stereospecific cleavage of hydantoin ring by so-called hydantoinase and gives N-carbamoyl-L-phenylalanine. Second reaction is a hydrolysis of N-carbamoyl-L-phenylalanine by so-called carbamoylase and gives L-phenylalanine. The advantage of using DL-5-benzylhydantoin is that it is very easily racemized under mild conditions. Consequently, enzymatic hydrolysis and chemical racemization occur simultaneously, so that DL-5-benzylhydantoin is completely converted to L-phenylalanine. If DL-5-benzylhydantoin can be directly converted to L-phenylalanine by microorganism, cost down of production is expected, because the cost of DL-5-benzylhydantoin is lower than that of another raw material for L-phenylalanine production. So, we screened microorganisms having the ability of L-phenylalanine production from DL-5-benzylhydantoin.

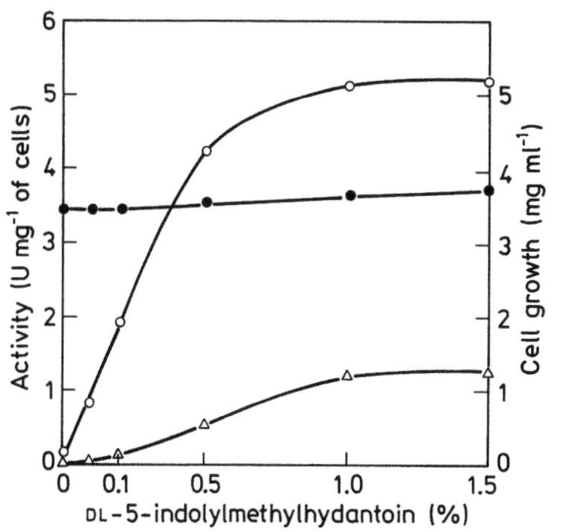

Fig. 8. Induction of hydantoinase and carbamoylase in *F.* sp. strain I-3, DL-5-Indolylmethylhydantoin was added to the medium, and cultivation was carried out at 28 °C for 40 h. Symbols: ○, Hydantoinase activity; △, carbamoylase activity; ●, cell growth

5.1 Induction of Both Enzymes

As the results of screening, we found a strain belonging to the genus *Flavobacterium* having both hydantoinase and carbamoylase from soil. The strain was named as *F.* sp. I-3. Both enzymes of strain I-3 formed inducibly as shown in Fig. 8. At the concentration of 1% or higher, highest activities of both enzymes were obtained.

5.2 Conditions for Conversion

Figure 9 shows a typical time course of L-phenylalanine production from DL-5-benzylhydantoin by two-step enzyme reaction using *F.* sp. strain I-3.

In the initial stage of the reaction time, the concentration of N-carbamoyl-L-phenylalanine in the reaction mixture increased in proportion to consumption of DL-5-benzylhydantoin, and finally L-phenylalanine could be produced in very high yield about 100%.

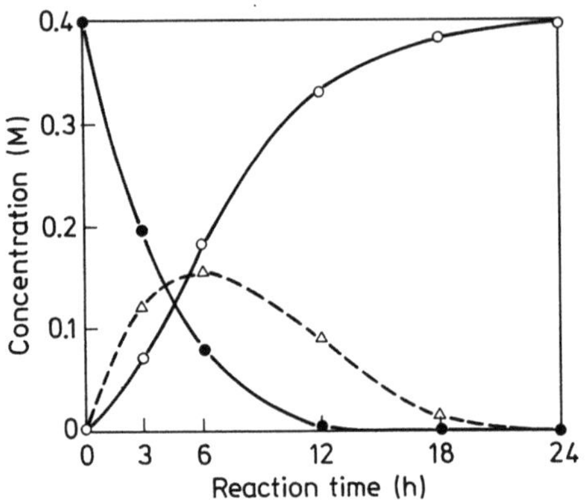

Fig. 9. Production of L-phenylalanine from DL-5-benzylhydantoin fed in cultured medium of *F.* sp. strain I-3. Symbols: ○, L-Phenylalanine; △, N-carbamoylphenylalanine; ●, DL-5-benzylhydantoin

6. Conclusions

As mentioned above, we have explained some examples of our studies on the enzymatic production of L-phenylalanine. It seems that the enzymatic production of L-phenylalanine using these raw materials as stated above is very promising for the industrial process. In the near future, we expect that one of these methods will be commercialized replacing the immobilized aminoacylase method of Tanabe Seiyaku Co., Ltd.

Acknowledgements

The author is indebted to Dr. I. Chibata, Vice President and Dr. S. Yamada, General Manager of Research Planning and Investigation Division, Tanabe Seiyaku Co., Ltd., for their helpful discussions and suggestions. These studies have been done by Dr. K. Nabe, Mr. Y. Nishida, and Mr. K. Nakamichi, the expert co-operative researchers. The author heartily expresses his thanks to all of them.

7. Abbreviations

C. sp. *Corynebacterium* species
F. sp. *Flavobacterium* species
P. sp. *Paracoccus* species
R. sp. *Rhodosporidium* species
ACA acetamidocinnamic acid
NTG N-methyl-N′-nitro-N-nitrosoguanidine
PAL phenylalanine ammonia-lyase

8. References

1. Chemical marketing Reporter, Oct. 22 1984
2. Chibata, I., Tosa, T.: Adv. Appl. Microbiol. *22*, 1 (1977)
3. Ogata, K., Uchiyama, K., Yamada, H.: Agric. Biol. Chem. *30*, 311 (1966)
4. Ogata, K., Uchiyama, K., Yamada, H.: Agric. Biol. Chem. *31*, 200 (1967)
5. Japan Patent No. 53-96388 (23 August 1978)
6. Japan Patent No. 56-26197 (13 March 1980)
7. Japan Patent No. 58-86082 (23 May 1982)
8. Yamada, S., Nabe, K., Izuo, N., Nakamichi, K., Chibata, I.: Appl. Environ. Microbiol. *42*, 773 (1981)
9. Nakamichi, K., Nabe, K., Yakada, S., Chibata, I.: Appl. Microbiol. Biotechnol. *18*, 158 (1983)
10. Chemical Engineering, November 25, 1985, p. 21
11. Hamilton, B. K., Hsiao, H.-Y., Swann, W. E., Anderson, D. M., Delente, J.: Trends in Biotechnol. *3*, 64 (1985)
12. Florkin, M., Stotz, E. H.: Comprehensive Biochemistry, p. 156, Amsterdam, Elsevier 1973
13. Ziehr, H., Kula, M.-R.: J. Biotechnol. *3*, 19 (1985)
14. Bulot, E., Cooney, C. L.: Biotechnol. Letters *7*, 93 (1985)
15. Calton, G. J., Wood, L. L., Updike, M. H., Lantz, L., Hamman, J. P.: Bio/Technol. *4*, 317 (1986)
16. Mavrides, C., Orr, W.: J. Biol. Chem. *250*, 4128 (1975)
17. Powell, J. T., Morrison, F.: Eur. J. Biochem. *87*, 391 (1978)
18. Lee, C.-W., Desmazeaud, M. J.: J. Gen. Microbiol. *131*, 459 (1985)
19. Higashijuma, M., Matsunaga, N., Nishimura, S., Kitamura, T., Kawaguchi, T., Tsuji, M.: In "The Proceeding of Annual Meeting of Agricultural Chemical Society of Japan" p. 540 (1986)
20. Nakamichi, K., Nishida, Y., Nabe, K., Tosa, T.: Appl. Biochem. Biotechnol. *11*, 367 (1985)
21. Nishida, Y., Nakamichi, K., Nabe, K., Tosa, T.: Enzyme Microbiol. Technol. *9*, 479 (1987)
22. Nakamichi, K., Nabe, K., Yamada, S., Tosa, T., Chibata, I.: Appl. Microbiol. Biotechnol. *19*, 100 (1984)
23. Nakamichi, K., Nabe, K., Tosa, T.: J. Biotechnol. *4*, 293 (1986)
24. Tsugawa, R., Okumura, S., Ito, T.: Agric. Biol. Chem. *30*, 27 (1966)
25. Yamada, H., Oishi, K., Aida, K., Uemura, T.: Agric. Chem. Soc. Jpn. *43*, 528 (1969)
26. Sano, K., Yokozeki, K., Eguchi, C., Kagawa, T., Noda, I., Mitsugi, K.: Agric. Biol. Chem. *41*, 819 (1977)
27. Takahashi, S., Ohashi, T., Kii, Y., Kumagai, H., Yamada, H.: J. Ferment. Technol. *57*, 328 (1979)

3. Genetic Engineering

3.1 Genetic Engineering in Coryneform Bacteria

M. T. Follettie and A. J. Sinskey

Massachusetts Institute of Technology, Department of Applied Biological Sciences, Cambridge, Massachusetts 02139, USA

Abstract

Recently, we have established recombinant DNA technology for *Corynebacterium glutamicum*. Recent advances in the molecular biology of the phenylalanine biosynthetic pathway will be presented. Specifics include the gene sequence for prephenate dehydratase and findings about promoter structure and function. In addition, the utility of a recently isolated *C. glutamicum* restriction minus strain will be demonstrated and discussed.

Bioproducts and Bioprocesses
Editors: Fiechter, Okada, Tanner

3.2 Formation of Jumbo Yeast Cells by Introduction of Methylglyoxal Resistant Genes, Development of New Vector for Wild or Industrially Used Cells

Akira Kimura

Research Institute for Food Science, Kyoto University, Uji, Kyoto 611, Japan

Contents

The two genes responsible for methylglyoxal (MG) resistance were cloned onto YEp13 and designated as pYMG14 and pYGL7, respectively. The former was proved to code the gene for MG reductase, which converted MG to lactaldehyde. The latter was for glyoxalase I, which converted MG to S-lactoylglutathione in the presence of glutathione. Introduction of either one of these plasmids into yeast cells made the transformants not only resistant to high concentration of MG, but also 2 times bigger in size of diameter and 8 times in cubic content than control yeast cells. This is the first report that jumbo yeast cells are breeded by introduction of some genes. These jumbo yeast cells are not only interesting from the point of basic microbiology, but also industrially useful. The physiological properties of them are now underinvestigated.

One of MG resistant gene pYMG14 can be used as a new vector for wild and industrially used strains of yeasts, for transformants containing it become resistant to various toxic compounds, to which control yeast cells are sensitive.

1. Outline

Only Trp and/or Leu vector have been available for the breeding of yeast cells. Moreover, to use these vectors host cells should be mutated in advance. Therefore, these vectors can not be used for wild or industrially used cells of yeasts. The yeast killer factor [1] and an antibiotic, gentamycin [2], have become available recently, but their utilization is limited because of their narrow spectrum of action and high cost.

Transformants of the mutated yeasts, although dosed with multiple genes, usually showed still weaker activity than wild or industrially used yeasts. Therefore,

Bioproducts and Bioprocesses
Editors: Fiechter, Okada, Tanner

we had looked for new vectors, which could be used for wild or industrially used yeasts.

We cloned genes responsible for methylglyoxal (MG)- and metalresistance in *Saccharomyces cerevisiae* onto YEp13 and designated pYMG14, pYGL7, pYZN, and pYCD, respectively [3,4,5]. We also cloned the resistant gene for nucleotide analogues [6].

In our study on glutathione, we noticed that methylglyoxal (MG) is a toxic metabolite and the resistance to MG can be used as the most useful new marker. Out of the two MG resistance genes cloned, one (pYMG14) was proved to be a gene for MG reductase, which catalizes the conversion of MG to lactaldehyde, whereas the other (pYGL7) was a gene for glyoxalase I, which converts MG to S-lactoylglutathione in the presence of glutathione. The *S. cerevisiae* cells dosed with the gene (pYMG14) acquired higher resistance not only to MG, but also to various toxic compounds. In the transformant containing pYMG14 the activity of MG reductase increased. Since wild and/or industrially useful yeasts are usually sensitive to MG or metal compounds, the introduction of hybrid plasmids containing these resistance genes can be used as a new vector for the breeding of wild and/or industrially used cells [7,8]. Besides the transformants containing either plasmid pYMG14 or PYGL7 become 2 times larger in diameter and 8 times in cubic content.

The reaction, which is catalized by glyoxalase I, is said quite active in cancer cells, so we thought we may, by the introduction of the gene for glyoxalase I, be able to breed yeast cells, which can grow as rapidly as cancer cells.

The jumbo yeast cells containing either pYMG14 or pYGL7 might be industrially quite useful for they are easily filtered and/or immobilized in polymers.

2. Methods

Strains: Strains used in this were was *Saccharomyces cerevisiae* DKD-5D-H (a, *leu2-3*, *leu2-112*, *trp1*, *his3*), *Escherichia coli* C600 (F^-, *recA*, *thr*, *leu*, *thi*, *lacY*, *supE*, *tonA*).

Isolation of Gene Responsible for MG Resistance of Yeasts: Chromosomal DNA was extracted from *S. cerevisiae* DKD-5D-H cells grown on a nutrient medium (YPD: 2.0% glucose, 1.0% yeast extract, 2.0% peptone, pH 5.0) by the method of Saito and Miura [9] and was digested with restriction endonuclease *Sau*3AI. Vector plasmid YEp13 was also digested with restriction endonuclease *Bam*HI. These two DNA digests were mixed annealed, and ligated with T4 DNA ligase. Restriction endonuclease and ligase reactions were carried out under the conditions recommended by the vendor. The whole ligation mixture was used to transform *S. cerevisiae* DKD-5D-H cells. Transformation of yeast cells was done by our method with lithium acetate [10], and the transformed cells were directly spread on minimal agar plates (SD medium) supplemented with appropriate nutritional requirements and 5.0 mM MG. The plates were incubated at 30 °C for about 3 d, and the colonies that appeared on the plates were selected as the transformants of the gene responsible for MG resistance. The largest colony among the transformants that appeared on the plates was picked up, and the cells were cultured on SD medium, harvested, and lysed with lytic enzyme (Zymolyase). Crude DNA extracts were prepared from the lysate and were used to transform *E. coli* C600 cells treated with calcium. The preparation of crude DNA extracts from yeast cells and the transformation of *E. coli* were done by our method [10], and Cohen et al. [11], respectively. *E. coli* C600 cells transformed by the crude DNA extracts were selected as colonies resistant to ampicillin on agar plates of L broth supplemented with 20 μg of ampicillin per ml. A hybrid plasmid in one of the *E. coli* C600 transformants was isolated and designated as pYMG14. The details of the isolation and purification and others were described in our previous papers [5, 12].

3. Construction of a New Host-Vector System to Improve Wild or Industrially Used Yeast Strains

A commonly used route for the introduction of new genes into microbial cells by gene cloning is, first, to isolate an auxotrophic mutant and then to use it as a host for a vector which can complement the deficiency. This method is simple and convenient when a mutation is lethal under appropriate growth conditions. However, it is sometimes difficult (especially in polyploid strains) to isolate a mutant for a special enzyme. In addition, the mutagenesis procedure might affect genes other than the intended one, so that the metabolic pathways of the host cell are seriously disturbed and it cannot be used for actual industrial production. It is necessary and important, therefore, to develop vector plasmids that can be introduced into wild-type or industrially used strains. For this purpose good dominant markers are required, which distinguish transformants from their wild-type host cells. We have chosen resistance to high amounts of MG and certain metals, as they are agents to which most yeast cells are usually sensitive. It is known that wild-type yeast cells do contain biochemical detoxification systems which can metabolize these chemicals and confer resistance to low concentrations. We therefore reasoned that if the level of the enzymes inolved in these processes could be increased, then the cells should become resistant to higher concentrations. Overproduction of an enzyme protein can be brought about is by increasing the dosage of the structural gene, for example, by cloning it onto a multicopy vector. We therefore thought that it should be possible to isolate genes encoding enzymes involved in these resistance mechanisms by the selection of clones which can grow on media containing the compounds at concentrations normally inhibitory to yeast growth after transformation of the cells with a yeast gene library constructed on a multicopy plasmid vector. Once they have been isolated, the hybrid plasmids could be used as vectors for the introduction of foreign DNA into any senstive strain of yeasts by use of the resistance genes encoding enzymes inolved in resistance to MG and certain metal ions by an approach based on this rationale [13].

4. The Metabolism of Methylglyoxal (MG)

Methylglyoxal (MG) is a toxic metabolite usually found in all organisms at extremely low concentration and is thought to be a regulator of cell division, for example, MG can arrest the growth of *E. coli* at millimolar concentrations. It is synthesized via

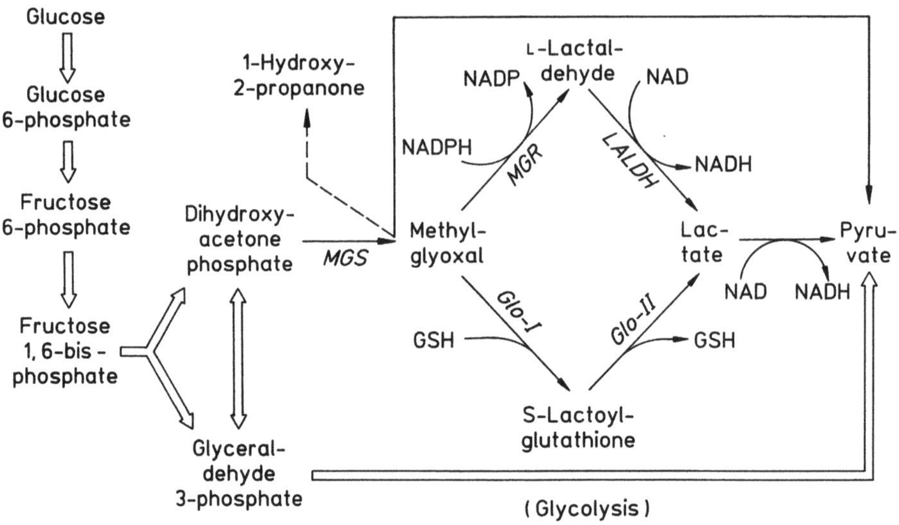

Fig. 1. Glycolytic bypass of *Saccharomyces cerevisiae*. *MGS*, Methylglyoxal synthase; *Glo-I*, Glyoxalase I; *Glo-II*, Glyoxalase II; *MGR*, Methylglyoxal reductase; *LALDH*, L-Lactaldehyde dehydrogenase

a bypass of the glycolytic pathway and/or aminoacetone cycle. As the compound is toxic, its synthetic and degradative pathways must be strictly controlled to avoid overaccumulation in cells. In *E. coli* cells, this is achieved by inhibition of MG synthase (EC 4.2.99.11), which catalyzes the synthesis of MG from dihydroxyacetone phosphate (Fig. 1) [14].

Several enzymatic reactions are involved in the degradation of MG. Among them the glyoxalase system, consisting of glyoxalase I (lactoylglutathione lyase; EC 4.4.1.5) and glyoxalase II (hydroxyacylglutathione hydrolase; EC 3.1.2.6) (Fig. 1), is thought to have an important role, not only in the degradation of MG, but also in conferring resistance to MG. These reactions have been shown to transform MG or its derivatives into lactate or corresponding alpha-hydroxy acids in the presence of glutathione in a wide variety of organisms. Thus the biosynthetic and degradative pathways of MG metabolism have been studied in detail [13-15].

The glyoxalase system is found also in bacteria (*P. putida*) [16]. On the other hand, MG is also oxidized directly to pyruvate by the enzyme from *P. putida* [17].

MG is supposedly involved in the regulation of cell dividion [18,19]. Like other alpha-ketoaldehydes, MG interacts with NH- or SH-groups, which may react with polyamines and function in cell division [20]. If alpha-ketoaldehydes are really involved in the regulation of cell division, the glyoxalase system could also be involved in the regulation of cell division and release of cells from the inhibitory action of MG and its derivatives. Using CDC (cell division cycle) mutants of yeasts, Dudani and coworkers demonstrated an important role of the glyoxalase system (glyoxalases I and II), as an indicator of cell growth [21].

It is possible that the unregulated growth of cancer cells may be attributable in part to the inactivation of alpha-ketoaldehydes by the glyoxalase system. For this reason, glyoxalase system indicators have been thought promising in screening for anticancer agents. The carcinostatic activities of various glyoxal derivatives towards certain tumours and leukaemias support this possibility.

5. Cloning of the Gene Responsible for MG Resistance

The yeast cell of *S. cerevisiae* DKD-5D-H could not grow on plates containing MG at concentrations above 1.25 mM for at least 3 to 4 d, because of the toxicity of MG. For the isolation of a gene responsible for MG resistance, the yeast cells were transformed with a mixture containing recDNAs, and spread on SD agar plates containing MG of 5.0 mM, a concentration sufficient to arrest the growth of the host cells. Colonies that appeared on the plates were isolated as the transformants harboring the gene responsible for MG resistance. About 10 transformants were obtained when the cells were transformed with 400 μg of DNA. No colonies were obtained when the parent yeast cells were spread and incubated on the same plates containing 5.0 mM MG. All the transformants obtained showed the same nutritional requirements as *S. cerevisiae*.

To extract the hybrid plasmid, the largest colony of transformants was picked up. The cells from it were grown on SD medium containing 5.0 mM MG, and were harvested and lysed with zymolyase. The crude DNA extracts were prepared and used to transform *E. coli* C600 cells treated with calcium. The *E. coli* C600 transformants were selected as the colonies resistant to ampicillin on agar plates of L broth. The hybrid plasmid in one of the *E. coli* C 600 transformants was extracted, purified, and used to retransform *S. cerevisiae* DKD-5D-H. Two plasmids containing genes for MG resistance were isolated and designated as pYMG14, and pYGL7, respectively. Restriction maps of these genes are shown in Fig. 2.

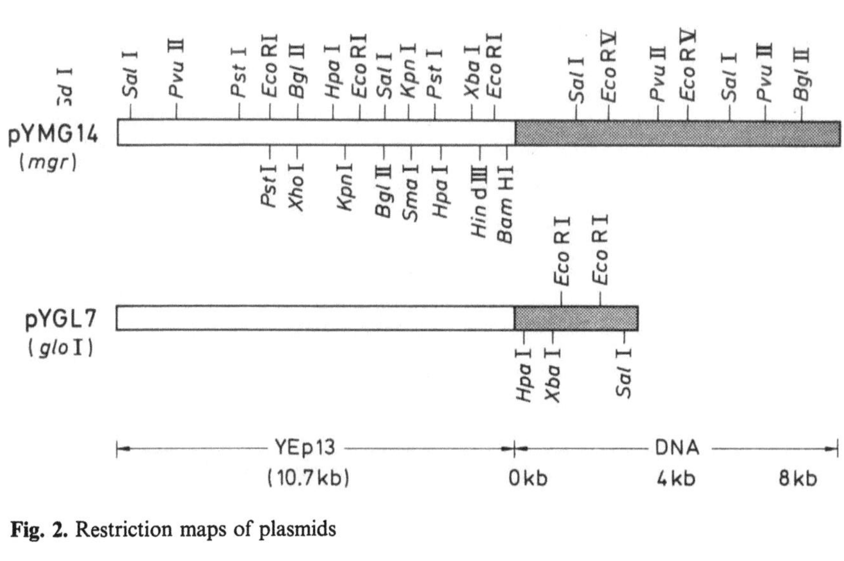

Fig. 2. Restriction maps of plasmids

6. Resistance of Yeast Transformants to Various Chemicals

S. cerevisiae carrying the hybrid plasmid pYMG14 showed marked resistance to heavy metal ions and some organic chemicals. Its resistance to Cd^{++} and Co^{++} was 100–50 fold higher, respectively, when compared with that of the original strain. Resistance to TMTD (tetramethylthiuram disulfide) and iodoacetamide also increased remarkably. In addition to them it showed the increased resistance also to other metals, phenylglyoxal, N-ethylmaleimide, and 8-hydroxyquinoline etc. The mechanism of this resistance to various compounds is not understood yet and now underinvestigated. Interestingly this plasmid gene pYMG14 was expressed in both eukaryote and prokaryote as well.

7. Formation of Jumbo Yeast Cells by MG Resistant Genes

Recently, the study of relationship between oncogene and cell division has been greatly advanced, because some homology in base sequence was found between oncogene and yeast gene, which compensates the deficiency of CDC (cell division cycle) mutation [21].

As pointed out by Szent-Gyoergyi MG have been thought to inhibit cell growth [18, 19], so it should be scavenged quickly. MG is converted by glyoxalase I to S-lactoylglutathione in the presence of glutathione. Since this reaction is active

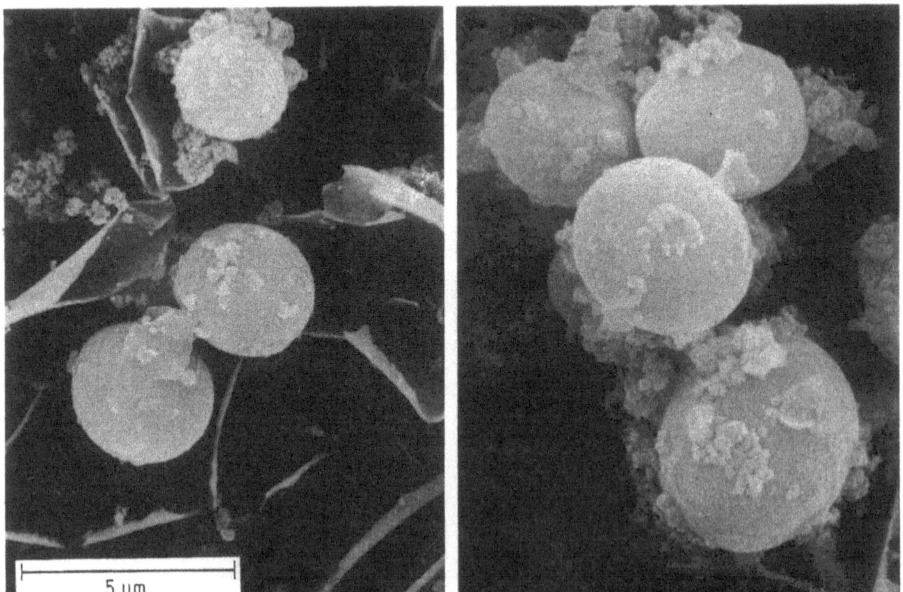

Fig. 3. Effect of Glyoxalase I gene on yeast cell size. *Left*: DKD-5D-H/YEp13; *right*: DKD-5D-H/pYGL7

especially in cancer cell, we thought that by the introduction of MG resistant genes (encoding MG reductase or glyoxalase I) we might be able to breed yeasts which can grow as rapidly as cancer cells. We tried to clone the gene for glyoxalase I, which is closely involved in cell growth. We cloned two genes (pYMG14 and pYGL7), and introduced them into yeast cells. Introduction of either genes make yeast cells 2 times bigger in diameter, and 8 times bigger in cubic content. The comparison of the transformant and control cells are shown in the photos (Fig. 3).

8. Discussion

Methylglyoxal (MG), a toxic compound, is one of the biological metabolites usually found in all organisms at extremely low concentration and is thought to function as a regulator of cell division. Genetic and enzymatic analyses of the biosynthetic and degradative pathways of MG, therefore, are of importance for better understanding of the functions of MG in cell division. In this regard, analysis of the characteristics of MG-resistant and MG-sensitive cells may give significant information. We first isolated genes responsible for the MG resistance of *S. cerevisiae* and characterized the phenotypic properties of the genes. One of them, the gene for MG reductase, can be used as new vector for breeding of wild or industrially used yeasts.

Generally for the isolation of genes from microbial cells it is preferable to use mutants having a deficiency in the gene product. This method is simple and convenient if the mutation is lethal. For the isolation of genes responsible for MG resistance of yeast, this was not applied. We assumed that the transformants with the gene could grow on the plates containing MG at the concentration that inhibited the growth of original strain (a host cell), since the gene recombined with a vector plasmid could multiply in the cells and produce a large amount of the protein responsible for the resistance. By this method, the gene responsible for MG resistance was easily obtained. This method seems to be applicable to the isolation of other genes resistant to toxic compounds.

Results of the enzyme assays also suggested that the hybrid plasmid pYMG14 contained the gene for MG reductase. Another interesting point with pYMG14 is the fact that the transformant with pYMG14 overaccumulates glutathione in the medium.

Although we should not generalize the reason for the multiresistance of yeast cells with pYMG14 to various toxic compounds, these phenotypic characteristics have a practical use as markers for the selection of transformants of wild-type yeast strains. In fact, the transformants of wild-type strains such as baker's, sake, and beer yeasts were effectively selected. The yeast strain utilized industrially are all diploid and have no genetic markers effective in the breeding of these strains. Only the yeast killer factor [1)] and antibiotic, gentamicin [2)], are now available for this purpose, but their utilization is limited because of their narrow spectrum of actions. In contrast, since the gene cloned here has a wider spectrum among various yeasts, it can be generally applied to the breeding of industrially important yeast strains.

MG is metabolized not only by MG reductase, but also by glyoxalase I, which converts MG to S-lactoylglutathione in the presence of glutathione. This latter reaction is active especially in cancer cell. We also succeeded in the cloning of the gene (pYGL7) for glyoxylase I. Introduction of either this gene (pYGL7) or another gene (pYMG14) makes cells 2 times bigger in size of diameter, and 8 times bigger in cubic content.

In conclusion we think it is quite interesting to investigate the relationship between proliferation of cancer cells and growth of yeast cells. We may be able to construct yeast cells, which can grow as rapidly as cancer cells. The recombinant DNA technique is quite powerful and promising tools for cell breeding.

9. Conclusions

Two methylglyoxal resistant genes were cloned. One is a gene for methylglyoxal reductase, and the other for glyoxalase I. The former can be used as a new vector for wild or industrially used yeast cells. Introduction of either gene made yeast cells 2 times bigger in diameter, and 8 times in cubic content. The jumbo yeast cells thus formed will be interesting not only microbiologically, but also industrially, for these jumbo cells can be easily filtered.

10. References

1. Hara, S.: J. Ferment. Technol. *62*, 325–339 (1984)
2. Jimenez, A., Davis, J.: Nature (London). *287*, 869–871 (1980)
3. Murata, K., Fukuda, Y., Shimosaka, M., Watanabe, K., Saikusa, T., Kimura, A.: Eur. J. Appl. Microbiol. Biotechnol. *151*, 631–636 (1985)
4. Fukuda, Y., Saikusa, T., Watanabe, K., Shimosaka, M., Murata, K., Kimura, A.: Agric. Biol. chem. *50*, 1341–1343 (1986)
5. Murata, K., Fukuda, Y., Shimosaka, M., Watanabe, K., Saikusa, T., Kimura, A.: Appl. Environ. Microbiol. *50*, 1200–1207 (1985)
6. Sato, N., Shimosaka, M., Fukuda, Y., Murata, K., Kimura, A.: Agric. Biol. Chem., *50*, 1339–1340 (1986)
7. Kimura, A.: Adv. Biochem. Eng. *33*, 29–51 (1986)
8. Kimura, A.: Biotechnol. Genet. Eng. Reviews. *4*, 39–47 (1986)
9. Saito, H., Miura, K.: Biochim. Biophys. Acta *72*, 619–629 (1963)
10. Ito, H., Fukuda, Y., Murata, K., Kimura, A.: J. Bacteriol. *153*, 163–168 (1983)
11. Cohen, S. N., Chang, A. C. Y., Hsu, L.: Proc. Natl. Acad. Sci. USA. *49*, 2110–2114 (1972)
12. Murata, K., Kimura, A.: Appl. Environ. Microbiol. *44*, 1444–1448 (1982)
13. Murata, K., Fukuda, Y., Shimosaka, M., Watanabe, K., Saikusa, T., Kimura, A.: J. Ferment Technol. *50*, 1200–1207 (1985)
14. Murata, K., Fukuda, Y., Shimosaka, M., Watanabe, K., Saikusa, T., Kimura, A.: Eur. J. Biochem. *151*, 631–636 (1985)
15. Murata, K., Inoue, T., Saikusa, T., Watanabe, K., Fukuda, Y., Shimosaka, M., Kimura, A.: J. Ferment. Technol. *64*, 1–4 1984
16. Inoue, Y., Watanabe, K., Shimosaka, M., Saikusa, T., Fukuda, Y., Murata, K., Kimurar, A.: Eur. J. Biochem. *153*, 243–247 (1985)

16a. Rhee, H., Murata, K., Kimura, A.: Biochem. Biophys. Res. Commun. *147*, 831–838 (1987)

17. Rhee, H., Watanabe, K., Murata, K., Kimura, A.: Agric. Biol. Chem. *51*, 1059–1066 (1987)
18. Szent-Gyoergyi, A.: Science. *149*, 34–37 (1965)
19. Egyued, L. G., Szent-Gyoergyi, A.: Proc. Natl. Acad. Sci. *55*, 388–393 (1966)
20. Russel, D. H. and McVicker, T. A. Biochim. Biophys. Acta, *259*, 247–258 (1972)
21. Dudani, A. K., Srivastava, L. D. and Prasad, R.: Biochem. Biophys. Res. Comm., *119*, 962–967 (1984)

3.3 Cloning and Over-Expression of the *E. coli* D-Xylose Isomerase Gene in Various Microorganisms

Nancy W. Y. Ho

Laboratory of Renewable Resources Engineering, Purdue University,
West Lafayette, IN 47907, USA

Contents

D-Xylose isomerase, also known as D-glucose isomerase, catalyzes the conversion of D-xylose to D-xylulose and D-glucose to D-fructose. Both reactions have important industrial applications. Recently, we have succeeded in the identification of a ColEl-hybrid plasmid containing the *Escherichia coli* xylose isomerase gene (*xylA*) from the Clarke and Carbon *E. coli* gene bank. Subsequently, through subcloning and restriction mapping, *xylA* was found to be located on a 1.6 kb BglII fragment. We found that the intact *E. coli xylA* gene was not expressed in yeast (*Saccharomyces cerevisiae*) or *Bacillus subtilis*. Fusion of the *xylA* structural gene to the yeast *TRP5* or *ADH1* promoter did not result in the expression of the hybrid gene in yeast. However, through *lacZ* fusion, we have been able to demonstrate that the *TRP5-xylA* hybrid gene did contain the necessary genetic elements to direct the synthesis of a protein in yeast. On the contrary, fusion of the *xylA* structural gene to a promoter functional in *B. subtilis* resulted in the expression of the *xylA* structural gene in the latter organism.

Overproduction of the xylose isomerase (or glucose isomerase) in *E. coli* by recombinant DNA techniques has also been extensively studied. We found that high levels of overproduction of the enzyme cannot be accomplished by cloning the intact *xylA* gene on high copy-number plasmids alone. However, we have been able to demonstrate that this difficulty can be overcome by the removal of the natural promoter from the *xylA* gene and fusion of the resulting promoterless *xylA* with other strong promoters which are functional in *E. coli*. For example, we have recently constructed several high copy-number plasmids, and each of them contains the *xylA* structural gene fused to a different strong promoter (*lac*, *tac*, γP_L, or γP_R). *E. coli* transformants bearing these plasmids can all overproduce 20- to 50-fold more xylose isomerase than the wild-type *E. coli* under the same conditions.

1. Introduction

The recent advances in recombinant DNA and gene cloning techniques have provided new stimuli to biotechnology. These new techniques allow the cloning and overexpression of a gene from one organism in another desired organism. In this paper the cloning and overexpression of the *E. coli* xylose isomerase gene in various

Bioproducts and Bioprocesses
Editors: Fiechter, Okada, Tanner

microorganisms will be discussed and used as examples to demonstrate the powerfulness and limitations of these new techniques. Wood and agricultural by-products are considered to be an abundant, virtually inexhaustible resource materials in nature. The biomass in these renewable materials is composed mainly of cellulose, hemicellulose, and lignin with a ratio approximately 4.5 to 3.5 to 2.0 [1]. It is obvious that these renewable materials could become the most valuable raw materials of the future. D-Xylose is a major constituent of hemicellulose. One of our recent research interests is to study the mechanism and regulation of xylose metabolism in various microorganisms including *E. coli*, yeast (*Saccharomyces cerevisiae*), and *Bacillus subtilis*.

Yeast (*S. cerevisiae*) cannot utilize D-xylose as a sole carbon source, but it can metabolize xylulose. D-Xylose can be converted to xylulose in vitro by the enzyme xylose isomerase. Both *B. subtilis* and *E. coli* can readily utilize D-xylose as a sole carbon source for growth.

In *E. coli* the initial metabolism of xylose involves the conversion of xylose to xylulose by the enzyme xylose isomerase, followed by the conversion of xylulose to xylulose-5′ phosphate by xylulokinase. The latter is then metabolized via the pentose pathway.

Genetic study provided evidence for four genes responsible for the initial metabolism of xylose in *E. coli*: *xyl*A, the xylose isomerase gene; *xyl*B, the xylulokinase gene; *xyl*T, a gene specifying a xylose transport component; and *xyl*R, a gene encoding a regulatory element necessary for the transcription of the *xyl* genes [2].

2. Cloning the *E. coli* Xylose Isomerase Gene in *E. coli* and Yeast

Recently we have cloned and characterized the *xyl*A and *xyl*B gene from *E. coli* [3,4]. We have attempted to subclone the intact *xyl*A gene on various high copy-number *E. coli* plasmids to study the gene dosage effect on the production of the enzyme. We found that *E. coli* was unable to produce significantly higher levels of D-xylose isomerase by increasing the gene dosage alone. This indicates that the expression of *xyl*A in *E. coli* is highly regulated.

We have also attempted to express the intact *E. coli xyl*A in yeast (*S. cerevisiae*) and found that the intact *E. coli* gene could not be expressed in the latter host. In order to express *xyl*A in yeast, we have constructed a *Trp*5-*xyl*A fusion, using the yeast *Trp*5 promoter to express *xyl*A in yeast [5]. The *Trp*5-*xyl*A fusion was cloned in plasmid pLX10–14, and the fusion was constructed in such a way that a XhoI linker was used to join the *Trp*5 promoter and the promoterless *xyl*A structural gene. The promoterless *xyl*A can be separated from the *Trp*5 promoter and the rest of the plasmid as a XhoI-EcoRI fragment. The construction of such a portable promoterless *xyl*A gene has been very helpful for accomplishing much of the research described below. Nevertheless, the *Trp*5-*xyl*A fusion was still not able to be expressed in yeast. Since *Trp*5-*xyl*A fusion only contains a yeast promoter while still retaining its original *E. coli* non-coding sequences immediately upstream to the *xyl*A structural gene including the ribosomal binding site, it is, therefore, desirable to determine whether the *Trp*5-*xyl*A hybrid gene contains all the necessary genetic signals, particularly the ribosomal binding site, for gene expression in *S. cerevisiae*.

3. Construction of a *xyl*A-*lac*Z Fusion

In *E. coli*, a cluster of three genes constitutes the lactose operon. The first two gene products are required for the utilization of lactose as a carbon source. The *z* gene codes for the structures of the enzyme β-galactosidase, which catalyzes the cleavage of lactose into glucose and galactose. The *y* gene directs the synthesis of a permease required to transport lactose into the cell. A third gene, the α gene, codes for the structure of thiogalactoside transacetylase [6]. Muller-Hill and Kania [7] have shown that the substitution of up to about 30 amino-terminal residues of *E. coli* β-galactosidase with a variety of other protein sequences results in the construction of various hybrid proteins which still retain substantial β-galactosidase activity. Since then, fusions of a defective *lac*Z gene (missing its control signals and the nucleotide sequences near the 5′ coding region) with other genes has been a powerful tool for defining the genetic elements responsible for the expression of numerous genes.

Furthermore, the *E. coli lac*Z gene was reported able to be expressed in *S. cerevisiae*, and *lac*Z fusion has also been frequently used to study gene expression in the latter organism [8]. Hence, we believe that the construction of a *lac*Z fusion between the *Trp*5-*xyl*A gene and a defective *lac*Z gene should be able to provide the crucial information on whether *Trp*5-*xyl*A contains a suitable ribosomal binding site

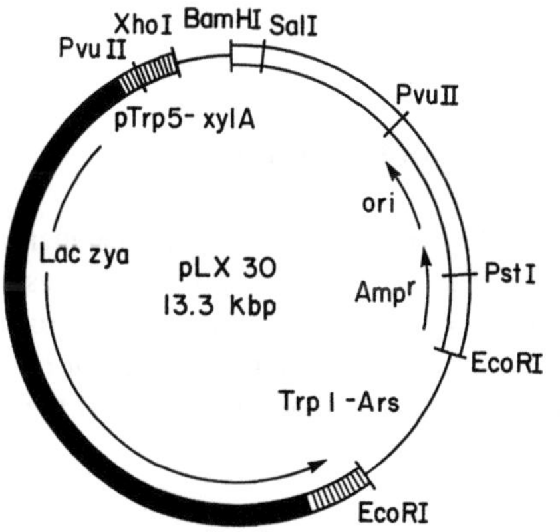

Fig. 1. Construction of the yeast-*E. coli* shuttle plasmid pLX30 containing the *Trp*5-*xyl*A-*lac*Z fusion. The fusion was constructed by linearizing plasmid pLX10-14 (11, also see Fig. 2) and trimming the linearized plasmid with Bal31 nuclease. The resulting mixture of fragments was then ligated with the SmaI-SalI fragment (*laczya*) containing the defective *lac*Z gene, which was isolated from plasmid MC1403 (8) and subsequently treated with DNA polymerase I to fill in the cohesive ends prior to ligation. The resulting plasmid containing the fusion, designated pLX24 (not shown), was used to construct the shuttle plasmid pLX30 by inserting an EcoRI fragment containing the Trp1-Ars, isolated from plasmid YR_p7 (15), in the EcoRI site of pLX24. The plasmid was used to transform *S. cerevisiae* Trp1 mutant, DBY 747, and the Trp1$^+$ transformants were found able to produce functional β-galactosidase

for gene expression in *S. cerevisiae*. A *Trp5-xyl*A-*lac*Z fusion, cloned on an *E. coli*-yeast shuttle vector, was therefore constructed (Fig. 1) and the resulting fusion was able to be expressed in yeast. This proves that the *Trp5-xyl*A fusion contains the necessary genetic elements upstream to the structural gene for gene expression in yeast. Thus, our results described above demonstrate that the fusion of *xyl*A to a functional expression system alone is not sufficient enough to make yeast producing functional *E. coli* xylose isomerase.

4. Expression of the *E. coli xyl*A Gene in *B. subtilis*

We also studied the expression of *E. coli xyl*A gene in *B. subtilis* and found that the intact *E. coli* gene was again not expressed in this foreign host. In order to express *xyl*A in *B. subtilis*, we constructed two *pen*P-*xyl*A fusions, subjecting the promoter-less *xyl*A under the control of *Bacillus* promoter from the pennicillinase gene. We found that both fusions could be expressed very well in *B. subtilis*. In one fusion, *xyl*A was inserted downstream to the leader sequence; in the other, *xyl*A was inserted into the structural gene of *pen*P [9,10].

Although it is not difficult to manipulate *E. coli xyl*A suitable to be expressed in *B. subtilis*, we found that the cloned *E. coli xyl*A gene cannot be stably maintained on the plasmid in *B. subtilis* under non-selective conditions. Under the latter conditions, the cloned *xyl*A gene gradually lost its activity, which was coincided with the formation of mutant plasmids containing partially deleted *xyl*A. Hence, in order to overproduce xylose isomerase via plasmid mediated gene cloning techniques in *B. subtilis*, it has to be carried out under selective conditions with the use of an *xyl*A mutant as the host.

5. Over-expression of the *E. coli* Xylose Isomerase Gene

As described before, we found that *E. coli* was unable to produce significantly higher levels of D-xylose isomerase simply by increasing the gene dosage. This indicates that the expression of *xyl*A in *E. coli* is highly regulated. One approach for overcoming this difficulty is to place the *xyl*A structural gene under the control of other promoters. As reported before, we had already succeeded in the removal of the

---

Fig. 2. Construction of pLX10-14 E-2 plasmids. The BamH1-BglII fragment containing the *xyl*A gene (F2) was ligated with BamH1 linearized pCQV2.4, constructed by Queen (16). The BamH1 linearized plasmid was treated with phosphatase to favor the insertion of the BamH1-BglII fragment. The resulting plasmid pLX10-14 E-1 was first linearized by XhoI, followed by trimming with Bal31 nuclease. The resulting DNA molecules were subsequently cut with BamH1 to completely remove the rest of the *Trp5* promoter. The DNA fragments were then treated with S_1 nuclease to generate ligatable even ends for the subsequent ligation to create a bank of pLX10-14 E-2 plasmids. Those pLX10-14 E-2 plasmids which can transform *E. coli* wild-type and make the resulting transformants produce white colonies on MacConkey-xylose-Amp indicator plates are xylose isomerase overproducers (12), and they were selected for further characterization ▶

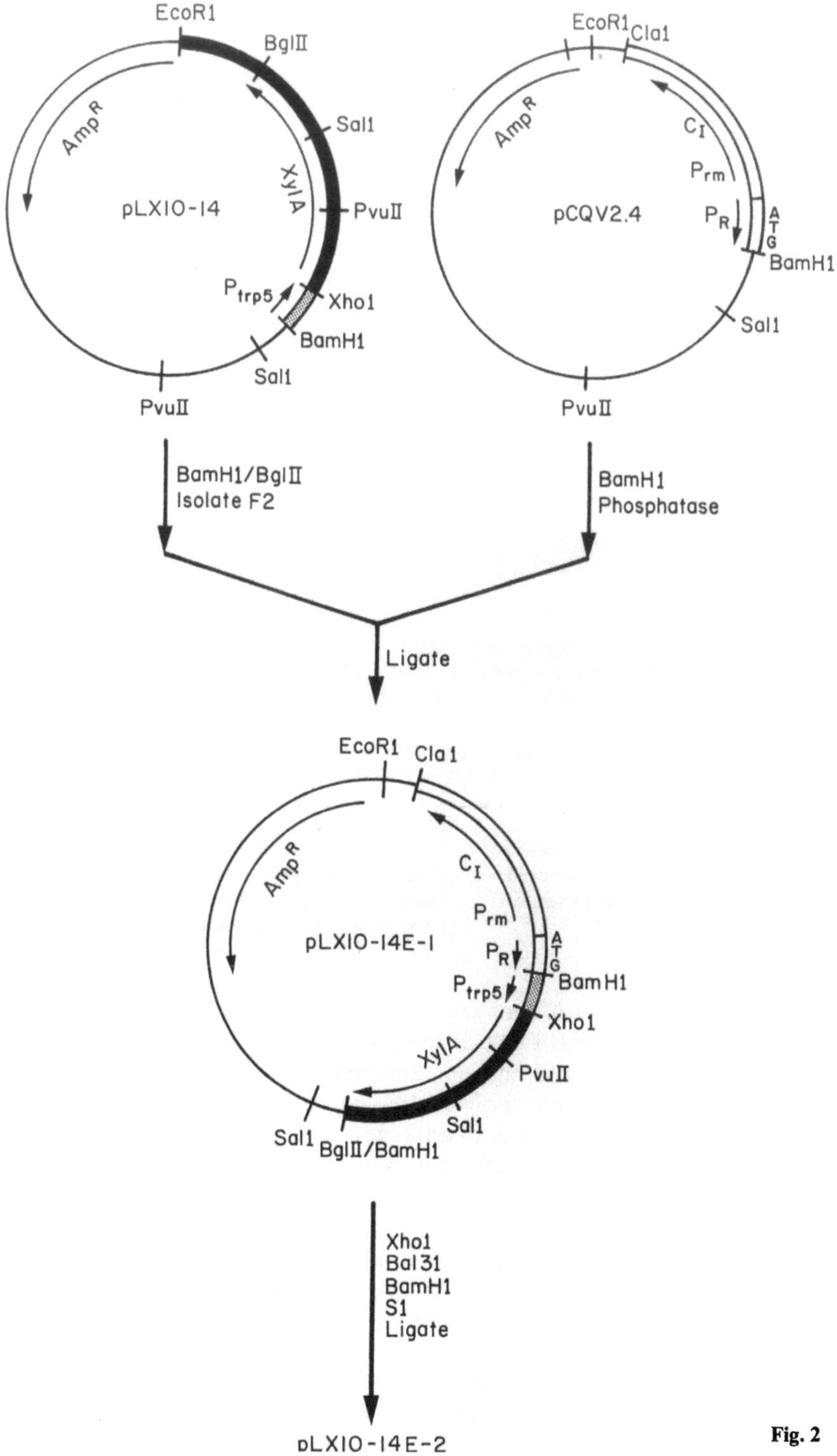
EcoR1
BglII
Sal1
XylA
PvuII
pLXIO-14
AmpR
Ptrp5
Xho1
BamH1
Sal1
PvuII
EcoR1
Cla1
CI
Prm
PR
ATG
BamH1
pCQV2.4
Sal1
PvuII
BamH1/BglII
Isolate F2
BamH1
Phosphatase
Ligate
EcoR1
Cla1
CI
Prm
pLXIO-14E-1
PR
ATG
BamH1
Ptrp5
Xho1
XylA
PvuII
Sal1
Sal1
BglII/BamH1
Xho1
Bal31
BamH1
S1
Ligate
pLXIO-14E-2

Fig. 2

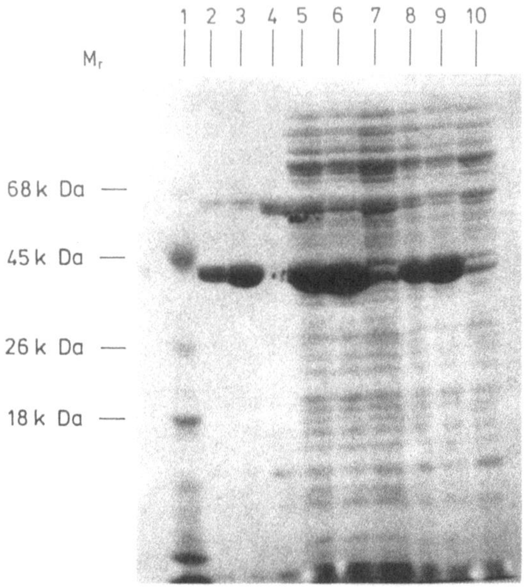

Fig. 3. SDS-PAGE analysis of the total protein from *E. coli* transformants containing pLX10-14 E-2 plasmids. Transformants A and B contain pLX10-14 E-2 plasmids, capable of overproducing *xyl*A protein product. Transformant C contains pCQV2.4 plasmid, unable to overproduce *xyl*A protein product, to serve as the control. The expression of *xyl*A in this system is regulated by the λ P_R promoter and the λ temperature sensitive repressor (16). Due to the fact the *xyl*A protein product, xylose isomerase, is heat-stable, one set of the proteins was heat-shocked at 65 °C for 5 min prior to SDS-gel analysis. Lanes 5–7 contain twice as much protein as Lanes 2–4 and 8–10.

Lane 1 Protein molecular markers
Lane 2 Total protein from A, heat-shocked
Lane 3 Total protein from B, heat-shocked
Lane 4 Total protein from C, heat-shocked
Lane 5 Total protein from A
Lane 6 Total protein from B
Lane 7 Total protein from C
Lane 8 Total protein from A
Lane 9 Total protein from B
Lane 10 Total protein from C

natural promoter from the *xyl*A gene and also succeeded in the construction of a portable promoterless *xyl*A structural gene with suitable restriction sites at its ends, convenient for recombinant manipulation. Subsequently, we have subjected the promoterless *xyl*A under the control of the *lac* and *tac* promoter and found that *E. coli* transformants containing such fused genes, cloned on various high copy-number plasmids, could overproduce 20 times more xylose isomerase activities than the wild-type *E. coli* [11)]. Recently we have constructed other high copy-number plasmids containing *xyl*A under the control of the P_R promoter of λ-phage. The construction of these plasmids are shown in Fig. 2. Analysis of the *E. coli* transformants containing pLX10–14 E-2 indicate that a 40- to 50-fold overproduction of the enzyme had been accomplished. SDS-PAGE analysis of the total protein of the *E. coli* transformants containing pLX10–14 E-2-type plasmids is shown in Fig. 3.

Our results demonstrate that the removal of the original *xyl*A promoter and the fusion of the resulting promoterless *xyl*A with other strong promoters, coupled with cloning the fused gene on high copy-numer plasmids have made it possible to overproduce xylose isomerase activity in *E. coli.* During these studies, we have also observed that the presence of high levels of xylose isomerase in *E. coli* will inhibit xylose degradation in *E. coli*[4,11]. Hence the overproduction of xylose isomerase in *E. coli* has to be carried out under nonselective conditions such as using wild-type *E. coli* strains as the host and using other sugars rather than xylose as the carbon source. Recently, we have been able to exploit this phenomena for the development of two useful vector systems (a positive selection system and a direct selection system) for gene cloning[12-14].

6. Conclusion

Cloning and overexpression of the *xyl*A gene in *E. coli* yeast, and *B. subtilis* provide ideal model systems to demonstrate the powerfulness of these techniques and the difficulties that may be encountered during manipulation of the microorganisms with these techniques. Proper application of these techniques will undoubtedly lead to dramatic improvement of the productivity of microbial based biotechnology. In general, most of the difficulties encountered during cloning and expression of a specific gene can be overcome by a very careful study of the problem.

7. Acknowledgements

This work was supported in part by NSF Grant MBB-8305043 and USDA grant 85-CRCR-1-1713. Dr. S. Rosenfeld was responsible for the construction of the *lacZ* fusion and Dr. J. Huang was responsible for the expression of *E. coli xyl*A in *B. subtilis* while they were working in my laboratory. Mr. P. Stevis was responsible for the over-expression of *xyl*A in *E. coli.*

8. References

1. Jeffries, T. W.: Adv. in Biochemical Engineering/Biotechnology *27,* 1 (1983)
2. Shamana, D. K., Sanderson, K. C.: J. Bacteriol. *139,* 64 (1979)
3. Ho, N. W. Y., Rosenfeld, S., Stevis, P., Tsao, G. T.: Enzyme Microb. Technol. *5,* 417 (1983)
4. Rosenfeld, S., Stevis, P., Tsao, G. T., Ho, N. W. Y.: Mol. Gen. Genetics *194,* 410 (1984)
5. Ho, N. W. Y., Stevis, P., Rosenfeld, S., Huang, J. J., Tsao, G. T.: Biotechnol. Bioeng. Symp., No. 13, 245 (1983)
6. Zabin, I., Fowler, A. V.: β-Galactosidase, the lactose permease protein and thiogalactoside transacetylase in: The Operon, p. 89, Cold Spring Harbor Laboratory, Cold Spring Harbor, NY 1980
7. Muller-Hill, B., Kania, J.: Nature *249,* 561 (1974)
8. Casadaban, M. T., Martinez-Arias, A., Shapiva, S. K., Chou, J.: Methods Enzymol. *100,* 293 (1983)
9. Huang, J. J., Ho, N. W. Y.: Biochem Biophys. Res. Comm. *1126,* 1154 (1985)

10. Huang, J. J., Ho, N. W. Y.: Expression of the *E. coli* xylose isomerase gene in *B. subtilis* using a *Bacillus* Promoter, in: Plasmids in Bacteria, Basic Life Science (A. Hollaender, General Ed.) Vol. 30, p. 953 Plenum Press 1985
11. Stevis, P. E., Ho, N. W. Y.: Enzyme Microbial Technology *7*, 592 (1985)
12. Stevis, P. E., Ho, N. W. Y.: Gene *55*, 67 (1987)
13. Stevis, P. E., Ho, N. W. Y.: Fed. Amer. Soc. Exp. Biol., Proceedings *45*, 1851 (1986)
14. Ho, N. W. Y., Feldman, D., Stevis, P. E.: Fed. Amer. Soc. Exp. Biol., Proceedings *45*, 1851 (1986)
15. Tschumper, G., Carbon, J.: Gene *110*, 157 (1980)
16. Queen, C.: J. Mol. and Appl. Genetics *2*, 1 (1983)

3.4 Production of Useful Proteins by *Bacillus brevis*

Shigezo Udaka

Department of Food Science and Technology, Faculty of Agriculture, Nagoya University, Chikusa, Nagoya 464, Japan

Contents

1. Introduction

Proteins are informational macromolecules with numerous structures and functions, and play essential roles in all living things for activities of life. From the viewpoint of usefulness to our human life, there are many proteins important as food proteins, enzymes, hormons, etc.

In 1971, I started to search the way to produce such useful proteins by using microorganisms, since we knew that some organisms have extraordinary cellular activities. Although it has long been known that various microorganisms secrete enzymes and proteinaceous toxins, the amount of such proteins is usually small [1]. I have therefore attempted to screen efficient protein-producing bacteria. The reasons I was interested in protein-secreting bacteria were two-fold. A large amount of limited kinds of proteins could be produced extracellularly so that we can purify the secreted protein rather easily. The yield of protein secretion could be higher than that of protein accumulated inside the cells because the protein content of cells is limited. The other interesting problem was the mechanism of passage of large molecular proteins across cell membrane which might be clarified by such a protein producer.

As a result of a screening program, we found several excellent protein-producing bacteria. Employing such bacteria, we are now able to carry out protein production in a very high yield. The present review describes how we developed the host-vector system suitable to the production of useful proteins.

2. Screening for Protein-producing Bacteria

The initial screening of protein-producing bacteria was done by selecting bacteria whose colonies formed an opaque area after flooding the plate with a protein-denaturing agent such as perchloric acid [2]. The logical basis for such a sreening

Bioproducts and Bioprocesses
Editors: Fiechter, Okada, Tanner

method is as follows. A large number of different bacteria could be tested in a very simple and efficient manner on agar plates. Many bacteria were spotted or replicated at one time on several plates having media of various compositions and could be cultured under different conditions (e.g., at different temperatures). It was inferred that a more efficient producer of protein gives a denser turbid area beneath the colony with the addition of acid, and autolysis of bacteria usually does not occur on an agar medium, so that the formation of an opaque area by acid is a measure of true protein secretion rather than the consequence of cellular lysis.

The bacteria thus selected were cultured in liquid medium with shaking to measure the amount of extracellular protein accumulated. About 1200 isolates from nature (mainly, soil samples) were tested on agar plates [2]. As shown in Table 1, over 80% of bacterial isolates gave little sign of protein secretion by tests on the two kinds of agar plates. About 15% of bacteria showed a certain degree of accumulation of macromolecules by the agar plate test. In general, much more secretion was observed when bacteria were cultured in medium T_2 than in T_1, regardless of whether the medium was solid or liquid. The inorganic nitrogen compound in T_1 medium might be responsible for suppressing the protein secretion.

Among the bacteria producing protein to some extent, 5 strains produced extracellular proteins in exceedingly large amounts. These bacteria share many common microbiological features and were found to belong to *Bacillus brevis* or its closely related new species.

Recently, we carried out a very extensive screening of protein-producers by the method similar to that described above [3]. From 10^5 bacterial strains isolated from about 2000 samples, about 80 strains showed a productivity of over 1.2 g l^{-1} of protein in liquid culture. Thirty seven strains produced more than 5 g l^{-1} of protein when cultured in liquid medium with an appropriate composition and appeared to

Table 1. Distribution of protein-producing bacteria

Culture medium	Formation of turbid area[a]	Protein accumulated (mg ml^{-1})	Number of strains found	% of total number
T_1	—		ca. 1150	97
T_1	+	<0.2	26	3
		0.2–0.5	6	
T_2	—		ca. 1000	85
T_2	+	<0.2	59	7.7
		0.2–1	31	
	++	<0.2	23	4.2
T_2		0.2–1	26	
T_2	+++	<0.2	11	2.9
		0.2–1	18	
		3–5	5	

Note: The medium T_1 was composed of 2% glucose, 0.2% urea, 0.2% $(NH_4)_2SO_4$, 0.3% K_2HPO_4, 0.1% KH_2PO_4, 0.02% $MgSO_4 \cdot 7\,H_2O$, 0.3% polypepton, and 0.2% meat extract. T_2 contained 1% polypepton, 0.5% meat extract, 0.2% yeast extract and 1% glucose (pH was adjusted to 7).

[a] +, detectable; +++, either dense or large opaque area; ++, the middle of + and +++.

be mostly *B. brevis* or its related bacteria. Several of them showed no detectable protease activity in the culture supernatant.

3. Characteristics of Protein Production by *B. brevis* 47

3.1 Optimal Cultural Conditions for Protein Production

B. brevis 47 strain was found to be one of the best protein-producing bacteria isolated. We therefore used this strain in most of the subsequent studies. A large amount of protein can be produced relatively easily in nutrient rich media. However, the efficiency of protein production varied greatly depending on the composition of the culture medium [4]. *B. brevis* cells grow well in a medium containing yeast extract or meat extract, but significant amounts of protein are not accumulated. On the other hand, polypeptone which is a commercial product and equivalent to peptone was

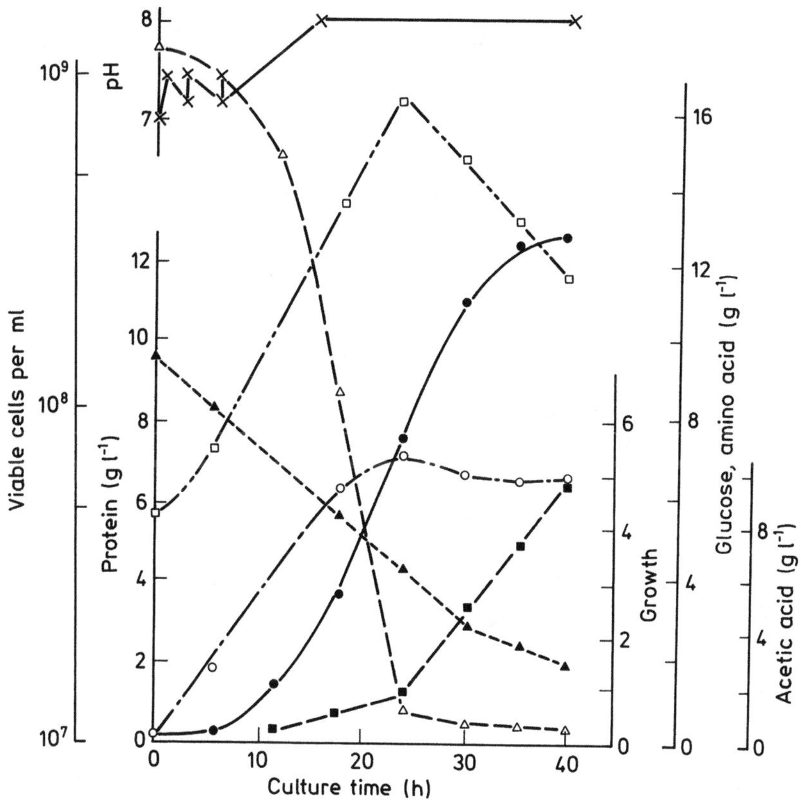

Fig. 1. Time course of protein production by *B. brevis* 47 in nutrient rich medium. The experiment was carried out in a stirred jar. The pH was controlled by acetic acid or ammonia gas. ●, extracellular protein; ○, growth; □, viable cells; ▲, glucose; ■, acetic acid; △, amino acids; ×, pH

found to be a nutrient suitable for protein production. It appeared to contain factors which stimulate protein secretion, since it showed a marked effect on protein production but only a little effect on growth when added to a medium containing other nutrients. We tried to characterize the stimulatory factors and found that it presumably consisted of polypeptide with a molecular weight of a few thousand [4].

Among factors affecting the protein production, glycine and L-isoleucine were found to be prominent in stimulating protein production in nutrient rich medium [5]. The simultaneous addition of appropriate amount of these amino acids resulted in the large accumulation of proteins, 12 $g\,l^{-1}$ (Fig. 1) which was about twice the amount accumulated in the medium without those amino acids. The mode of action of glycine and isoleucine appeared different. Isoleucine stimulated the synthesis of both extracellular and intracellular proteins, while glycine caused a considerable increase of extracellular protein accumulation with a concomitant decrease in the amount of intracellular protein. Therefore, glycine may have a function in stimulating protein secretion.

A simple chemically defined medium was devised for protein production by *B. brevis* 47 [6]. About 2 $g\,l^{-1}$ of proteins were secreted in the synthetic medium which contained 4% glucose, 1% ammonium sulfate, and some of inorganic salts such as phosphate, Mg^{2+} and Ca^{2+}. An essential component of the medium was

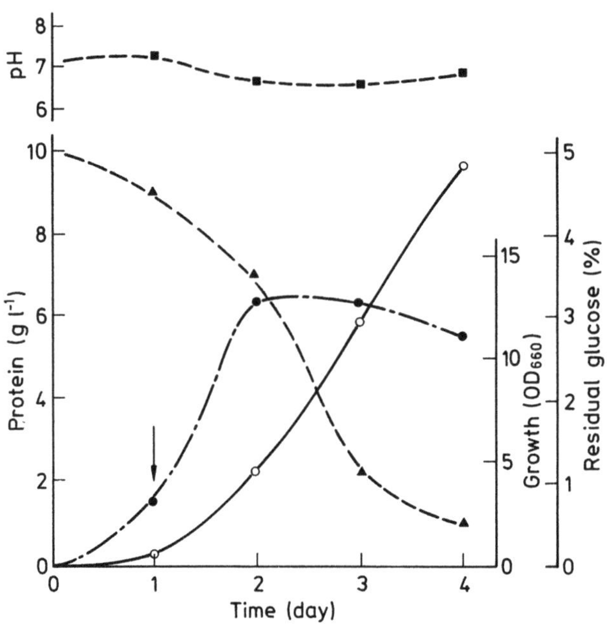

Fig. 2. Protein production by *B. brevis* 47 in synthetic medium. The culture was carried out in shake flask at 34 °C on a shaker. The medium contained 5% glucose, 1% $(NH_4)_2SO_4$, 0.1% KH_2PO_4, 0.02% $MgSO_4 \cdot 7\,H_2O$, 0.03% L-glutamic acid, trace amounts of Fe^{2+}, Mn^{2+}, Zn^{2+}, and 2% $CaCO_3$. pH was adjusted to 7.8 with KOH. Bacitracin was added to the culture at the time indicated by arrow. ○, extracellular protein; ●, growth; ▲, residual glucose; ■, pH

Ca salt which is required by this organism for assimilating glucose. Studies on the effects of various chemicals and antibiotics on protein production revealed that inhibitors of cell wall (peptidoglycan, teichoic acid, etc.) synthesis, such as bacitracin or β-lactam antibiotics, were effective in greatly increasing the accumulation of proteins [7].

When 60 mg l^{-1} of bacitracin was added to the culture at the early exponential growth phase, about 9 g l^{-1} of proteins were produced, compared to 3 g l^{-1} of protein production without bacitracin (Fig. 2) [7]. In this case, total protein production was 12 g l^{-1} from 50 g l^{-1} of glucose: 9 g l^{-1} of extracellular protein plus 3 g l^{-1} cellular protein. Since generally cell yield (dry weight basis) from glucose is known to be at most about 50% and average protein content of bacterial cells is about 60%, we may expect about 15 g l^{-1} as total protein yield from 5% glucose. Therefore, the yield of the protein production was about 80% on the basis of glucose consumption.

3.2 Mechanism of Protein Production

Proteins secreted were analyzed by sodium dodecyl sulfate polyacrylamide gel electrophoresis and found to be consisted mainly of two proteins with approximate molecular weights of 130 kDa and 150 kDa. Examination of the bacterial cell surface under electron microscope revealed a characteristic three-layered structure [8]. The three layers are designated, from the outermost to the innermost layer, as the outer wall (OW, 4.2 nm), the middle wall (MW, 8.5 nm), and the inner wall (2.1–3.7 nm) [9,10]. The inner wall may be a peptidoglycan layer, since the purified inner wall had a chemical composition characteristic of peptidoglycan. The cell wall structure was morphologically altered depending on the growth period (Fig. 3). At the early stationary phase of growth, cells began to shed the outer two layers of the surface. This shedding was complete after further cell growth. The morphological alterations in the cell wall occurred concomitantly with a prominent increase in protein secretion. The synthesis and secretion of extracellular proteins continued after cell growth ceased. When protein secretion was severely inhibited by growing cells with Mg^{2+}, morphological alterations in the cell wall were not observed, even at the late stationary phase of growth. This was also the case with a nonprotein-producing mutant [8].

When cells at the exponential phase were incubated in weakly alkaline buffers (pH 7.5–8.5), the outer two layers of the cell wall were specifically removed, leaving cells surrounded only by the peptidoglycan layer. The two surface layers were found to be composed mainly of two proteins with the same molecular weights as those secreted into the medium. Based on the results of analysis of amino acid composition, limited proteolysis followed by electrophoresis, and the cross-reactivity of antisera, we concluded that the 130 kDa and 150 kDa extracellular proteins were derived from the respective cell wall proteins [11]. Therefore, we propose the following scheme as the mechanism of the protein production by *B. brevis*. Cells synthesize at the exponential phase the paracrystalline cell wall proteins to form hexagonal arrays in the cell surface, and at stationary phase continue to synthesize and secrete the wall proteins, which do not stay on the cell surface and accumulate in the medium as extracellular proteins. This is a novel mechanism of protein production by bacteria hitherto unknown.

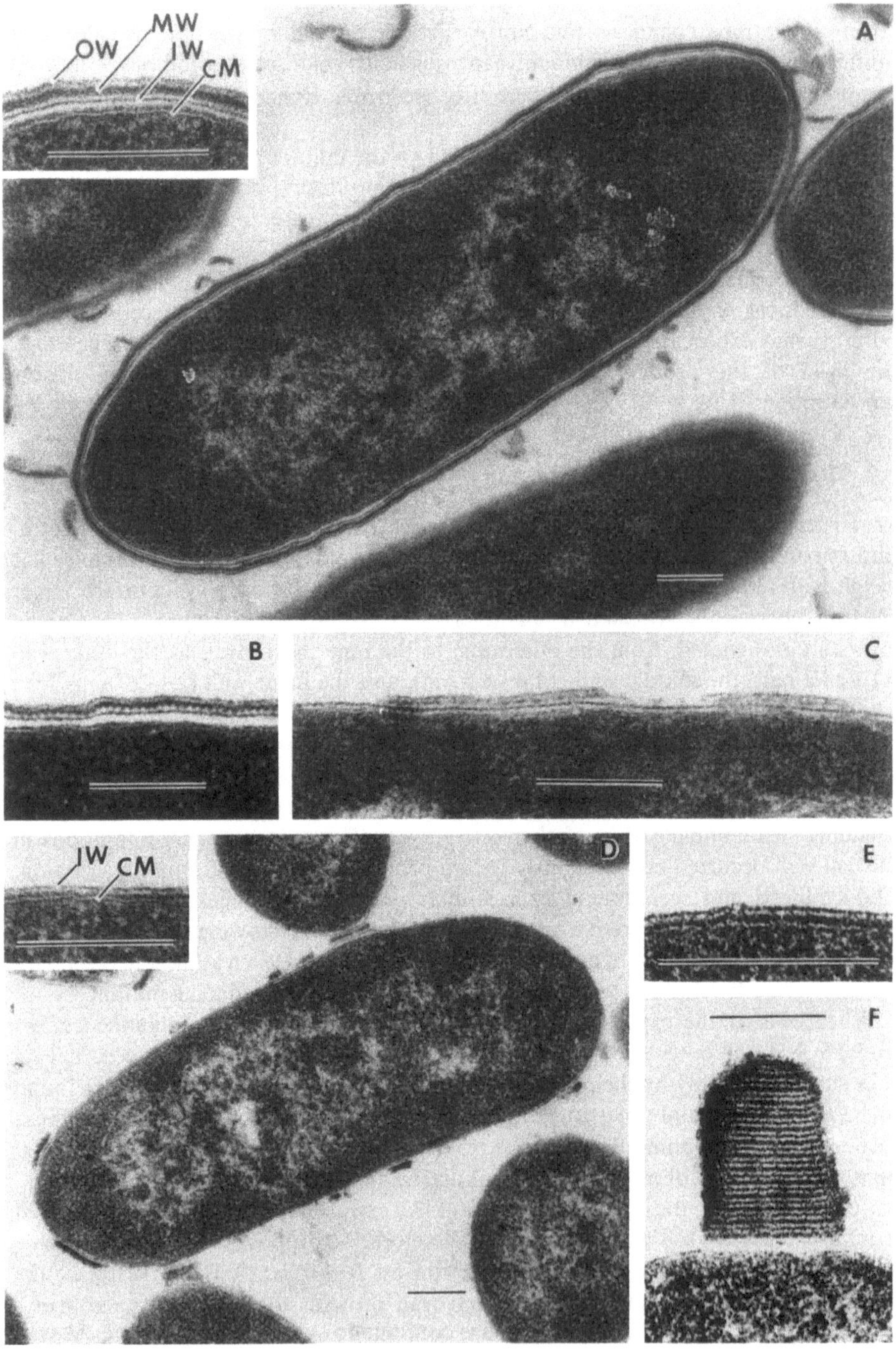

Fig. 3A–F

4. Gene Manipulation in *B. brevis*

To utilize the excellent protein-producing ability of *B. brevis* for the production of various useful proteins, we attempted to develop a method to engineer the genes in *B. brevis*. For that purpose, it was essential to establish a method of transformation and to find appropriate vectors available in this bacterium. Protoplast transformation has been successful in many bacteria in the presence of polyethylene glycol, but the method could not be applied directly to *B. brevis* due to the difficulty in regenerating its protoplast. We therefore devised a novel method for introducing plasmid DNA into *B. brevis* [12]. The method includes treatment of cells with tris-HCl buffer of alkaline pH that removes the outer two protein layers of the cell wall (as described in Sect. 3.2), followed by introduction of DNA-uptake with polyethylene glycol. The competent cells incorporated plasmids with a frequency of as high as 10^5–10^6 transformants per μg of DNA.

Two plasmids originated from *Staphylococcus aureus*, a low copy-numbered pHW1 and a high copy-numbered pUB110, have been useful as the vector in *B. brevis*. A cryptic plasmid found in *B. brevis* 481, one of the protein-producing bacteria, was used to construct a low copy number vector, pHY481, by combining an erythromycin resistant gene of pHW1 [13,14]. This plasmid when introduced into *B. brevis* 47 was maintained very stably even in the absence of selective antibiotics. The copy number of pHY481 could be amplified about 10-fold by conferring a mutation either on the host, *B. brevis* 47, or on the plasmid, pHY481 [14]. In this way, we could establish a novel host-vector system suitable for the production of proteins coded by foreign genes.

5. Expression of Foreign Genes and Secretion of Useful Proteins

A recombinant DNA, pBAM101, was prepared by ligating the pUB110 DNA and a thermophilic α-amylase gene which was cloned from *B. stearothermophilus* into *Escherichia coli* [15]. This was introduced into *B. brevis* as well as *B. subtilis*. These bacteria carrying pBAM101 synthesized the same α-amylase as the donor strain as determined by the enzyme's thermal stability and amino acid sequence of amino-terminal region [16]. Regardless of the host, the 34-amino acid signal peptide of the enzyme was processed at exactly the same site between two alanine residues.

◀ **Fig. 3A–F.** Alterations of fine cell wall structure of *B. brevis* 47 during growth. All cells were withdrawn at various growth stages. At the exponential phase, cells were surrounded by a three-layered cell wall **(A)**. A highly magnified picture of the surface **(inset)** shows clearly three layers above the cytoplasmic membrane (CM). At the early stationary phase, the MW layer changed morphologically to contain spikes **(B)**, and the outer two layers began to be shed **(C)**. At the late stationary phase, most of the cells shed the outer two layers and were surrounded only by a thin layer **(D)**. The *inset* shows only one layer above the CM. After 24 h of growth, shedding of the two layers was complete in most of the cells **(E)**, and cell wall fragments adhering to the surface were seen in some cells **(F)**. *Bars* represent 200 nm

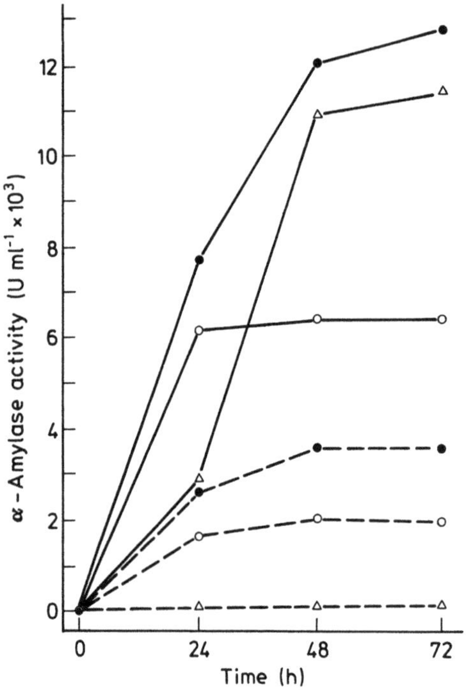

Fig. 4. Time courses of α-amylase production in *B. brevis* 47 (pBAM101) and *B. subtilis* 1A289 (pBAM101). Cells were grown at 37 °C in T_2 medium containing 60 μg of neomycin per ml, with 1% glucose (●), soluble starch (○), or glycerol (△) as carbon sources. Solid and broken lines indicate cultures of *B. brevis* 47 (pBAM101) (———) and *B. subtilis* 1A289 (pBAM101) (– – –), respectively

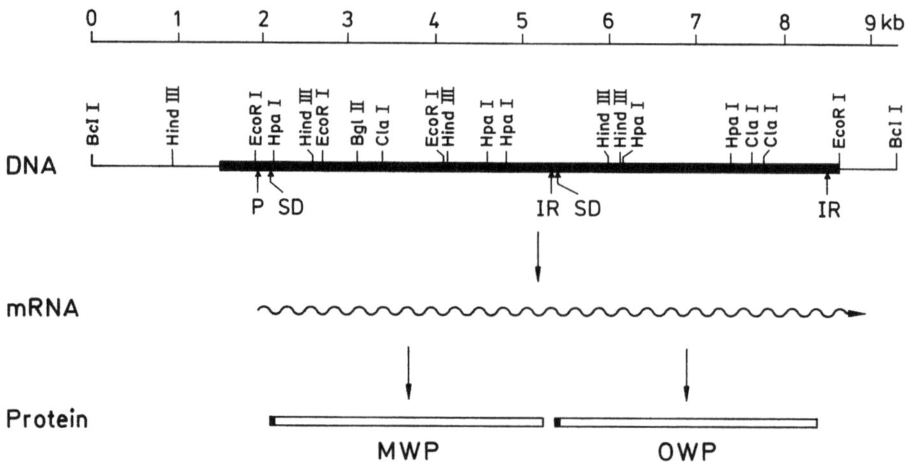

Fig. 5. Schematic representation of the expression of the two cell wall protein genes in *B. brevis* 47. In the genomic restriction map, the bold line represents the cell wall protein genes. The mRNA of the cell wall protein genes is depicted as a *wavy line* with a length of 7 kb. The *open bars* represent the entire coding regions for the two cell wall protein. P, Promoter region; IR, Inverted repeat structure; SD, Shine-Dalgarno sequence

B. brevis 47 with pBAM101 secreted the enzyme most efficiently of the hosts examined, 100, 15, and 5 times more than *B. stearothermophilus*, *E. coli*, and *B. subtilis*, respectively. A comparison of α-amylse productivity of *B. brevis* 47 (pBAM101) with *B. subtilis* 1A289 (pBAM101) in the media containing different carbon source is shown in Fig. 4 [16]. These results supported the presumption that *B. brevis* is an excellent host for protein produc on.

To express a gene efficiently and facilitate its product to secrete, it is necessary to have a powerful promoter, appropriate ribosome binding site (so called SD sequence), and signal sequence at the 5′ region of the gene. We tried to utilize the 5′ region of the cell wall protein gene in *B. brevis* as a controlling region of gene expression, since vast amounts of the cell wall proteins, i.e., extracellular proteins, are synthesized and secreted into the medium as mentioned above.

After an intensive work, we succeeded to clone the genes for the cell wall proteins with their flanking region into *E. coli* or *B. subtilis* [17,18]. Nucleotide sequence analysis of DNA fragment which directed the synthesis of the cell wall proteins revealed that it contains two open reading frames in tandem. The upstream gene corresponds to the middle wall protein (MWP) and the downstream gene codes for the outer wall protein (OWP), constituting a cotranscriptional unit, an operon (Fig. 5). The nucleotide sequence of the 5′ region of this cell wall protein operon suggested the existence of several promoter like sequences and two translation initiation sites on the upstream of the MWP gene. MWP and OWP are synthesized in precursor forms with signal peptides of 23 and 24 amino acid residues, respectively.

When the promoter region was inserted to the upstream of a foreign α-amylase gene, the extracellular production of the α-amylase in *B. brevis* became very efficient. The 5′ region of the MWP gene was fused to the *B. licheniformis* α-amylase gene, of which nucleotide sequence analysis was described previously [19], as follows. pHY483 was constructed by inserting the α-amylase gene into the HpaII site of pHY481. A 600 basepair AluI-AluI fragment containing the 5′ promoter region of the MWP gene was inserted into the PstI site within the signal sequence region of the α-amylase gene on pHY483. The resulting plasmid, pHY4831, containing the promoter region of the cell wall protein gene fused with the α-amylase gene (Fig. 6), was introduced into *B. brevis* 47 [17]. *B. brevis* carrying pHY4831 produced 50 times more of the fusion amylase (15,500 units ml^{-1}) than did *B. brevis* carrying pHY483 (300 units ml^{-1}), which contains the authentic *B. licheniformis* α-amylase gene. The extracellular amylase activity reached 44,000 units ml^{-1} (this corresponds to 0.7 g l^{-1} of the amylase protein), when pHY4831 was introduced into *B. brevis* 47 Cop11, a mutant in which the copy number of the plasmid increases about 10-fold [14].

The 5′ promoter region of the MWP gene is about 300 nucleotides long with a complex sequence structure. S1 mapping analysis of the transcripts from the cell wall protein gene operon suggested that the transcription starts from the 5 different sites within the 300 nucleotides [17]. The -35 and -10 regions of the promoters corresponding to each of the transcription start sites were assigned. The first putative promoter resembled the consensus recognition sequence of the σ^{43} RNA polymerase of *B. subtilis* [20] with only one base difference in the -35 region. The other promoters are not homologous to the consensus sequence. Interestingly, the first promoter (the most distal to the translation start site) was almost inactive and the other promoters were strong, when promoter activities were tested by inserting the fragment of each

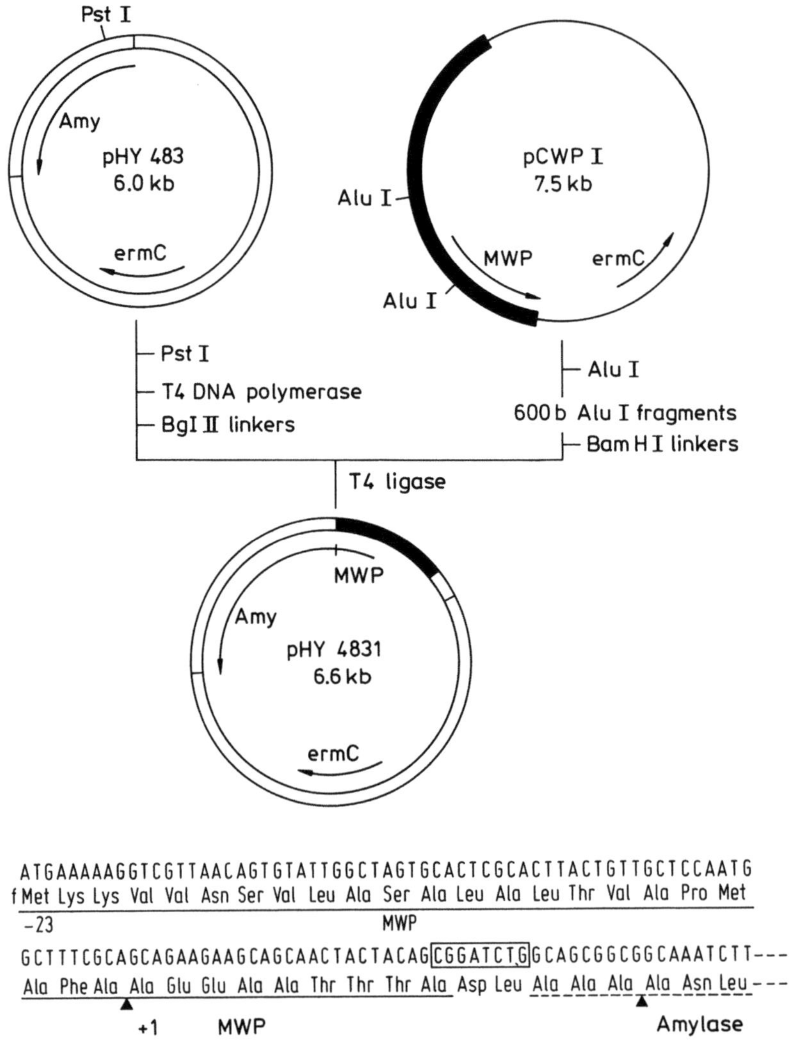

Fig. 6. Construction of pHY4831 carrying the MWP-α-amylase fusion gene. *Closed bars* denote *B. brevis* 47 DNA. The *arrows* indicate the direction of transcription. Amy, *B. licheniformis* α-amylase gene; ermC, erythromycin resistance gene. The DNA sequence of the junction region of the fusion gene is shown at the bottom. The linker DNA sequence is boxed. The amino acid sequence from the second translation initiation site deduced from the DNA sequence is numbered from −23. The amino acid sequence of the MWP and α-amylase genes are underscored with *solid* and *broken lines*, respectively. The signal peptide cleavage sites of MWP and α-amylase are indicated by *arrowheads*

promoter region into the 5′ region of a promoterless α-amylase gene and measuring the amylase activity [21].

Based on these findings, we could construct an excellent secretion vector containing the 5′ region of the cell wall protein gene operon. The vector and the host, *B. brevis*, have been very useful for efficient heterologous protein production [21].

Besides the high productivity, a prominent feature of this system is that protease is not significantly produced by the protein-hyperproducing *B. brevis*, thereby proteins once produced are stable even after prolonged culture.

6. References

1. Glenn, A. R.: Ann. Rev. Microbiol. *30*, 41 (1976)
2. Udaka, S.: Agric. Biol. Chem. *40*, 523 (1976)
3. Takagi, H., Kadowaki, K., Udaka, S.: ibid. in press
4. Shaku, M., Koike, S., Udaka, S.: Agric. Biol. Chem. *44*, 99 (1980)
5. Miyashiro, S., Enei, H., Hirose, Y., Udaka, S.: ibid. *44*, 105 (1980)
6. Tsuchida, T., Miyashiro, S., Enei, H., Udaka, S.: ibid. *44*, 2291 (1980)
7. Miyashiro, S., Enei, H., Takanami, K., Hirose, Y., Tsuchida, T., Udaka, S.: ibid. *44*, 2297 (1980)
8. Yamada, H., Tsukagoshi, N., Udaka, S.: J. Bacteriol. *148*, 322 (1981)
9. Tsuboi, A., Tsukagoshi, N., Udaka, S.: ibid. *151*, 1485 (1982)
10. Tsukagoshi, N., Yamada, H., Tsuboi, A., Udaka, S., Katsura, I.: Biochim Biophys Acta. *693*, 134 (1982)
11. Ohmizu, H., Sasaki, T., Tsukagoshi, N., Udaka, S., Kaneda, N., Yagi, K.: J. Biochem. *94*, 1077 (1983)
12. Takahashi, W., Yamagata, H., Yamaguchi, Y., Tsukagoshi, N., Udaka, S.: J. Bacteriol. *156*, 1130 (1983)
13. Yamagata, H., Takahashi, W., Yamaguchi, K., Tsukagoshi, N., Udaka, S.: Agric. Biol. Chem. *48*, 1069 (1984)
14. Yamagata, H., Nakagawa, K., Tsukagoshi, N., Udaka, S.: Appl. Env. Microbiol. *49*, 1076 (1985)
15. Tsukagoshi, N., Ihara, H., Yamagata, H., Udaka, S.: Mol. Gen. Genet. *193*, 58 (1984)
16. Tsukagoshi, N., Iritani, S., Sasaki, T., Takemura, T., Ihara, H., Idota, Y., Yamagata, H., Udaka, S.: J. Bacteriol. *164*, 1182 (1985)
17. Yamagata, H., Adachi, T., Tsuboi, A., Takao, M., Sasaki, T., Tsukagoshi, N., Udaka, S.: ibid. *169*, 1239 (1987)
18. Tsuboi, A., Uchihi, R., Tabata, R., Takahashi, Y., Hashiba, H., Sasaki, T., Yamagata, H., Tsukagoshi, N., Udaka, S.: ibid. *168*, 365 (1986)
19. Yuuki, T., Nomura, T., Tezuka, H., Tsuboi, A., Yamagata, H., Tsukagoshi, N., Udaka, S.: J. Biochem. *98*, 1147 (1985)
20. Doi, R. H., Wang, L.: Microbiol. Reviews. *50*, 227 (1986)
21. Udaka, S., Tsukagoshi, N., Yamagata, H., Tsuboi, A.: unpublished result

3.5 Extracellular Secretion of Cloned *Serratia marcescens* Protease through the Outer Membrane from *Escherichia coli* Cells

Teruhiko Beppu

Department of Agricultural Chemistry, Faculty of Agriculture, The University of Tokyo, Yayoi 1-1-1, Bunkyo-ku, Tokyo 113, Japan

Outer membrane structure of the Gram-negative bacteria plays as a barrier against penetration of exogenous molecules as well as against proteins secreted from the cells through the inner cell membrane. Well developed cloning and expression systems of *Escherichia coli* for heterologous proteins have enabled to achieve their effective production, however, in most cases the products form insoluble inclusion bodies in the cells which may cause some difficulties for subsequent extraction and refolding to obtain the correct tertiary structure of the products. Several systems by using the signal sequences of the periplasmic enzymes of *E. coli* have been developed to make secretion of the fused heterologous proteins into the periplasm, however, the outer membrane barrier prevents its release into the media. Secretion of proteins through the outer membrane of the Gram-negative bacteria will be an interesting problem for effective production of heterologous proteins by recombinant DNA technology.

Serratia marcescens is a Gram-negative bacteria belonging to the *Enterobacteriaceae*, the same family of *E. coli*. The organism has been selected as a producer of an extracellular metalloprotease which is utilized as an antiinflammatory drug. During the course of cloning the extracellular protease gene of *S. marcescens* in *E. coli*, we obtained a clone producing another extracellular protease of serine type which was different from the known metalloprotease of *S. marcescens*. A hybrid plasmid containing a 5.8-kb DNA fragment of *S. marcescens* was recovered from the clone. *E. coli* carrying pSP11 produced the serine protease (Mr, 41,000) into the medium during the exponential growth but not in the periplasm nor the cell. No concomitant release of the intracellular or periplasmic enzymes of the host was observed.

Subcloning and sequencing of the cloned fragment revealed a single open reading frame of 3,135 base pairs coding a protein of 1,045 amino acids (Mr, 112,000). Comparison of the 5′ nucleotide sequence with the N-terminal amino acid sequence of the protease indicated the presence of a typical signal sequence of 27 amino acids. The C-terminal amino acid of the enzyme was found at position 408, as deduced from the nucleotide sequence. Artificial frameshift mutations introduced into the coding sequence for the assumed distal polypeptide after the C terminus of the protease caused complete loss of the enzyme production. It was concluded that the *Serratia* serine protease is produced as a 112-kilodalton proenzyme and that its N-terminal signal peptide and a large C-terminal part are processed to cause excretion of the mature protease through the outer membrane of *E. coli* cells. The *Serratia* protease and its gene will provide a new concept for mechanism of protein secretion and possibly a new secretion vector in *E. coli.*

Bioproducts and Bioprocesses
Editors: Fiechter, Okada, Tanner

3.6 Cumulative Effect of Intragenic Amino Acid Replacements on Thermostability of a Protein

Shuichi Aiba

Department of Fermentation Technology, Osaka University, Faculty of Engineering, Suita-shi, Osaka 565, Japan

Abstract

The marginal net stability of a folded protein is considered to rest upon a small difference among large compensating, individual forces. Therefore, the net free energy of stabilization of proteins is unexpectedly small, ~ 40 kJ mol^{-1}. Meanwhile, the contribution of individual forces such as hydrogen bonds and salt bridges to the stabilization are evaluated as 4 to 12 kJ mol^{-1}, and several additional forces are thought to be sufficient to account for the extra thermostability of thermophilic proteins. The native conformation of a protein is determined by the totality of interatomic interactions and hence, by the amino acid sequence. If a few amino acid residues which individually contribute to the stabilization could be implemented concurrently into the sequence, the multiple replacement would enhance an overall stability of the protein molecule. Herein a report on concrete evidence in favor of this argument will be presented. Thermal inactivation kinetics and proteolytic resistance for mutants of a kanamycin nucleotidyltransferase revealed that a few intragenic amino acid replacements stabilized the protein cumulatively. This work demonstrates not only the feasibility of elevating a protein's thermostability but also a way to better understanding of forces and interactions that are responsible for the protein stability.

Bioproducts and Bioprocesses
Editors: Fiechter, Okada, Tanner

4. Down Stream and Integrated Processes

4.1 Separation and Purification in Bioprocess

Kenichi Suga and Shuichi Aiba

Department of Fermentation Technology, Faculty of Engineering, Osaka University, Japan

Abstract

A variety of methods for recovery, separation, and purification in biotechnology have been reported in recent years. These methods can be accomplished by different principles: chromatographic separations, membrane separation, purification by supercritical fluid, aqueous two-phase extraction systems and electrically driven separation process.

In this presentation, a part of recent advances in separation and purification of bioproduct studied in Japan, will be introduced.

The fermentation system with continuous separation will be presented in this seminar.

Immuno-affinity chromatography is a powerful separation method. The effectiveness of utilization of antibodies for adsorption and the adsorption capacity of the immobilized antibodies affect the economical feasibility of immuno-affinity chromatography. Kato et al. studied the effects of immobilization method and antibody concentration on the molar ratio of adsorbed antigen to immobilized binding sites on support beads.

The continuous separation of a glucose/fructose mixture was experimentally performed using a simulated moving-bed adsorber packed with the Ca^{2+} ion form of Y zeolite by Hashimoto et al. They presented two mathematical models for calculating the concentration profiles of glucose and fructose in the simulated moving-bed adsorber.

An isotachophoretic separation of protein was simulated, using amino acid as spacer. Firstly, the analysis of transient state in the separation presented a clue of how to estimate the column length and the time required, if an aqueous solution comprising a mixture of proteins were to be separated completely and continuously into the single component. Secondly, a simple graphic method to estimate values of pH, concentration and migration order with regard to constituents of a given sample undergoing a steady state isotachophoresis was presented. The proposal coming from the analysis would be useful for a quick and appropriate selection of spacers.

It was evident from the stimulation work that a constituent of the lower mobility became remarkably concentrated in the separation. Profiles of proteins and pH during the transient period were experimentally determined and the results were compared with the simulation. The agreement between calculation and experimentation was satisfactory. An increase of leading ion concentration was found to reduce the column length required for the separation as well as the concentrating effect of ionic constituent.

Bioproducts and Bioprocesses
Editors: Fiechter, Okada, Tanner

4.2 Bioseparations of Milk Proteins

M. R. Ladisch, S. R. Rudge[1], K. W. Ruettimann[1], and J. K. Lin[2]

[1] Laboratory of Renewable Resources Engineering, School of Chemical Engineering, and Department of Agricultural Engineering, Purdue University, West Lafayette, IN 47907, USA
[2] U.S. Food and Drug Administration, Food Engineering Branch, 1090 Tusculum Avenue, Cincinnati, OH 45226, USA

Contents

Milk is a complex biological fluid consisting of lipids, phospholipids, carbohydrates, proteins, sugars, salts and vitamins. Casein proteins exist in the milk serum (non-fat milk) as micellar structures stabilized by colloidal calcium phosphate and protein surfactant. Raw milk also contains microorganisms and somatic cells, which contribute small amounts of proteolytic enzymes and nucleic acids to milk in the course of processing. Since milk is a suitable carrier for a diverse range of biological molecules, its fractionation is of considerable interest. Knowledge of these fractionation methods may further facilitate recovery of proteins of therapeutic value or of enhanced nutritional qualities from milk. The purpose of this chapter is to show separation strategies aimed at recovering selected molecules found in milk serum, and to indicate possible approaches for larger scale fractionations.

Production of milk and milk products is a large industry in the United States, with production of 142.5 billion pounds in 1987 [1]. Cultured products (cheese, yoghurt, 9.6%) frozen products (8.5%) and dried products (4.2%) account for a large portion of dairy product consumption [2] (by weight, not milk equivalents). The processes leading to these products often produce waste streams of high BOD which need to be treated further and may contain useful chemicals to be recovered. Improvements in fractionation techniques for these streams may help improve the economics of the overall process itself.

1. Introduction

In this article, we emphasize the handling and processing of milk in the context of traditional chemical engineering. The properties of proteins and protein solutions

Bioproducts and Bioprocesses
Editors: Fiechter, Okada, Tanner

will be addressed, in terms of their effect on separation interactions. The phase and charge properties are important to consider when developing an appropriate fractionation scheme, as are the kinetics of various reactions or interactions upon which the separation depends. In this context, we will address fractionation of caseins, whey proteins, calcium chloride, and lactose.

2. Casein Protein Separations

The majority of the proteins in milk are caseins. The major casein proteins are α_{s_1}-casein (12–15 g l^{-1}), α_{s_2}-casein (3–4 g l^{-1}), β-casein (9–11 g l^{-1}) and $\varkappa$-casein (2 to 4 g l^{-1}) [3]. Two other caseins, γ and λ, are common by-products of casein processing. Another peptide, protease peptone, or glycomacropeptide, is a cleavage product of $\varkappa$-casein, produced during cheese making. Most of the caseins in milk exist as casein micelles, which are spherical aggregates of the casein proteins, stabilized by colloidal calcium phosphate.

These casein micelles are the important kinetic and structural units in the cheese making process. Cheese making begins with the addition of a bacterial culture to milk which provides part of the characteristic flavor. This step is followed by the addition of a milk-clotting enzyme (rennet) to milk which attacks the Phe_{105}-Met_{106} bond of $\varkappa$-casein, destabilizing the micelles and resulting in their coagulation. The coagulated micelles form a gel, which is cut, resulting in curd and whey. The curd is further processed to the final cheese product. A review of milk proteins is given by Swaisgood [4], while a review of casein micelles can be found in Ruettimann and Ladisch [3].

Separations techniques have been used to help deduce the structure and composition of the micelles as well as to study their behavior under different environmental conditions. Casein micelles found in milk have a broad size range of 20–600 nm and electron microscopy studies have shown that the micelles themselves are composed of smaller units known as submicelles [3]. The micelles disintegrate into the submicelle units when the colloidal calcium phosphate (CCP) is removed by dialysis, indicating the importance of CCP in stabilizing the micelle structure. Additionally, the micelles are not uniform in their casein composition with the smaller micelles having a larger proportion of $\varkappa$-casein, indicating that $\varkappa$-casein resides mainly on the micelle surface [3]. The hydrophobicities of α_{s_1}-, β-, and $\varkappa$-caseins have been calculated to be 1.24, 1.32, and 1.20 kcal $mole^{-1}$, respectively [5]. The combination of such high hydrophobicities with relatively low molecular weights (23,600, 24,000, and 19,000) allows for strong hydrophobic interactions among the proteins. These hydrophobic interactions favor the association of the caseins, which can complicate separation procedures.

2.1 Precipitation Techniques

The casein proteins can be separated from milk by isoelectric precipitation at pH 4.6 and 20 °C [4]. Acid precipitation of casein is an important process in itself, which leads to manufacture of sodium caseinate and cottage cheese and is usually the first step in casein and whey fractionation. A second method is to fractionate the casein micelles from milk using ultracentrifugation at 45,000 g for 90 min at 37 °C [4]). The caseins can also be separated into two protein fractions based on their stability in the presence of calcium. Milk is adjusted to 0.25 M Ca^{2+} at pH 7 and 37 °C. The insoluble fraction (calcium-sensitive) contains α_{s_1}-, α_{s_2}- and β-caseins. The soluble fraction (calcium-insensitive) contains $\varkappa$-, β-, and λ-caseins [4]).

2.2 Chromatography

The casein proteins undergo strong protein-protein interactions, as evidenced by the micelle and submicelle formation and by the self-association of caseins. These protein associations must be disrupted in order to separate the individual caseins. Urea is normally used to dissociate the casein complexes. Differential solubility, of casein proteins in urea can be used to fractionate whole casein into β-caseins, γ-caseins, and a mixture of α_s-, $\varkappa$-, and λ-caseins [4]). The mixture of α_s-, $\varkappa$-, and λ-caseins can then be separated by differential solubility in acid solutions of urea using trichloroacetic acid or sulfuric acid. The caseins have also been separated using differential solubility in ethanol [4]).

The casein proteins have been fractionated using ion exchange chromatography on DEAE-cellulose using urea as the dissociating agent with the proteins being eluted using a sodium chloride gradient [4]). Chromatography on hydroxyapatite uses a phosphate buffer to elute the proteins, which desorb in order of their increasing phosphate content ($\varkappa$-, β-, α_s-) for an increasing gradient [6]). Fast protein liquid chromatography (FPLC) using ion exchange columns has been an effective method to separate and quantify the casein proteins. Barrefors et al. [7]), used FPLC with a Mono-Q (Pharmacia) column. The caseins were dissolved in a 0.01 M imidazole buffer at pH 7.0 containing 3.3 M urea and 0.01 M 2-mercapto-ethanol and eluted from the column with a 0.0–0.5 M NaCl gradient. Detection of protein was via absorbance at 280 nm using a uv monitor. These conditions resulted in very good resolution of the caseins, with the proteins eluting in order of increasing phosphate content: $\varkappa$-, β-, α_{s_2}- and α_{s_1}-caseins.

Dalgleish [8]) used FPLC over Mono-Q and Mono-S to study the effects of $\varkappa$-caseins with differing degrees of glycosylation on rennet kinetics as well as to determine the distribution of glyco-$\varkappa$-casein from different sized micelles. Dalgleish found that the glycosylation of $\varkappa$-casein did not affect its distribution in differently sized micelles and also that enzymatic hydrolysis of $\varkappa$-casein was independent of micelle size. Chromatography has also been used to follow the enzymatic hydrolysis of $\varkappa$-casein during renneting. Van Hooydonk and Olieman [9]) used HPLC over TSK 2000 SW columns to separate and quantify the amount of glycomacro-peptide (GMP) released during various stages of renneting. GMP is the peptide (MW $\approx$ 8000) cleaved from $\varkappa$-casein by enzymatic attack. Possible interference from whey proteins was avoided by removing them by adjusting the solutions to 8% trichloroacetic acid. About 40 min were required to complete the separations.

Recently, reverse phase chromatography has been shown to effectively separate the casein proteins. Barrefors et al. [7]), used a C8 HR 5/10 column and eluted the proteins with a 30–40% acetonitrile gradient. The proteins eluted in order of their hydrophobicities, except for κ-casein (due to the presence of glycosylated residues): β-casein > κ-casein > α_{s_1}-casein > α_{s_2}-casein. Carles [10] also used reverse phase HPLC to separate the caseins with an octadecylsilyl stationary phase and a phosphate-buffered aqueous propan-2-ol mobile phase containing sodium dodecyl sulfate. This system not only resolved the four major caseins, but also separated two versions of the genetic variant β-casein A^1. The previously unidentified version had an amino acid substitution of proline for leucine. This separation shows the ability of RP-HPLC to resolve similar proteins that differ slightly in hydrophobicity.

The previous methods have been developed for separations of small samples. Larger quantities (gram amounts) of the caseins can be prepared using a method described by Wei and Whitney [11]. The researchers adsorbed the proteins on DEAE cellulose and separated γ-, κ-, β-, α_{s_2}-, and α_{s_1}-caseins by successive extractions with increasing concentrations of sodium chloride. The process was performed batchwise in 3.3 M urea at pH 7.4.

Casein micelles themselves have been fractionated using chromatography on controlled pore glass [12,13,14,15]. The fractionation of the micelles has aided in determining micelle size distributions and in determining the composition and structure of micelles of different sizes. Gel chromatography has been used to separate the submicelles which make up the casein micelles. The study by Ono and Takagi [16] found that the submicelles could be fractionated by size into four distinct classes, showing that the submicelles are not homogeneous. The two major fractions of submicelles had molecular weights of 500,000 and 200,000 or 100,000 (dependent upon concentration).

2.3 Summary

It is clear that several separation techniques have been used to separate the casein proteins. These techniques have played an important role in determining the structure and composition of casein micelles, the properties of the casein proteins and how they interact with themselves and their environment. Continued development of these techniques may lead to the separation of the individual genetic variants of each of the caseins. The separation of these very similar proteins can be carried out now using gel electrophoresis. Further refinement of column-based separation technology may allow the use of faster and more convenient methods for separating casein components.

Chromatographic and precipitation techniques are useful for separating the very hydrophobic caseins. Most chromatographic techniques rely on the differing amounts of phosphorylation each molecule has. This is because such strong dissociating conditions are required to treat casein molecules individually (3.3 M urea, mercaptoethanol, sodium dodecyl sulfate) that normal hydrophobic adsorption is not useful.

The precipitation technique, in which Ca^{++} ion is used to separate the caseins, gives insight into the role of calcium in stabilizing casein in milk. The results indicate that the more phosphorylated caseins are more sensitive to Ca^{++}. This corresponds with the biological importance of colloidal calcium phosphate in micelle stabilization [3].

3. Whey Protein Separations

Whey is the fluid which remains unincorporated in cheese after casein micelles have been coagulated. The major constituents of whey are water, lactose, salts, and globular proteins. Of the proteins, β-lactoglobulin makes up the largest proportion ($2\text{–}4\ g\ l^{-1}$ skim milk), and α-lactalbumin ($1\text{–}1.5\ g\ l^{-1}$) and bovine serum albumin ($0.1\text{–}0.4\ g\ l^{-1}$) make up most of the rest [4]. Small amounts of immunoglobulins ($0.6\text{–}1\ g\ l^{-1}$, mostly IgG) and ferritin ($<0.1\ g\ l^{-1}$) also exist in whey [4]. In colostrum (milk taken in the first day after calving) immunoglobulin is approximately 75% of the total non-casein protein, but this proportion quickly decreases after the first day [17].

There are several methods for recovering or concentrating proteins in whey. The choice of a technique usually depends on the intended end use of the protein and the amount of sugar in the whey. The proteins may be separated from each other, free of salts and sugars, by chromatography or ultrafiltration. If a whole protein fraction is to be recovered in an active state and in a concentrated solution free of sugar and salt, adsorption (or ion exchange) or filtration may be used. Alternatively, the protein, sugar, and salt may be desired in a dry bulk form (for livestock feed) in which case the water is evaporated, although filtration can also be used to complement evaporation in dewatering processes because of energy efficiency [18]. The strategies are shown in Fig. 1. A review of industrial preparations and applications of whey protein concentrates has been given [19].

Knowledge of the whey composition gives clues to the composition of the rest of the milk. For instance, β-lactoglobulin (β-lg) has two genetic variants in North American herds, called A and B, which differ by two out of 162 amino acids [20]. However, cows homozygous in one variant or the other have been shown to produce milk with different cheese processing properties, as shown in Table 1 [21–27].

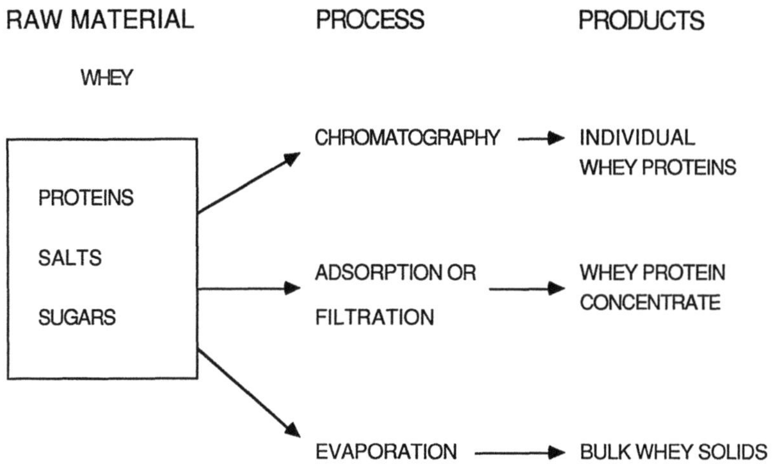

Fig. 1. Separation strategies for whey constituents. The operation used (*center*) depends on the product desired (*right*)

Table 1. The effects of genetic polymorphism of β-lactoglobulin on processing properties of milk. Animals homozygous in variant A are represented by AA, homozygous in B are represented BB.

Property	Increased by
β-lg concentration	AA
Heat processing stability	AA
Fat concentration	BB
Casein concentration	BB
Casein clotting time	BB
Curd firmness	AA

β-Lactoglobulin also binds and transports retinol (vitamin A), a photo-oxidizable vitamin important in vision [28], and has interesting heat and acid stable properties, which make it attractive in certain food additive applications as an emulsifying or foaming agent [29,30].

Many alternatives are available for the separation of whey proteins using chromatography or electrophoresis. Nearly every mode of chromatographic separation has been used successfully to separate the four (β-lg A and B, α-lactalbumin (α-la) and bovine serum albumin (BSA)) major proteins of whey, including ion exchange, reverse phase, size exclusion, and chromatofocusing. Native gel electrophoresis is also effective. In comparison, SDS-PAGE and isoelectric focusing have not been as widely reported for this type of separation.

3.1 Electrophoresis

Aschaffenberg and Drewry [31] made the initial discovery of β-lactoglobulin genetic variants on starch gel electrophoresis. Since then, Bell [32] has developed a starch gel technique whereby 4 variants of β-lg, two variants of α-la, BSA, Ig and ferritin are all simultaneously resolved. Current analytical techniques are based on the

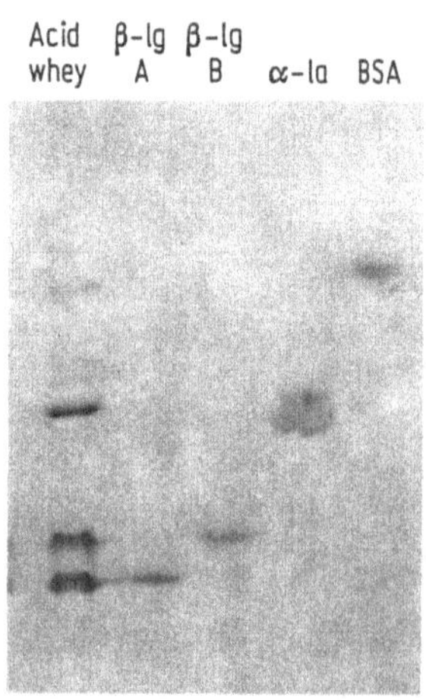

Fig. 2. Native polyacrylamide gel (13%) electrophoresis of whey proteins. Acid whey prepared from skim milk by adjusting the pH to 4.6 (for left) is run next to β-lactoglobulin A, β-lactoglobulin B, α-lactalbumin and bovine serum albumin standards (from left to right). The standards correspond to the major bands in acid whey

polyacrylamide technique of Darling and Butcher[33], which gives resolution of A and B variants of β-lg, α-la, BSA and Ig. These polyacrylamide techniques offer better resolution of smaller samples in shorter times than the DHA starch techniques of Bell. A native polyacrylamide gel is shown in Fig. 2. β-Lactoglobulin A migrates the furthest in the polyacrylamide gels, followed closely by β-lg B, whose mobility is about 5% less. α-Lactalbumin has a mobility about half that of the β-lg's. Bovine serum albumin has a very low mobility in the gels, but it is not clear whether this is due to its larger size or its smaller charge, or both. Immunoglobulins are effectively excluded from polyacrylamide gels of this concentration.

3.2 Selective Precipitation

Proteins can often be separated by varying the temperature, pH, or ionic strength of the solution. Selective enzymatic hydrolysis is also an effective means of eliminating or precipitating some proteins from a solution. As mentioned earlier, whey is formed by one such precipitation reaction, where casein micelles are precipitated by either a change to mildly acidic conditions, or with coagulation by a rennetting enzyme. McKenzie[20] gives a protocol for separating the whey proteins by salting out with ammonium sulfate. A method which uses trichloroacetic acid is also popular, especially for β-lg separation. Trichloroacetic acid is added to acid whey (pH 4.6) to a 3% solution[35]. The precipitate contains α-la and BSA, while β-lg's remain soluble. Polyacrylamide gel electrophoresis of the precipitate and supernate is shown in Fig. 3. Heat treatment has also been used to separate β-lg from α-la, resulting in a very pure α-la precipitate, but a supernate with α-la contamination[36].

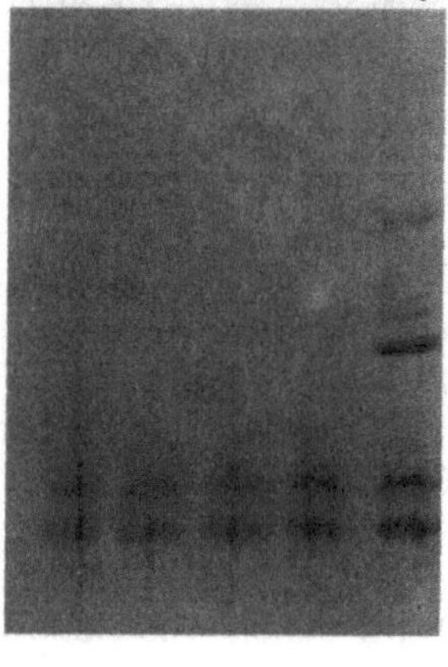

Fig. 3. Acid whey which has been further precipitated with trichloroacetic acid is compared to acid whey by polyacrylamide gel electrophoresis. Trichloroacetic acid treatments of 3.0%, 2.5%, 2.0%, and 1.5% (from left to right) all appear to be free of albumin fraction compared to acid whey and whey protein standards (β-lg A, β-lg B, α-la, BSA, left to right). β-Lactoglobulins appear stable across the acid range

3.3 Chromatography

As with the casein fraction of milk serum, the whey fraction may be fractionated by chromatographic methods. Chromatographic separations rely on a host of protein properties, such as size, charge, and hydrophobicity, to affect separations. Some examples follow.

Size Exclusion Chromatography

Size exclusion chromatography may be used to separate whey proteins. Poor resolution is usually achieved, except when analytical type inorganic resins are used. Bican and Blanc [37] were able to resolve Ig (100 kDa), BSA (67 kDa), β-lg (36 kDa), and α-la (14 kDa) (in that order, as expected) using TSK 3000SW (Toya Soda). Gupta got better resolution with the same method but a longer column [38]. A typical size exclusion result from a gel resin (Sephadex G-75) [39] is shown in Fig. 4. Immunoglobulin and BSA elute together in the first peak, while the β-lg's and α-la elute together in the second. Riboflavin is in the third peak.

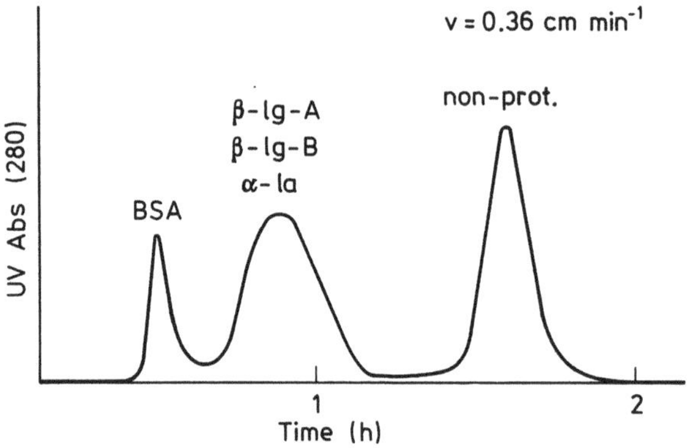

Fig. 4. Size exclusion chromatographic separation of acid whey proteins on Sephadex G-75. A 30 cm column resolves BSA, but does not separate the β-lactoglobulins from α-lactalbumin

Ion Exchange Chromatography

Resolution of four major whey proteins (BSA, α-la, β-lg A and B) has been demonstrated by Manji et al. [40], on a Mono Q ion exchange column. A stepwise gradient was run from distilled water to 0.7 M sodium acetate, with each step eluting a protein. α-Lactalbumin eluted at the lowest step (0.14 M sodium acetate), BSA eluted at 0.24 M sodium acetate, β-lg B at 0.39 M and β-lg A at 0.7 M. The separation is achieved over 55 min, but the time could be shortened with a faster responding column. Earlier, Humphrey and Newsome [41] had resolved these four proteins in piperazine buffer (pH 6.0) with a 0–0.5 M NaCl buffer. The elution order was the same as that obtained by Manji et al.

The chromatographic results indicate a stronger binding, and thus a larger concentration of positive charge on the β-lg's than on BSA and α-la around neutral pH. Skudder [42] also confirms this with frontal chromatography of these four proteins on Spherosil QMA. In this experiment, β-lg A served as the ultimate displacer causing concentration "roll up" in the other three protein elution profiles. Skudder's experiment showed that at pH 6.5, β-lg A is the most strongly adsorbed protein of the four. This is the opposite of the order these proteins run in alkaline (pH 8.4) electrophoresis gels, with the exception of the transposition of BSA and α-la. This reflects that total charge must be highest for the β-lactoglobulins, and that the mobility of BSA might be reduced more than the mobilities of other proteins by sieving effects in the gel matrix.

Chromatofocusing (where a pH rather than a salt gradient is used to elute protein) has been reported by Pearce and Shanley [43]. All four peaks are resolved, over a gradient of 5.2 to 4.2, with BSA eluting first, followed by β-lg B, α-la, and β-lg A. This would seem to indicate that the isoelectric point of α-la is between that of the two β-lg's. However, the values as reported by Swaisgood [4] indicate that α-la has a lower isoelectric point (4.8) that β-lg A (5.26) or β-lg B (5.34), and should have eluted later than either β-lg. The results from chromatofocusing indicates some mechanism of binding other than ion exchange for all three proteins.

Reverse Phase Chromatography

An example of reverse phase chromatography of whey proteins has been reported in the literature. In this, Pearce [44] used a 7.5 cm column packed with Spherisorb S5 C6 to separate whey proteins using a gradient from 0.15 M NaCl/HCl (pH 2.1) to 36% acetonitrile in 0.15 M NaCl/HCl. Again, β-lg A and β-lg B were the most strongly retained proteins, followed by α-la and BSA. This result is not in conflict with the result for ion exchange. The relative hydrophobicity of a protein is not related to the relative charge, which is pH dependent. It also suggests that some hydrophobic interaction between the lactoglobulins and ion exchange resins (nonspecific adsorption) may have contributed to the retention times of these proteins. Protein hydrophobicity appears to be most strongly affected by aromatic, rather than aliphatic amino acid groups (Hayakawa and Nakai) [45].

3.4 Summary

It has been shown that the various constituents of whey protein may be fractionated. Crude separations are attainable with salt, pH, and heat precipitation techniques, while more refined separations are possible when electrophoretic or chromatographic techniques are employed. In developing these techniques, important insights into the properties of the molecules can be made. The β-lactoglobulins appear to be more hydrophobic in general, and have higher charge around neutral pH (6–9) than the other major whey proteins. Bovine serum albumin appears to be slightly more charged than α-la, but its size becomes a factor when separating BSA from the much smaller lactalbumins and lactoglobulins. β-Lactoglobulin is also the most acid stable protein, which may be part of its physiological function as a retinol (vitamin A) carrier. β-Lactoglobulin likely keeps retinol soluble as it passes through the highly acidic environment of the stomach, so it may be adsorbed in the intestine.

4. Lactose/Calcium Chloride Separation

Lactose is a disaccharide of glucose and galactose present in milk serum. Its presence in whey contributes to high BOD values that make whey costly to dispose of. An essentially protein free, lactose rich liquor is produced in the manufacture of whey protein concentrate by ultrafiltration (see Fig. 1). Therefore an economic method for handling the lactose is a final consideration in milk protein purification and utilization. Lactose is responsible for indigestion in some individuals who develop lactose intolerance due to a deficiency of the lactase enzyme in their intestines [46]. This limits lactose's value as a sweetener. However, lactose is valuable in pure form, as USP lactose for the pharmaceutical industry. Lactose may also be hydrolysed by treatment with β-galactosidase to yield glucose and galactose, which are valuable as sweeteners [47]. Glucose may be further processed to fructose [48] for additional sweetness. The enzymes required for these sugar conversions are sensitive to the other components of the lactose mixture, especially ions, so methods of purifying lactose are important as the last step in processing these components of milk.

Lactose is usually purified by a crystallization step. This step is complex; it involves pH adjustment and centrifugation of very fine particles (0.5 to 1.5 mm) [49]. Chromatographic steps may also be taken to deash whey.

We have studied ion exclusion chromatography as a method for purifying lactose from inorganic salts such as calcium chloride, which constitute the major part of ash in lactose liquors. Ion exclusion chromatography utilizes resins in the same ionic form as the ion to be excluded, and works on the basis of a chemical and electrical field gradient which excludes ionic species, while its effect on non-ionic solute partitioning is small.

A resin which has been derivatized with negatively charged groups (such as a sulfonyl group) will exchange ions with an electrolyte solution. Negatively charged ions are largely excluded from the resin, according to a phenomenon known as Donnan equilibrium. The higher the charge on the negative species, the more strongly it will be excluded from the resin. Uncharged, or non-ionic species, may still partition between the resin and electrolyte phases.

4.1 Comparison of Ion Exclusion to Ion Exchange

The lactose/$CaCl_2$ system gives a good example for ion exclusion separation. A cation exchange resin, initially in the Ca^{++} form (Ca^{++} ions are in equilibrium with the resin's bound negative charge) is used to adsorb lactose, while excluding Cl^-. If a sample of lactose/$CaCl_2$ is injected into a column of this resin in a dilute buffer (or distilled water), Ca^{++} ions may exchange freely with Ca^{++} previously on the resin, lactose can distribute freely between the phases, and Cl^- ions will be excluded by the immobilized negative charges in the resin, and elute at an earlier time. Since Ca^{++} is exchanging with Ca^{++} in this case, Cl^- will elute as $CaCl_2$. This is known as ion exclusion. If the resin were initially in a H^+ or Na^+ form, Ca^{++} would displace these ions, and Cl^- would elute as HCl or NaCl. If this is the desired result, the column would have to be periodically regenerated with acid or sodium salt in high concentration to displace the absorbed calcium as is the usual case for ion exchange.

The choice of a buffer or eluent for ion exclusion is critical for an efficient separation. Excluding an ion depends on an electric repulsive force which can be screened or compressed with increasing ionic strength. Therefore, the efficiency of exclusion will decrease with increasing buffer ionic strength [50]. By contrast, partitioning of non-electrolytes like lactose are unlikely to be affected greatly by buffer ionic strength.

The screening effect of electrolyte on ion exclusion can be seen in the shape of the isotherm for $CaCl_2$ with cation exchange resin in the Ca^{++} form. This shape is generally referred to as concave upward. Solutes with this shape isotherm give a characteristic elution profile with a sharp tail (right hand side of peak) and a diffuse front (left hand side of peak). Solutes whose partitioning is unaffected by concentration have Gaussian shaped chromatographic elution profiles. These two shapes can be seen for $CaCl_2$ and lactose, respectively, in Fig. 5 for a 2.54 cm i.d. × 119 cm long column at 57 °C.

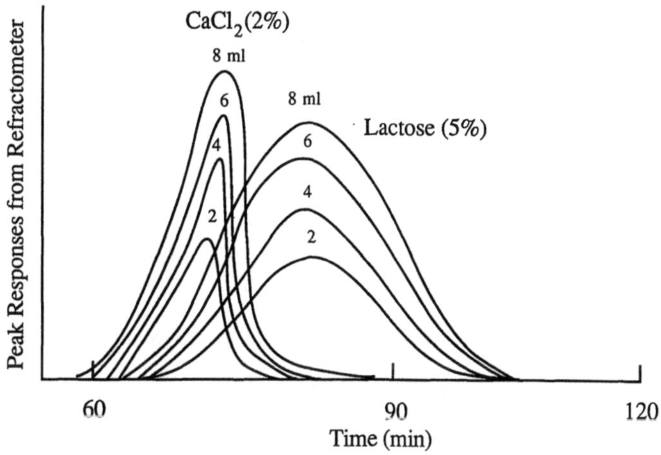

Fig. 5. Ion exclusion chromatography of lactose and $CaCl_2$ of different sample sizes (2, 4, 6, and 8 ml). The $CaCl_2$ peak shows a sharp tailing edge, indicating a concave upwards (unfavorable) isotherm. The lactose peak is essentially Gaussian, indicating linear equilibrium with the resin, IR-118.

5. Conclusions

The major milk serum molecules may be purified to varying degrees by a number of techniques. These techniques are developed based on a working knowledge of the varied properties of the different milk constituents. Casein micelles may be destabilized either by pH adjustment or enzymatic hydrolysis, to separate the caseins from the serum. The individual caseins may then be dispersed using urea and separated based on phosphorylation or glycosylation. Casein fractionations will help elucidate the role of casein in milk, as well as the cheese making properties of individual constituents.

The remaining serum may be fractionated by electrophoretic or chromatographic techniques associated with soluble and globular proteins. The behavior observed helps characterize the whey proteins, and suggests applications in which they might be used as food additives, either individually or as a whey protein concentrate. Sugars from the whey fractionation could be further processed to yield USP grade lactose by crystallisation or adsorption processes.

Milk is a source of a diverse set of biological molecules. The strategy of fratio-nation procedures for milk serum reflects general strategies available for fractionation of biological solutions. In this context the milk protein separation techniques briefly reviewed here may have parallels in other areas of the biotechnology industry.

6. Acknowledgements

The material in this work was supported by NSF Grant CBT8351916, and Kraft, Inc. (Glenview, IL). We thank Lisa Smith for assistance with the electrophoresis and Rick Hendrickson for assistance with ion exclusion separations. We also thank Jill E. Porter and Belinda F. Roettger for helpful suggestions in preparing this manuscript.

7. References

1. Agric. Outlook; Research Services, U.S. Dept. Agric., April (1988)
2. Jacobson, R. E.: J. Dairy Sci. 69, 1447 (1986)
3. Ruettimann, K. W., Ladisch, M. R.: Enzyme Microb. Technol. *9*, 578 (1987)
4. Swaisgood, H. E.: In: "Developments in Dairy Chemistry": p. 1 (Fox, P. F., ed.). Appl. Sci. Publ., New York 1982
5. Schmidt, D. G., Payens, T. A. J.: In: "Surface and Colloid Science" *9*, p. 165 (Matijevic, E. ed.), Wiley & Sons, New York 1976
6. Aoki, T., Kako, Y., Imamura, T.: J. Dairy Res. *53*, 53 (1986)
7. Barrefors, P., Ekstrand, B., Fägerstam, L., Larsson-Raznikiewicz, M., Schaar, J., Steffner, P.: Milchwissen. *40*, 257 (1985)
8. Dalgleish, D. G.: J. Dairy Res. *53*, 43 (1986)
9. Van Hooydonk, A. C. M., Olieman, C.: Neth. Milk Dairy J. *36*, 153 (1982)
10. Carles, C.: J. Dairy Res. *53*, 35 (1986)
11. Wei, T.-M., Whitney, R. McL.: J. Dairy Sci. *68*, 1630 (1985)
12. McGann, T. C. A., Kearney, R. D., Donnelly, W. J.: J. Dairy Res. *46*, 307 (1979)
13. Larsson-Raznikiewicz, M., Almlöf, E., Ekstrand, B.: J. Dairy Res. *46*, 313 (1979)
14. Anderson, M., Griffin, M. C. A., Moose, C.: J. Dairy Res. *51*, 615 (1984)
15. Robson, E., Horne, D. S., Dalgleish, D. G.: J. Dairy Res. *52*, 391 (1985)
16. Ono, T., Takagi, T.: J. Dairy Res. *53*, 547 (1986)
17. Butler, J. E.: Veterin. Immunology Immunopath. *4*, 43 (1983)
18. Hiddink, J., deBoer, R. and Nooz, P. F. C.: Milchwissen. *36* (11), 657 (1981)
19. Matthews, M. E.: J. Dairy Sci. *67*, 2680 (1984)
20. McKenzie, H. A.: In: "Milk Proteins: Chemistry and Molecular Biology:" p. 257 (McKenzie, H. A., ed.), Academic Press, New York (1971)
21. Marziali, A. S., and Ng-Kwai-Hang, K. F.: J. Dairy Sci. *69*, 1793 (1986)
22. Marziali, A. S. and Ng-Kwai-Hang, K. F.: J. Dairy Sci. *69*, 2533 (1986)
23. Ng-Kwai-Hang, K. F., Hayes, J. F., Moxley, J. E., and Monardes, H. G.: J. Dairy Sci. *69*, 22 (1986)

24. Kroeker, E. M., Ng-Kwai-Hang, K. F., Hayes, J. F., and Moxley, J. E.: J. Dairy Sci. *68*, 1637 (1985)
25. Lin, C. Y., McAllister, A. J., Ng-Kwai-Hang, K. F., and Hayes, J. F.: J. Dairy Sci. *69*, 704 (1986)
26. Kroeker, E. M., Ng-Kwai-Hang, K. F., Hayes, J. F., and Moxley, J. E.: J. Dairy Sci. *68*, 1752 (1985)
27. Kroeker, E. M., Ng-Kwai-Hang, K. F., Hayes, J. F., and Moxley, J. E.: J. Dairy Sci. *68*, 1346 (1985)
28. Fugate, R. D. and Song, P.-S.: Biochim. Biophys. Acta. *625*, 28 (1980)
29. Harwalker, V. R. and Modler, H. W.: Milchwissen. *36* (10), 593 (1981)
30. Bech, A. M.: Dairy Ind. Int. Nov., 25 (1981)
31. Aschaffenberg, R. and Drewry, J.: Nature *176*, 218 (1955)
32. Bell, K.: Biochem. Biophys. Acta *147*, 100 (1967)
33. Darling, D. F. and Butcher, D. W.: J. Dairy Sci. *59* (5), 863 (1975)
34. Hillier, R. M.: J. Dairy Res. *43*, 259 (1976)
35. Fox, K. K., Holsinger, V. H., Posoti, L. P., and Pollansch, M. J.: J. Dairy Sci. *50* (9), 1363 (1967)
36. Pearce, R. J.: Austral. J. Dairy Technol. Dec., 144 (1983)
37. Bican, P. and Blanc, B.: Milchwissen. *37* (10), 592 (1982)
38. Gupta, B. B.: J. Chromatogr. *282*, 463 (1983)
39. Rudge, S. R.: Masters Thesis, Purdue University (1986)
40. Manji, B., Hill, A., Kakuda, Y., and Irvine, D. M.: J. Dairy Sci. *68*, 3176 (1985)
41. Humphrey, R. S. and Newsome, L. J.: N. Zeal. J. Dairy Sci. Technol. *19*, 197 (1984)
42. Skudder, P. J.: J. Dairy Res. *52*, 167 (1985)
43. Pearce, R. J. and Shanley, R. M.: Austral. J. Dairy Technol. Sept., 1120 (1981)
44. Pearce, R. J.: Austral J. Dairy Technol. Sept., 114 (1983)
45. Hayakawa, S., and Nakai, J.: J. Food Sci., *50*, 486 (1985)
46. Stryer, L.: In: "Biochemistry", p. 378, W. H. Freeman and Co., New York, 2nd edition 1975
47. Chiu, C. P. and Kosikowski, F. V.: J. Dairy Sci. *68*, 16 (1985)
48. Chiu, C. P. and Kosikowski, F. V.: J. Dairy Sci. *69*, 959 (1986)
49. Modler, H. W. and Lefkovitch, L. P.: J. Dairy Sci. *69*, 684 (1986)
50. Helfferich, F.: In: "Ion Exchange", p. 95, McGraw Hill Book Co., Inc., New York 1962

4.3 Design and Use of Enzyme Systems for Selective Product Release from Microbial Cells

J. A. Asenjo and B. A. Andrews

Biochemical Engineering Laboratory, Columbia University, New York, N.Y. 10027, USA
Biochemical Engineering Laboratory, University of Reading, P.O. Box 226, Reading, RG6 2AP, England (present address; where correspondence should be addressed)

Contents

This paper describes the characteristics, production and use of enzyme systems with lytic activity against yeast and bacterial cells. Factors affecting enzymatic lysis including the sequential protease and glucanase degradation of the two-layered yeast wall and the obtention of sufficiently high lytic activities are analyzed. The advantages of using enzyme systems as well as their design and actual processes for the controlled lysis and selective product release (concept of a "Biochemical Cell Refinery") are discussed.

1. Introduction

For the recovery of a number of valuable intracellular and subcellular materials from microbial cells, controlled methods of cell breakage and lysis to produce the selective release of products would be highly attractive. This is particularly relevant for a number of processes being developed by modern biotechnology and will also result in the generation of novel applications and the more widespread use of many valuable cell associated materials or cellular substructures. In the case of yeast cells such products would include recombinant proteins [1], specialized lipids [2], wall glucan polymers [3,4], and mitochondria [5].

Traditional methods of cell breakage, particularly at the pilot and large scale level, have relied mainly on mechanical principles which are highly non specific. The use of enzymes to produce the partial lysis of the cell wall and maybe other subcellular structures appears very attractive due to the specificity that can be obtained in enzyme reactions.

Two crucial questions that we have to answer are: How do lytic enzymes that degrade microbial cells work? and, How are they synthetized?. The second question

Bioproducts and Bioprocesses
Editors: Fiechter, Okada, Tanner

Table 1. Some potential uses of lytic enzyme systems

- Release of cloned intracellular proteins
- Manufacture of specialized wall glucan polymers
- Synergistic action with mechanical cell breakage
- Selective release of products from different locations in the cell

Table 2. Yeast-lytic enzyme systems

Yeast-lytic systems	
Essential enzymes	*Supporting enzymes*
Lytic β(1–3)glucanase	β(1–3)Glucanases
Lytic protease	Mannanase
	β(1–6)Glucanases
	Chitinase
	Protease

is particularly relevant if our intention is the large scale production and use of such enzymes. Some interesting uses of lytic enzyme systems are shown in Table 1.

Lytic enzyme systems are usually composed of more than one enzyme and have the ability to break or partially degrade whole microbial cells. Yeast-lytic systems for instance contain β(1-3)glucanases, proteases, mannanase, chitinase and β(1-6)-glucanase. Enzymes can hydrolyze the walls of microbial cells and when sufficient will has been removed the internal osmotic pressure can burst the periplasmic membrane thus allowing the intracellular content to be released into the surrounding medium.

It appears that the enzymatic degradation of the wall of whole yeast cells is better understood than that of bacterial cells. The wall of yeast is composed of two main layers: an external protein-mannan layer and an internal glucan layer. In order to produce the breakage of whole yeast cells, it is essential to have a specific wall-lytic protease to degrade the outside layer and a lytic β(1-3)glucanase to degrade the inside layer [6] (Table 2). These two enzymes are essential for the lysis of whole yeast cells. With bacteria, a single enzyme (e.g. lysozyme: N-acetylmuramidase) can be responsible for the lysis of Gram-positive bacteria, but a pretreatment

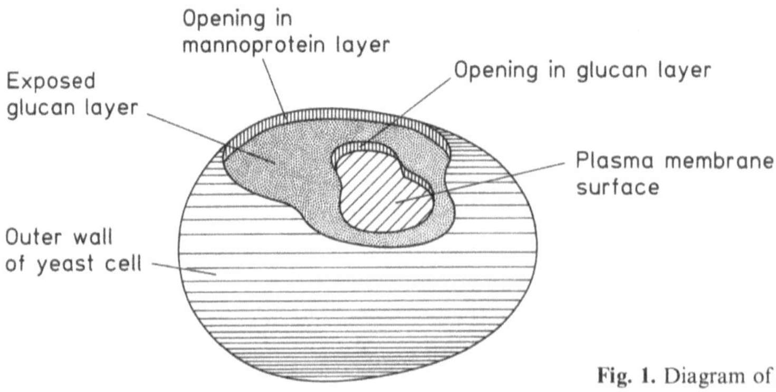

Fig. 1. Diagram of lysing yeast cell [5]

with a detergent (e.g. Triton X-100) is usually necessary for the lysis of Gram negative ones. The effect of the detergent is to rupture the outside membrane, which is a characteristic of Gram negative bacteria. Some proteases have also been shown to have bacteriolytic activity.

In the lysis of yeast cells, the lytic protease acts on the outside mannan-protein layer making a hole which allows the β(1-3)glucanase to degrade the inside structural glucan layer. Usually only one glucanase in a system is lytic and the lytic protease is very specific having little or no activity on the soluble or insoluble protein product. The other supporting enzymes of the system act synergistically and can strongly affect the distribution or "type" of product obtained (e.g. specialised glucan polymers, protein-free glucan. particulate or soluble intracellular proteins).

2. Materials and Methods

Proteolytic activity of enzyme preparations on yeast cell free extract, which largely corresponds to intracellular protein, was measured. Yeast cell free extract was prepared using a bead mill and separating the broken solids by centrifugation [7]. The supernatant was used as the cell free extract. Proteolytic activity of the undiluted *Oerskovia* and *Cytophaga* enzyme preparations was measured as described previously [8] but using the cell free extract as substrate instead of casein. Protein left in solution after TCA precipitation was measured by the Biorad assay.

3. Lytic Enzymes Towards Microbial Cells

A number of different microorganisms have been found to produce yeast-lytic enzymes; the most important ones are *Arthrobacter luteus/Oerskovia xanthineolytica*, *Cytophaga* sp., *Rhizoctonia* sp. and *Bacillus circulans*. The first three appear to be the most important from a commercial viewpoint with very high concentrations of activity of the extracellular enzyme system. The enzymes from *Bacillus* sp. have not been studied on whole yeast cells.

The most important source of bacteriolytic enzymes is lysozyme which is produced commercially from hen egg white. Other bacteriolytic enzymes include those from *Cytophaga* sp., *Staphylococcus* sp. and *Streptomyces* sp. which can all lyze $Gram^+$ bacteria. A lytic protease from *Micronospora* sp.can lyze lyophilized cells of $Gram^-$ bacteria and a lytic protease produced by *Bacillus subtilis* that can lyze cells of the bacterium *Escherichia coli* ($Gram^-$) apparently without the need of pretreatment has been reported [9].

Tables 3 and 4 show pH and temperature optima for yeast-lytic and bacteriolytic enzymes. pH and temperature optima varies considerably between different lytic systems. Usually at temperatures 5–10 °C above the optima the stability of the enzyme is low and hence the rate of denaturation is high.

In Table 3 the lytic systems of *Oerskovia xanthineolytica/Arthrobacter luteus* and of *Arthrobacter* GJM-1 are shown to include two enzymes [6, 10, 11]. The systems of *Oerskovia* CK and *Rhizoctonia* sp. have basically one enzyme with both glucanase

Table 3. Yeast-lytic enzymes

Source	Optimum pH	Optimum temperature (°C)
Oerskovia xanthineolytica/ Arthrobacter luteus (Zymolyase/Lyticase)		
Glucanase	6.5	45
Protease (alkaline)	10.0	35 (heat labile above 40 °C)
Whole yeast cell activity	7.5	35
Arthrobacter GJM-1		
Glucanase	5.0	30–40 (stable up to 55 °C)
Glucanase on walls		50
Protease (alkaline)	9.0	
Whole yeast cell activity	7.5	30–35
Oerskovia CK		
Glucanase with		35–40
Proteolytic activity		60
Whole yeast cell activity	9.0	35–40
Rhizoctonia sp.		
Glucanase	5.5	55–60
Protease	6.5	40 (at 50 °C 1/2 life = 30 min)
Whole yeast cell activity	6.0	40
Cytophaga NCIB 9497		
Whole yeast cell activity (also see Table 4)		55

and protease activity [12,13]. The system of *Cytophaga* NCIB 9497 has both activities but has not yet been sufficiently characterized.

Table 4 shows the characteristics of some bacteriolytic enzymes. Lysozyme is lytic towards a large number of both Gram$^+$ and Gram$^-$ bacteria. *Cytophaga* NCIB 9497 enzyme activity has been measured against *Bacillus subtilis* and *Corynebacterium lilium* [14] and recently against *E. coli*. *Cytophaga* B-30 enzyme was used against *Staphylococcus* and had little activity against other strains [15]. The enzymes of the three strains of *Staphylococcus* [16] were assayed against heat killed cells of *Micrococcus luteus*. The N-acetylmuramidase (Mutanolysin) is active towards *M. lysodecticus*, *Streptococcus* sp. and others [17]. The *Micronospora* lytic protease was tested against lyophilized cells of *Serratia marcescens* (100% activity), *Pseudomonas aeruginosa* (96%), *E. coli* (91%) and *B. subtilis* (187%) [18]. The *B. subtilis* 797 lytic protease was active against *E. coli* [9].

The cell walls of yeast and bacteria are distinctly different. Hence in general lytic systems are specific for different microorganisms. For instance the lytic system of *Oerskovia/Arthrobacter* is specific for yeast, and lysozyme is specific for bacteria. One exception to this is the lytic system of *Cytophaga* NCIB 9497 which has a broad activity spectrum. It has high activity on both yeast and bacterial cells.

Table 4. Bacteriolytic enzymes

Source	Optimum pH	Optimum temperature (°C)
Lysozyme (hen egg-white)	6–7	35
Cytophaga NCIB 9497	9.0	50 (30 min) 45 (60 min)
Cytophaga B-30 (Lysopeptase)	9.5	50
Staphylococcus sp.		
S. aureus	7.0	37
S. simulans	6.0	37
S. saprophiticus	6.5	37
Streptomyces globisporius (N-acetylmuramidase) (Mutanolysin)	6.5	50
Micromonospora sp. Nr. 152 (lytic protease)	11.0	60
Bacillus subtilis 797 (lytic protease)	7.8–8.5	30 (assay temp.)

Microbial production of high concentrations of lytic enzymes has been achieved using *Cytophaga* sp. [8] *Rhizoctonia* sp [13] and *Oerskovia/Arthrobacter* [19, 20]. Several fold increases in enzyme concentration and productivity has been achieved using continuous culture techniques [8, 20, 21] and by mutation [19]. Lysozyme is a commercial lytic enzyme that has been used for a number of years in industry [22] for several different purposes including enzyme extraction.

3.1 Sequential Treatment

Enzymatic lysis will introduce a considerable amount of specificity to the process of cell breakage and product release. This can be achieved by selecting or designing the lytic enzyme or enzyme system to be used. To release a bacterial protein located in the wall or membrane one would incubate the bacteria or yeast and lytic enzyme in an isotonic solution. This allows wall degradation and product release without disrupting the cytoplasm and thus facilitating later purification of the wall or membrane linked enzyme.

Yeast has a double layer wall. By choosing an enzyme system with a higher glucanase activity or higher lytic protease, or a combination of different glucanases (producing glucan oligomers of different sizes), the rate of release of final product and by-products as well as its qualities and composition can be engineered. Examples would constitute the release of invertase from cell walls, the production of protein-free glucan or glucan oligomers of prespecified characteristics. The strategy

Table 5. Abilities of lytic enzymes to lyse yeast cells

Enzyme	Amount (μg)[a]	% Decrease in OD	% Decrease in cell number
g	225	0	0
	450	0.2	0
	900	0.6	0
p	4.2	5.0	0
	8.3	7.0	0
	17	9.0	0
	34	11.0	0
	68	13.0	0
	135	13.0	0
g + p	g 5.9 p 8.3	78.8	100

[a] μg 10 ml^{-1} of reaction mixture.
(Agric. Biol. Chem., *46*, 967, 1982)

for intracellular protein release aiming at maximum product recovery is discussed later.

Table 5 shows the use of purified lytic glucanase (g) and purified lytic protease (p) by themselves, and mixed (g + p), in the lysis of whole yeast cells. It can clearly be seen that both enzymes are needed to produce cell breakage. Table 6 shows the effect of sequential treatment with glucanase (g) and protease (p), washing the cells in isotonic buffer in between. Cell breakage was only obtained when the cells were first incubated with protease and then with glucanase and not in the reverse order [10]. Mannanase could not substitute for the protease [11]. Another important impact could be in the selective lysis of mixed populations. This cannot be achieved by mechanical breakage.

Table 6. Stepwise treatment of yeast cells with lytic enzymes

Enzyme		Decease in OD 800 nm (%)
Pretreatment	Treatment	
g	None	0
g	p	9.4
p	None	—[a]
p	g	64.8
None	g	0
None	p	7.2
None	g + p	62.5

[a] Increase in OD.
(Agric. Biol. Chem. *46*, 967, 1982)

3.2 Factors

There are a number of factors that have to be considered when analyzing the possibility of using lytic enzymes for cell disruption and product release. Amongst these are several potential limitations. Some of the crucial questions that have to be answered are shown in Table 7. Apart from these questions, of which the first two are specific to enzymatic cell breakage, there are some other potential limitations which are either common to mechanical breakage (debris removal) or have to be addressed specifically, like the obtention of high activities or safety of the microbial strains.

Table 7. Important factors to be considered when analizing the possibility of using lytic enzymes for cell disruption and product release

1. Does the lytic enzyme or lytic enzyme system contain any protease, and if so, does it have any activity on the products?
2. Can the lytic enzymes be easily separated from the product (difference in molecular weight or pI) or does the addition of enzymes add a protein that will be tedious to separate from the product?
3. Are the activities of the lytic enzymes sufficiently high to obtain fast cell breakage and product release without allowing endogenous intracellular proteases to degrade the product?

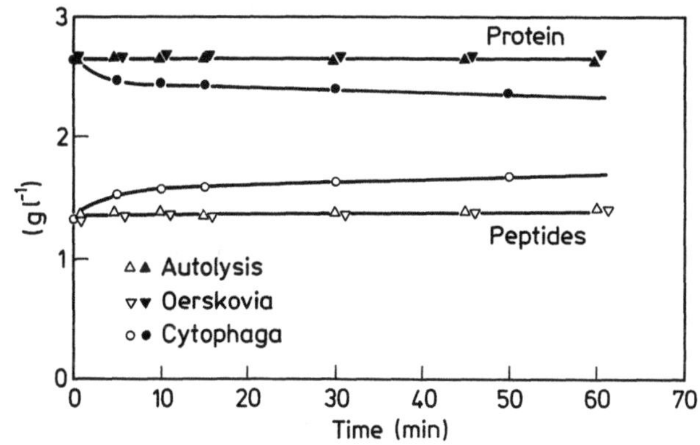

Fig. 2. Hydrolysis of intracellular protein from yeast in the presence of lytic enzymes from *Oerskovia xanthineolytica* and *Cytophaga* NCIB9497. Temp. = 37 °C, pH 7.0

Figure 2 shows an incubation of intracellular yeast protein with the two lytic enzyme systems. No degradation of the protein into TCA soluble peptides is observed for the *Oerskovia* enzyme and little hydrolysis occurs with the *Cytophaga* enzyme. The enzyme concentration used in this experiment was several fold the enzyme concentration used for most of the lysis experiments. This is a broad protein solubilization analysis, hence the denaturation and actitivity of individual proteins should be analized in each particular application. Protease can be very specific and hence the stability of the protein or proteins targeted for recovery will be the only real test in specific systems. Tables 3 and 4 give information of pH and temperature optima for lytic enzymes which are very important when choosing conditions for an enzyme lysis and product release process.

As has already been stated, recent work on the development of continuous culture processes [8,14,20,21,23] and mutation [19] has resulted in several fold increases in enzyme concentration and productivity of both yeast-lytic (*Oerskovia/Arthrobacter*, *Cytophaga*) and bacteriolytic enzyme systems (*Cytophaga*). Work on mutation has been limited to *Arthrobacter*, hence further work in this field could improve enzyme concentration and productivity very substantially. Cloning of lytic enzyme genes in other bacteria could also be explored.

3.3 Design and Use

Cell breakage is becoming a critical technique in large scale biotechnology since many of the new products are relatively large or even particulate cloned proteins that the cell cannot secrete. At present the only industrial scale cell breakage equipment available is the high pressure homogenizer, although there are reports that lytic enzymes have been used industrially with commercial success; bead mills have been used up to pilot plant scale. In places where mechanical breakage facilities exist, the addition of lytic enzymes to the mechanical disruption process can have a very substantial synergistic effect. This could result not only in decreasing energy costs of the disruption process but also in higher rate, yield and selectivity, thus minimizing intracellular product destruction. Figure 3 shows the basic problem, where a multienzyme system is produced in a bioreactor and subsequently used to lyze microbial cells. This can be achieved in the presence of lytic enzymes only, or in combination with mechanical means. A separation stage (S) is also shown. In the separation stage two products are shown as well as the possibility of enzyme (E) recycle. Enzymes can be used either in batch or in continuous reactors. Batch and tubular reactors would give a higher recovery of product released [24].

The cells to be lysed can be usually harvested from the bioreactor by centrifugation or maybe by ultrafiltration of microporous filtration. They are added to the lysis reactor together with the lytic enzymes and buffer.

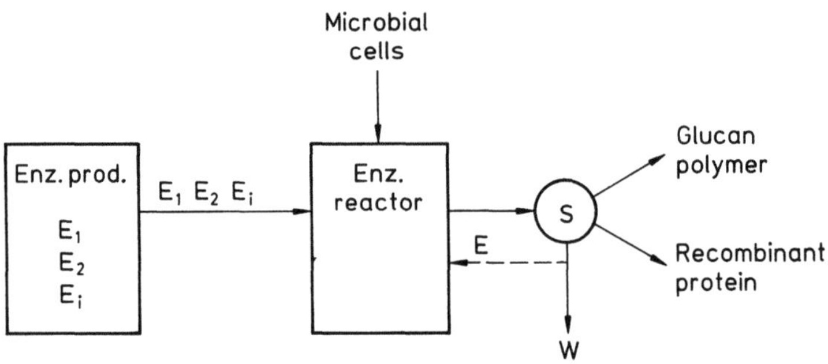

Fig. 3. Bioreactor for production of enzyme system (E_1, E_2, E_i) linked to a cell-lysis enzyme reactor for the release of cell products. S = separation stage; E = enzyme recycle; W = waste stream

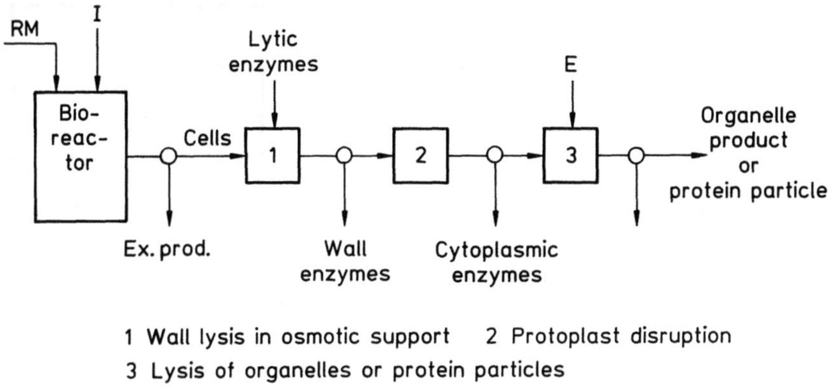

Fig. 4. Process for the lysis of whole microbial cells including sequential disruption for selective product release. R.M. = raw materials; I = inoculum; Ex. Prod. = extracellular product; E = enzyme or reagent used for lysis of organelle or renaturation of protein

Typical enzyme concentrations are 3–30% of crude enzyme supernatant or as low as 0.4% if the enzyme is produced in continuous culture [14]. These values are for the lysis of *B. subtilis* cells at a concentration of 9–13 g l^{-1} using the enzyme produced by *Cytophaga*. With yeast cells, concentrations of up to 110 g l^{-1} dry weight have been used in actual disruption reactors. With lysozyme it was estimated that using 4,000–5,000 U l^{-1} good protein extraction could be obtained from the cells in one hour. Commercial preparations contain 14,000–22,000 U g^{-1}.

The design and use of Lytic Enzyme Systems will depend on a number of important variables which include the microbial cells, the products to be harvested, the enzyme sytems available or desired, and the process conditions for its use. Typical examples of microbial cells to be used would include purposedly designed yeast cells for manufacture of specialized wall polymers or human serum albumin. The design of the enzyme system will be pursued by strain selection, use of inducers, growth rate control as well as enzyme fractionation and purification. Process design for use of the enzyme will be determined by the actual reaction, separations and product purification conditions. Two such processes are shown here.

Figure 4 shows a process where the extracellular product has been removed from the cells (beer and yeast). The lytic enzymes are added to a reactor where the microbial cells are in osmotic support. This would allow the separation of wall products before bursting the cells. Protoplast disruption in 2 can be achieved by means of gentle agitation. After 2, soluble cytoplasmic enzymes are separated from recombinant protein particles or mitochondria. The renaturation or lysis of the particles or organelles is achieved in 3. Figure 5 shows our structured model of cell lysis that we have used for process simulation [5].

Figure 6 shows a different process where the sequential use of lytic protease (for the release of wall protein) and lytic glucanase, is used to avoid the presence of protease when intracellular protein is released. In this particular process no osmotic support is used. It has been found however, that wall lytic proteases are very

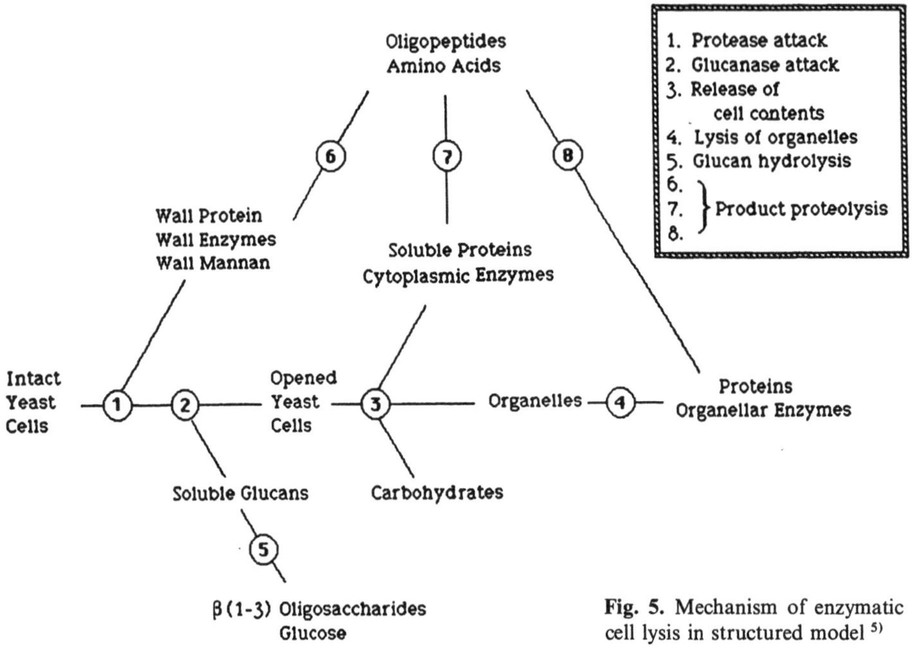

Fig. 5. Mechanism of enzymatic cell lysis in structured model [5]

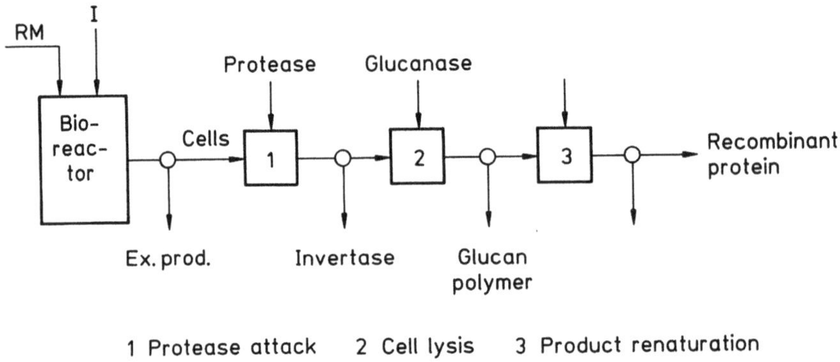

Fig. 6. Process for the lysis of whole microbial cells with sequential lysis of the cell wall in 2 stages; it includes selective product release. For abreviations see caption of Fig. 4.

specific showing little activity on intracellular protein. In many cases it has not been possible to substitute the specific wall-lytic protease by conventional proteases [6].

pH and temperature optima for the glucanase and protease can be very different (Table 3). Hence a pH or temperature shift can also be used to have a high protease activity in the initial stage of the lysis reaction and a high glucanase and low protease activity later. The use of protease inhibitors after the initial stages of reaction has also been considered. Mannan is a very strong inhibitor of the protease reaction.

The design of a particular lytic system will be largely determined by its specific use; not only the particular microbe to be lysed but also the specific product to be released from the cell. This is a good example of process synthesis where the recovery stage (use of the lytic system) is closely integrated to the process used for synthesis and production of the lytic enzyme system as well as with the fermentation used for intracellular product formation and later purification stages.

4. Conclusions

This paper has both reviewed and described enzyme systems for the lysis of yeast and bacterial cells. Production and design of a particular lytic system will be largely determined by the type of microbial cells to be lysed, product to be released and its location and the possibility of isolating other byproducts in the same process. Mechanical techniques for cell disruption are effective but highly non specific. The concept of a Biochemical Cell Refinery where enzymatic, physico-chemical and/or genetic techniques are used to selectively release products from the cell is vital (Figs. 4 and 6). The advantage of using enzymes in conjunction with other techniques is their lytic specificity.

5. Acknowledgements

This work was supported by a grant from the National Science Foundation, NSF to whom thanks are due. Support from Enzyme Biosystems is also gratefully acknowledged.

6. References

1. Asenjo, J. A., Andrews, B. A. and Pitts, J. M.: Design and Use of Microbial Lytic Enzyme Systems for Selective Product Release, Proc. IV European Congress of Biotechnology", Amsterdam, Holland, June 14–19, 1987
2. Hammond, E. G., Glatz, B. A., Choi, Y. and Teasdale, M. T.: in: New Sources of Fats and Oils (E. H. Pryde, L. H. Princess, K. D. Mukherjee, eds.) chapter 14, p. 171. Am. Oil Chem. Soc. 1981
3. Jamas, S., Rha, C. K. and Sinskey, A. J.: Biotech. Bioeng. *28*, 769, (1986)
4. Kobayashi, R., Miwa, T., Yamamoto, S. and Nagasaki, S.: Eur. J. Appl. Microbiol. Biotechnol. *15*, 14–19, (1982)
5. Hunter, J. B. and Asenjo, J. A.: in: Separation, Recovery and Purification in Biotechnology: Recent Advances and Mathematical Modeling, p. 9, (J. A. Asenjo and J. Hong, eds.) American Chemical Society, Washington 1986
6. Scott, J. H. and Schekman, R.: J. Bacteriol. *142*, 414, (1980)
   2. A Simple Model of Lysis Kinetics, Biotech. Bioeng. *30*, 481 (1987)
7. Hunter, J. B. and Asenjo, J. A.: Kinetics of Enzymatic Lysis and Disruption of Yeast Cells:
8. Andrews, B. A. and Asenjo, J. A.: Biotech. Bioeng. *28*, 1366, (1986)
9. Borovikova, V. P. et al.: Biokhimiya *45*, 1524, (1980)
10. Kitamura, K.: Agric. Biol. Chem. *46*, 963, (1982)
11. Vrsanska, M., Kratky, Z. and Biely, P.: Z. Allg. Microbiol. *17*, 391, (1977)

12. Obata, T., Iwata, H. and Namba, Y.: Agric. Biol. Chem. *41*, 2387, (1977)
13. Kobayashi, R., Miwa, T., Yamamoto, S. and Nagasaki, S.: J. Ferment. Technol. *58*, 319, (1980)
14. Le Corre, S., Andrews, B. A. and Asenjo, J. A.: Enz. Microb. Technol. *7*, 73, (1985)
15. Miles, Technical Information,: N-Acetylmuramidase (1983), Zymolyase (1984), Lysopeptase (1984)
16. Valisena, S., Varaldo, P. E. and Satta, G.: J. Bacteriol. *51*, 636, (1982)
17. Sigma Chemical Co., 1986 Catalog.
18. Suzuki, K., Uyeda, M. and Shibata, M.: Agric. Biol. Chem., *49*, 1719, (1985)
19. Kitamura, K.: Ph. D. Thesis, Kyoto University 1984
20. Andrews, B. A. and Asenjo, J. A.: Continuous Culture Studies of the Synthesis and Regulation of Extracellular B(1–3)glucanase and Protease Enzymes from *Oerskovia xanthineolytica*, Biotech. Bioeng., *30*, 628, (1987)
21. Andrews, B. A. and Asenjo, J. A.: Production of Enzyme Systems in Continuous Culture for the Controlled Lysis of Microbial Cells, Ann. N.Y. Acad. Sci. *506*, 637 (1987)
22. Scott, D., Hammer, F. E. and Szalkucki, T. J.: Bioconversions: Enzyme Technology, in: Food Biotechnology (Knorr, D., ed.) p. 413, Marcel Dekker, New York 1987
23. Asenjo, J. A., Andrews, B. A., Hunter, J. B. and Le Corre, S.: Process Biochem. *20*, 158, (1985)
24. Asenjo, J. A.: Adv. in Biotechnol. Vol. 3, p. 295, (C. Vezina and K. Singh, eds.), Pergamon Press 1981

4.4 Dynamic In-Situ Fractionation of Extracellular Proteins Produced in a Baker's Yeast Cultivation Process

William T. Effler Jr.[1], Robert D. Tanner[1], and George W. Malaney[2]

[1] Department of Chemical Engineering, Vanderbilt University, Nashville, Tennessee 37235, USA
[2] Department of Civil and Environmental Engineering, Vanderbilt University, Nashville, Tennessee 37235, USA

Contents

At the first Japan-U.S. Biotechnology Conference we presented results from a preliminary study of protein present within the medium of a baker's yeast cultivation process [1]. In that study, conducted with a tall beaker bioreactor, the concentration of proteins in the medium during the cultivation experiments (as measured by the Bradford Coomassie Blue Method) was found to vary in both magnitude, with respect to the vertical position within the bioreactor, as well as in time. The data from the non-aerated cultivation experiment in that work indicated that the cultivation medium total protein concentration could differ by as much as 5 times between the top of the cultivation vessel and the middle or bottom positions.

In a subsequent study, utilizing the same tall beaker bioreactor, a series of cultivation experiments were performed to determine the effect of various medium conditions on the total protein variation phenomena [2]. As much as a two-fold greater total protein concentration at the top as compared to the bottom of the bioreactor was observed in the non-aerated experiments of that study. In addition, the concentrations of individual polypeptides present in the cultivation medium (extracellular proteins secreted into the medium by the growing yeast cell population and cellular components from cell lysis) were measured using SDS-polyacrylamide gel electrophoresis. That analysis revealed as much as a 5-fold greater concentration of particular polypeptides (described by approximate molecular weight) at the top of the bioreactor as compared to the bottom. This

Bioproducts and Bioprocesses
Editors: Fiechter, Okada, Tanner

enhanced concentration of particular polypeptides near the top of the bioreactor was attributed to the phenomenon of bubble fractionation.

Portions of these earlier studies [1,2,3] have been replicated, employing a taller bioreactor with a modified aeration apparatus, and are reported here. Measurement of individual polypeptides found in the cultivation medium, using SDS-gel electrophoresis, indicated even further enhancement of the protein stratification effect: as much as a thirty five-fold greater concentration of particular polypeptides at the top as compared to the bottom of the bioreactor. These significant differences in molecular weight profiles in two different size bioreactors independently confirm that protein fractionation is occurring within the bioreactor column and at various stages of cell growth. The individual polypeptide concentration differences are particularly noticeable within the molecular weight range of 45 kDa and 95 kDa.

1. Introduction

1.1 Microorganisms and Selective Protein Production

Progress in the development of production methods for high value therapeutic proteins utilizing recombinant DNA techniques has proceeded rapidly in recent years. Yeast and bacteria currently are the preferred vectors. Commercial production employing these microorganisms offers many advantages over mammalian cell culture, and knowledge of their cultivation processes is extensive. In comparing the two, the strengths of yeast over bacteria include: the innate ability to secrete proteins (so that the cells need not be ruptured to retrieve the desired proteins, as is generally the case with bacteria), ease of growth in large numbers with minimal contamination, and production of proteins in a more active form (yeast are capable of glycosylating secreted proteins, whereas bacteria are not [4]). Bacteria, however, are much better characterized in terms of their gene structure. The introduction of synthesized DNA sequences into bacterial cells, such as *Escherichia coli*, has become a relatively routine procedure. In addition, recent advances by Japanese researchers at the Institute for Chemical and Physical Research on penicillinase synthesis by *E. coli* should enable bacteria to secrete other proteins in large quantities into the culture fluid, just as yeast cells do [5].

Recovery of the desired extracellular proteins secreted during a cultivation process, however, poses a significant problem in exploiting these genetically engineered microorganisms. Processing the large volumes of cultivation medium necessary for adequate recoveries by conventional methods of protein purification can become both time and cost intensive. A rapid, cost effective method of initial purification of these desired proteins, therefore, is an important consideration in determining the feasibility of applying any of these cultivation based technologies to commercial scale production.

1.2 Previous Work

In previous work this laboratory has studied the proteins present in the medium during a baker's yeast cultivation process, which are collectively referred to as the total extracellular protein concentration (the yeast cells are removed from the medium sample by centrifugation), in a series of experiments utilizing a 1 l tall beaker bioreactor [2]. The concentration of total extracellular protein, and that of particular polypeptides, present in medium samples taken from three vertical positions within the bioreactor (top, middle, and bottom) were measured at various time points during the cultivation experiments. As much as a two-fold higher total extracellular protein concentration was observed at the top of the bioreactor, as compared to the bottom, in that study. Separation of individual polypeptides present in those medium samples by gel electrophoresis, and subsequent quantification by laser densitometry analysis of the visualized protein bands, revealed as much as a 5-fold difference in concentration of particular proteins between the top and bottom of the bioreactor. This enhanced concentration of particular polypeptides near the top of the bioreactor was attributed to the phenomenon of bubble fractionation.

1.3 Bubble Fractionation and Its Use in Protein Purification

Bubble fractionation results when surface active substances, such as proteins, adsorb to the surfaces of rising bubbles (in our case, carbon dioxide bubbles produced by yeast cells during the cultivation or bubbles created from gas introduced into the medium near the bottom of the bioreactor) and then are carried upward on the buoyant bubbles to the top of the vessel. Usually at the top surface (the liquid-air interface), the bubbles break, releasing the protein solute back into the medium (or into the surface foam).

This protein transport mechanism can establish a vertical concentration gradient for these macromolecules which may, perhaps, be used as a basis for conducting a quick and effective separation scheme for cultivation systems. By removing a portion of the medium from the bioreactor column (during the cultivation process) at the point where the desired polypeptide is at its highest concentration (usually at the top of the vessel), and modifying the medium environment to enhance the fractionation of the specific protein of interest, the volume of medium to be purified could be greatly reduced. The cost and time involved in the protein purification process could thereby be lowered.

This phenomenon has been utilized in the past with separate operations: the cultivation process was conducted, the cultivation medium was collected and the cells removed, then the medium was stripped of the desired protein by injecting gas bubbles into the container and removing the surface foam (often called foam fractionation). Our approach is to combine these separate steps and utilize the gas bubbles produced naturally during the cultivation process (or augment them with a sparged gas stream) to strip the desired protein from the medium on-line, as it is produced. Thus, once synthesized and transported outside the cell, the residence time of the desired protein in the cultivation medium is minimized. This should: (a) lessen the chance of product inhibition in the synthesis of the desired protein, and, (b) reduce the opportunity for proteolysis of the desired protein (by the extracellular proteases present in the cultivation medium), both especially important considerations when utilizing a genetically engineered organism.

There is no inherent reason why the bubble fractionation mechanism could not be applied to the separation of excreted proteins produced in the submerged culture of yeast, protein secreting bacteria, or filamentous fungi [6] (and, possibly, in a tissue culture system arranged in a vertical column). It is, therefore, our contention that the separation method discussed here may apply to a wide spectrum of protein secreting microorganisms and cell cultures.

1.4 Research Objectives

With the promising applications of this phenomenon in mind, this study was undertaken to test the premise that bubble fractionation is the controlling mechanism underlying the protein stratification effect observed in our three previous cultivation studies [1, 2, 3]. Here, a series of experiments were conducted using a modified cultivation apparatus designed to enhance bubble fractionation effects within the medium environment. Should these modifications increase the protein concentration effect previously observed, it would tend to validate our hypothesis. The current system includes: (a) a modified system, utilizing a sparger, for introducing gas into the medium from the bottom of the bioreactor (to augment the bubbles naturally produced during the cultivation) which creates smaller diameter bubbles than those produced by our previous system (which consisted of a glass tube connected to a gas line, inserted into the bioreactor from the top, which released large bubbles (~ 1 cm) near the bottom of the cultivation medium), (b) a bioreactor with increased height to diameter ratio, to increase the effective contact time of the rising bubbles for a given volume of medium and minimize axial dispersion (promote plug flow), and, (c) a medium sampling method which minimizes disturbances of the stirred medium caused by the sampling procedure itself. In addition, a non-cultivation experiment was performed, utilizing a partially purified-protein (invertase, an enzyme derived from yeast) to confirm that the protein concentrating effect is indeed caused primarily by bubble fractionation and not metabolic or mixing phenomena.

2. Materials and Methods

2.1 Cultivation Organism

The cultivation organism studied was *Saccharomyces cerevisiae*, purchased at a local grocery store as Fleischmann's brand active dry yeast (Lot No.: Aug. 16, 85A-1). The inoculation concentration of yeast was 2.5 g to 1000 ml of medium (0.25% wt/vol).

2.2 Medium

The cultivation medium utilized was the Synthetic Medium C of Maxon and Johnson [7]. The carbon substrate was 10% (wt/vol) glucose. For the non-cultivation experiment (Experiment 5), 100 mg (100 $\mu g\ ml^{-1}$) of partially-purified invertase (Grade V: Practical, Sigma I 9253) was added to the normal 1 l of cultivation medium.

2.3 Bioreactor

The cultivation vessel used consisted of a 35 cm (height) × 6.2 cm (diameter) Nalgene polycarbonate 1 l graduated cylinder. The upper rim was modified so that the column could be sealed with a rubber headplate. The headplate was fitted with apertures for a pH probe, an automatic pH control feed line, a thermometer, and an off-gas reflux condenser. Sample ports fitted with rubber septa were inserted into the side of the column at the top position (1 cm below the 1000 ml mark), the middle position (500 ml mark), and the bottom position (1 cm above the bottom) of the column. The cylinder was filled to the 1000 ml mark with liquid cultivation medium and rested on a magnetic stirring plate. A 2.54 cm stirring bar rotated at 400 rpm at the bottom of the vessel. The lower half of the vessel was submerged in a constant temperature bath maintained at 32 °C. The temperature of the medium was measured at 10 cm intervals throughout the column and little temperature gradient within the bioreactor was found to exist (±1.5 °C). All of the cultivations were conducted at

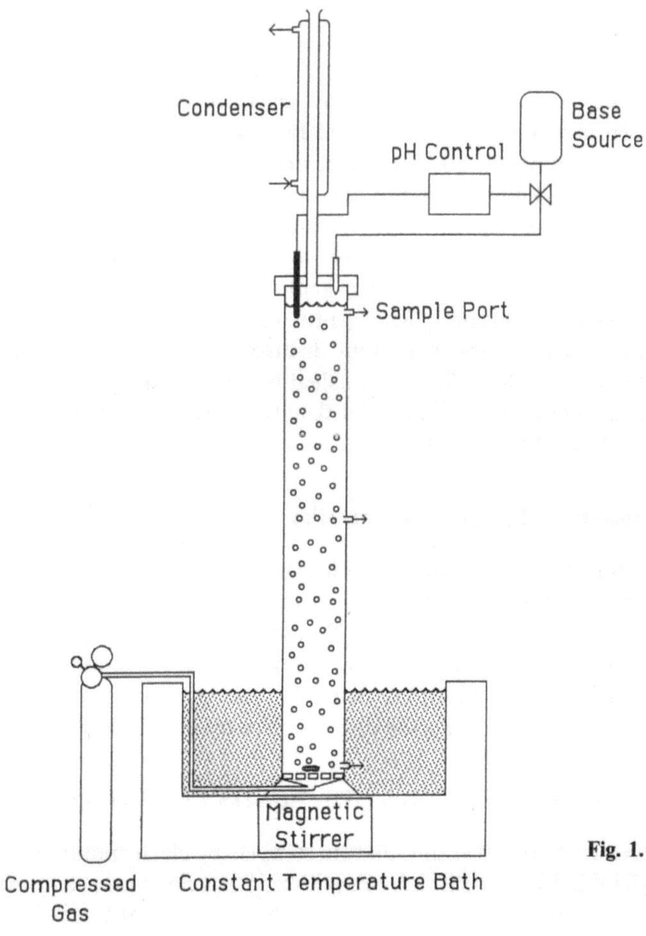

Fig. 1. Bioreactor setup

a constant pH (5.0 ± 0.25), maintained by an automatic titrimeter (Fisher Model 36, charged with a 10% NH_4OH solution, to neutralize the acid produced by the cultivation). The pH of the medium was measured at 10 cm intervals throughout the bioreactor column and no significant pH gradient was evident (5.0 ± 0.125). The water and ethanol vapors in the gas mixture, which evolved from the cultivation process, were cooled by a 5% ethanol-water cooled reflux condenser at 0 °C and returned to the cultivation medium as a liquid. The cooling solution was pumped by a Blue M constant temperature circulator, Model RBC-10A. All equipment and media (except the pH probe, which was sanitized with an absolute ethanol rinse) were sterilized in an autoclave with pressurized steam at 121 °C for 15 min. For the gas sparged experiments, filtered compressed air or CO_2 was introduced at 0.14 volumes of gas per volume of medium per min (0.14 vvm) through a 3.0 cm diameter fritted glass disc fitted flush to the bottom of the bioreactor. During the gas sparged experiments, the magnetic stirring bar spun directly over the top of the sparger. The whirling action helped to create very small bubbles (~100 µm in diameter). A diagram of the bioreactor setup is shown in Fig. 1.

2.4 Sampling Protocol

Samples of the medium were taken at the 0, 4, 7, and 10 h time points (the "0 hour" samples were taken just after the dried yeast cells had been added to the medium, before cultivation had begun). Using a 5 ml Becton-Dickinson disposable sterile syringe, with an 18 gauge needle (200 µm inner diameter), 2 ml samples were withdrawn from the top, middle, and bottom positions of the bioreactor by inserting the needle through the respective septum until it just entered the medium. The syringe was filled slowly (~60 s). The contents were slowly dispensed (so as not to burst the yeast cells) into 10 mm × 65 mm test tubes kept at 4 °C and quickly centrifuged (Beckman Model 2600) at 3100 rpm (1850 × g) for 15 min at 4 °C. The supernatant was removed with a Pasteur pipette without disturbing the yeast pellet and stored at 4 °C for various assays. The pellet was washed by resuspending in 2 ml of distilled water to remove entrained sugars. The pellet then was dried for 24 h at 40 °C for subsequent determination of dry cell weight.

2.5 Assay of Total Extracellular Protein Concentration

The Lowry microassay for proteins [8] was used for the determination of total extracellular protein in each medium sample. Each determination was repeated in triplicate and the results averaged. In all cases the standard deviation for the three samples was less than 5%. A comparison of the total protein concentration values derived from the Lowry and Bradford (used for the previous work [1,2,3]) methods was made for one experiment and the differences were all within 8%.

2.6 Assay for Individual Proteins Using Polyacrylamide-SDS Gel Electrophoresis

Polypeptides were resolved from the medium mixture samples by electrophoresis on 12.5% (wt/vol) polyacrylamide-SDS 10 cm slab gels with a 5% stacking gel according to the procedure of Laemmli and Favre [9]. The sample buffer (62.5 mM Tris base containing 20% sucrose, 10% SDS, 0.001% bromophenol blue, pH 6.8) and the medium sample (minus cells) were mixed in equal proportions and the solution heated at 100 °C for 5 min before loading 50 μl on the gels. Electrophoresis was performed at 25 mA in 25 mM Tris base — 190 mM glycine — 3.47 mM SDS — 9 mM B-mercaptoethanol at pH 8.5, until the dye marker had reached 2 cm above the bottom of the gel. The polypeptide bands were visualized by the silver staining procedure described by Wray [10]. The gels were purged in preparation for drying with three two-h swell-shrink cycles (soak gel: 1 h distilled water, 1/2 h 25% ethanol/water, 1/2 h 50% ethanol/water). The gels were then vacuum dried between two sheets of cellophane (Bio-Rad 165-0922) on a slab gel dryer (Bio-Rad Model 443 Slab Dryer). Bio-Rad standard molecular weight markers (high and low range) were used to calculate Rf values for approximation of molecular weights.

Individual polypeptide band densities in each lane were analyzed using a scanning laser densitometer (LKB 2202 Ultroscan). Each polypeptide peak was quantified by integrating the peak on a digital integrator (LKB 2220 Recording Integrator). The total area under each polypeptide band peak was used to represent the "concentration" of that polypeptide at that position. These individual polypeptide concentrations were used for comparisons of the same polypeptide at different sample positions (representing a relative concentration difference) and not for comparing concentrations between different polypeptides, since each separated polypeptide was likely to take up a different amount of dye (differential dye-binding [11]. For consistency, the integrator was "normalized" at the beginning of each gel scan by adjusting the attenuation function to give the same "standard size" peaks for the molecular weight marker bands. This compensated for staining differences and allowed for comparison of protein concentrations between different cultivation runs.

Most proteins present in the cultivation medium, mainly cellular components from cell lysis and secreted enzymes such as L-Asparaginase, Exo-1,3-B-glucanase, Endo-1,3-glucanase, and pheromone peptidase [12], are detected using our methods of SDS-polyacrylamide gel electrophoresis. Smaller polypeptides, however, such as pheromone or killer toxin [12], may not be detected by our system.

3. Experimental Conditions

Five experiments were conducted, all at a controlled pH of 5. To directly determine the effect of scale-up from the bioreactor used in our previous paper [2], which had a medium volume of 800 ml and a height/diameter ration, H/D, of 1.6, to the bioreactor used in these experiments, with a medium volume of 1000 ml and an H/D of 5.7, the first experiment was conducted without sparging. Comparison of the sparged experiments is not too meaningful, since in addition to the change in H/D, the bubble size also changed. The aerated experiments in our previous paper were performed with a simple open tube (gas was bubbled into the medium near the bottom of the bioreactor), producing large (~ 1 cm diameter) bubbles, as contrasted with the present setup, which used a flat porous disc sparger (implanted into the bottom of the bioreactor) that produced much smaller bubbles (~ 100 μm diameter).

To test the observation that the addition of moderate amounts of salt to the cultivation medium can reduce the degree of bubble fractionation (as it did in our previous study [2]), the 2nd experiment was performed with the addition of 0.6 M NaCl to the cultivation medium (and without sparging).

The third and fourth experiments employed sparging, the third being sparged with air, and the fourth with carbon dioxide, to investigate the effect of additional bubble surface area and differences between these gases on protein stratification. To eliminate the possibility of overlooking significant cycling in the protein concentration data, additional samples of the medium were taken, during the fourth experiment, both 5 min before and 5 min after the predetermined sample time. These additional samples were also analyzed for both total protein concentration and individual polypeptide concentrations (using gel electrophoresis) to check for significant perturbations about a smoothly evolving time trajectory.

To demonstrate the extent of protein stratification that can occur in a sparged column, without the dynamics of a cultivation process involved, Experiment 5 was conducted without inoculating the medium with yeast cells. In Experiment 5, partially-purified invertase was mixed in a beaker with 1 l of Maxon and Johnson cultivation medium (containing 10% glucose) to a concentration of 100 $\mu g\ ml^{-1}$, then stirred with a glass rod until dissolved. This 1 l mixture was poured into the bioreactor and the mixture was (in the following order): (a) allowed to stand for 20 min, followed by normal sampling, (b) stirred at 400 rpm for 20 min, followed by normal sampling, and, (c) stirred at 400 rpm and sparged at 0.14 vvm with carbon dioxide for 20 min, followed by normal sampling. A sparged but not stirred run was not conducted since the whirling action of the stirring bar was essential for the formation of small bubbles. In other words, removing the stirring bar would change the system, negating meaningful scale comparisons.

The experiments performed are summarized as follows:

Exp. 1 — Cultivation, non-sparged
Exp. 2 — Cultivation, non-sparged, 0.6 M added salt
Exp. 3 — Cultivation, sparged with air
Exp. 4 — Cultivation, sparged with carbon dioxide
Exp. 5 — Non-cultivation, partially-purified enzyme added to medium, sparged with CO_2

4. Results and Discussion

4.1 Influence of Cultivation Medium Conditions on the Total Extracellular Protein Concentration

The results of the total protein assays for Experiments 1 through 4 are shown in Figs. 2 through 5, respectively. Generally, top to bottom gradients in total protein concentration were established in all of the cultivation experiments once cultivation had begun. Virtually identical total protein concentrations were observed in the medium samples at all three positions along the vertical axis of the vessel when taken at the "0 hour" time point (just after adding the dried yeast to the culture medium, before the cultivation process or aeration had begun) in all the cultivation experiments. This substantiates the premise that the presence of bubbles is a key factor for

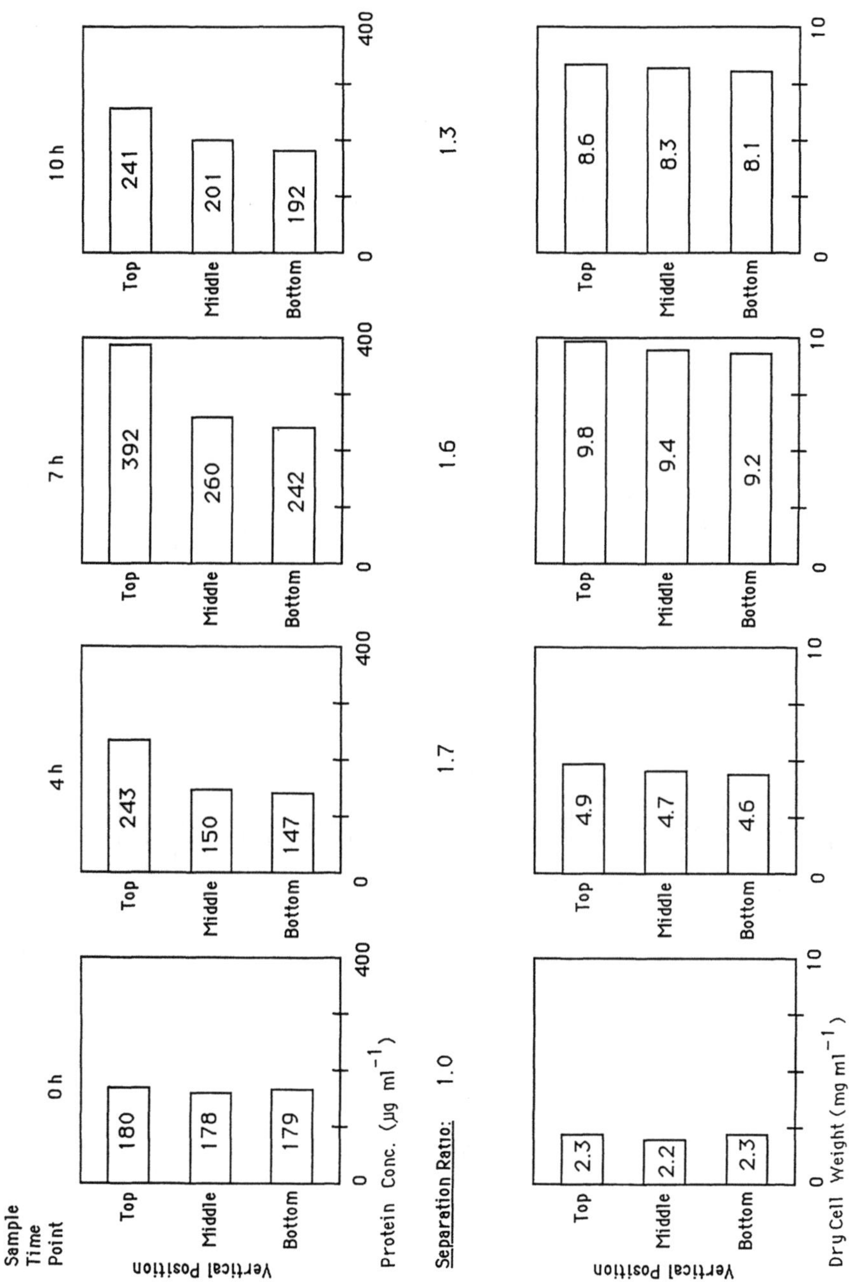

Fig. 2. Cultivation medium total protein assays for Experiment 1, non-sparged

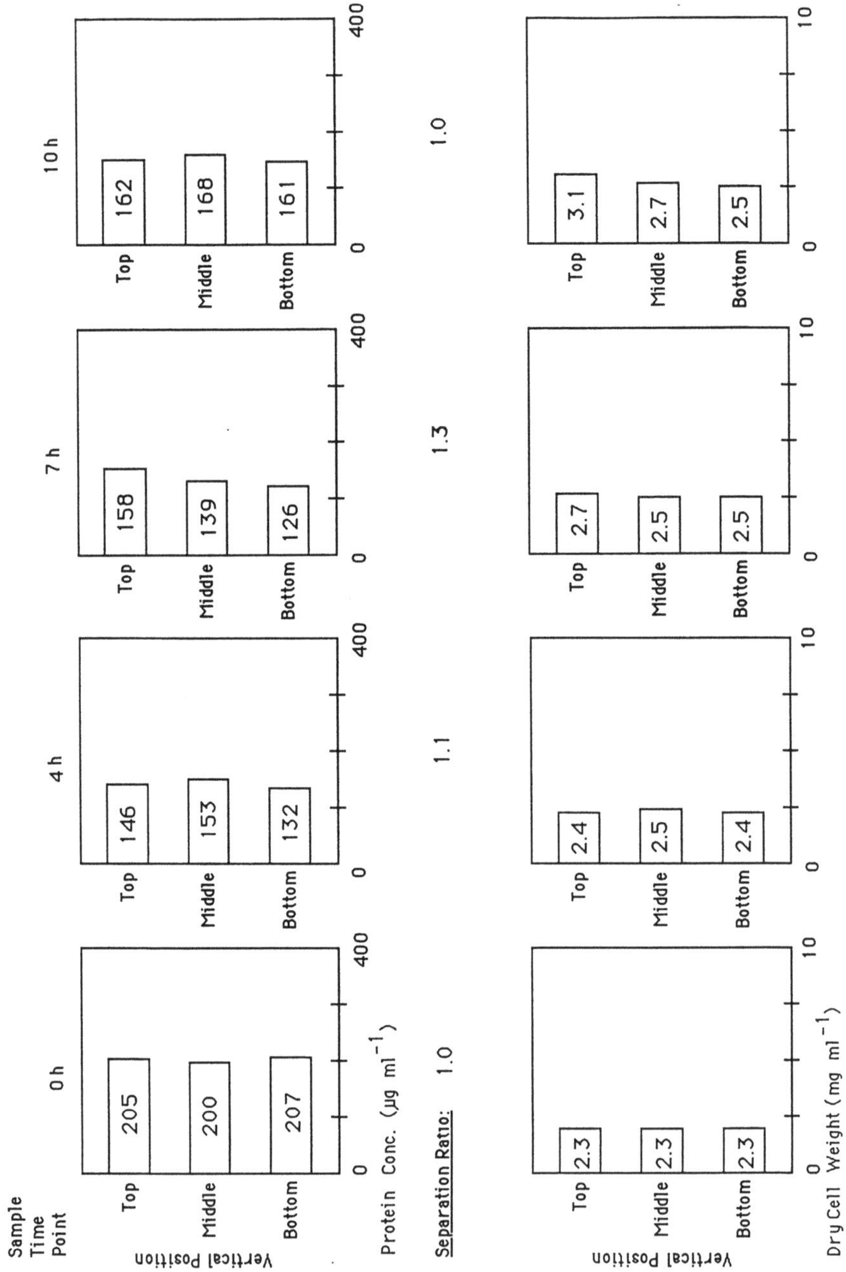

Fig. 3. Cultivation medium total protein assays for Experiment 2, non-sparged. 0.6 M NaCl added

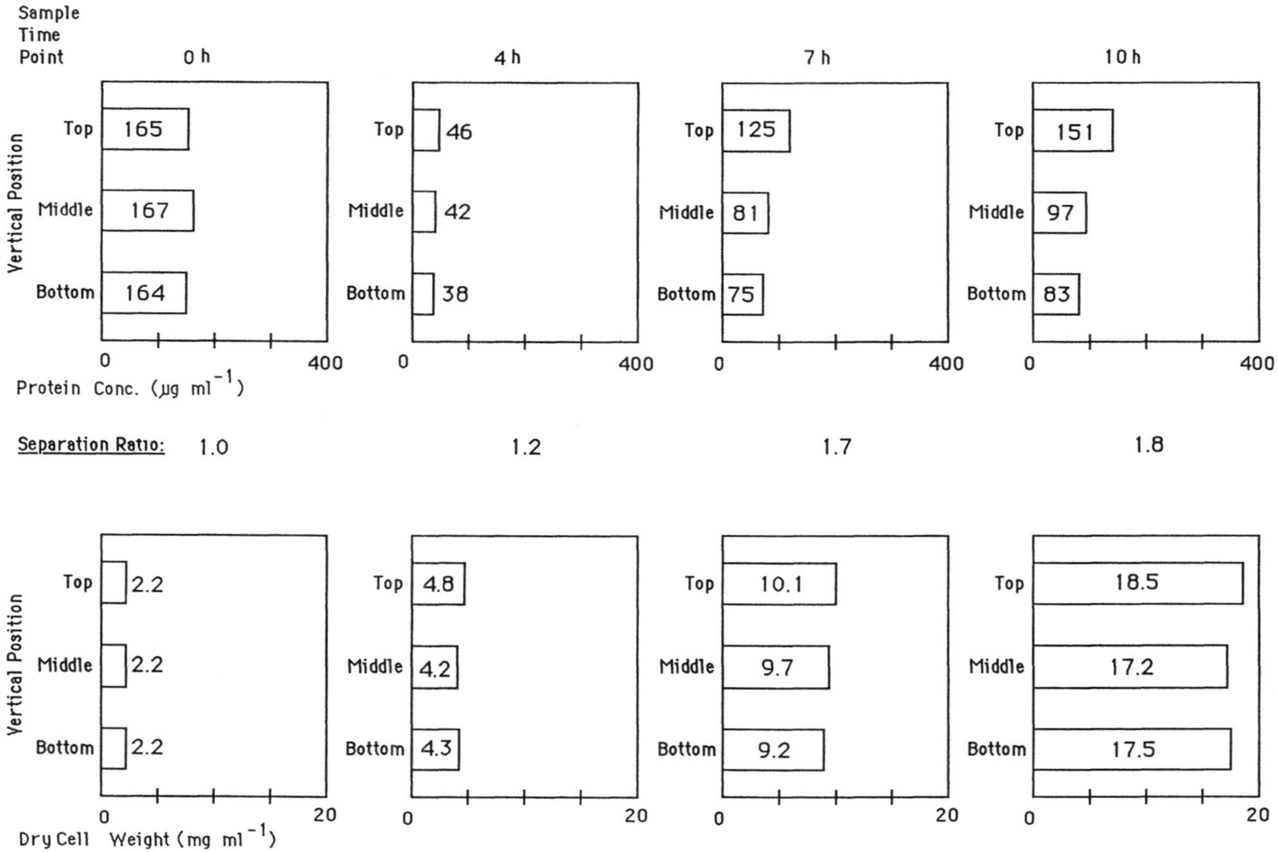

Fig. 4. Cultivation medium total protein assays for Experiment 3, sparged with air (0.14 vvm)

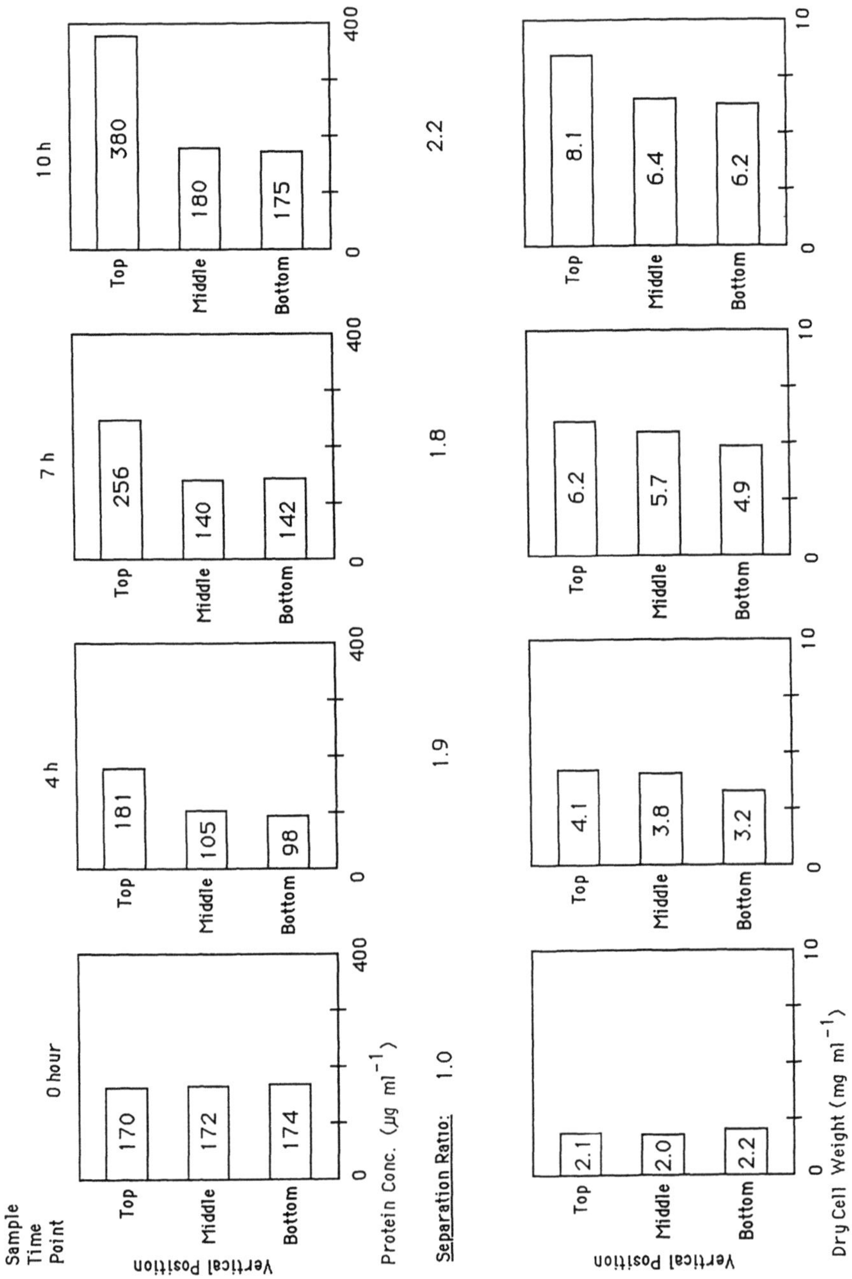

Fig. 5. Cultivation medium total protein assays for Experiment 4, sparged with CO_2 (0.14 vvm)

protein stratification to occur. The results of the dry cell weight analysis for each medium sample are shown in the lower portions of Figs. 2 through 5. Higher cell densities at the top of the bioreactor, as compared to the bottom, occurred in all experiments once cultivation had begun. Apparently, the yeast cells themselves tend to be carried upward in the rising bubble stream. The presence of a greater number of yeast cells near the top of the bioreactor column probably contributes to the greater total protein concentration at the top sample position observed in the cultivation experiments.

To illustrate the extent of protein fractionation occurring at the four time points during the cultivation experiments, we introduce the concept of a separation ratio for proteins. The separation ratio for the total extracellular protein concentration is defined here as the concentration of extracellular protein at the top position of the bioreactor divided by the concentration of extracellular protein at the bottom position.

In Experiment 1 (results depicted in Fig. 2), the cultivation was conducted without sparging. Rapid cell growth was exhibited, though not to the extent of Experiment 3 in which the medium was sparged with air. The overall total protein concentrations of the cultivation medium at the 4 and 7 h points were the highest of any at these times in the experiments performed in this study. Significant fractionation of total extracellular protein occurred, with a separation ratio of 1.7 at the 4 h time point and 1.6 at the 7 h time point. The cultivation had apparently reached its log phase prior to the 10 h sample time, as demonstrated by the reduction in extracellular protein levels and dry cell weights by 10 h. The degree of protein fractionation was also reduced by the 10 h point (with a separation ratio of 1.3), presumably due to the diminished amount of CO_2 bubbles resulting from the lower cell numbers and lowered cell metabolism.

The diminished separation ratios for total extracellular protein in Experiment 2 (1.0 at 4 h, 1.3 at 7 h, and 1.00 at 10 h, see Fig. 3) when compared with Experiment 1, corroborate our previous results [2] of reduced protein stratification (and reduced total extracellular protein) in yeast cultivation medium subjected to a moderate addition of salt. This "salting in" effect seems to result in the proteins within the medium having an increased solubility in water and reduced surface activity [13] and, consequently, the protein concentrations are less affected by bubble fractionation. The overall total extracellular protein concentration at the 0 h time point was the highest of any at the 0 h time point, indicating, perhaps, that the 0.6 M NaCl medium extracted more protein from the yeast cell interior and/or surface by increasing the protein solubility and/or by increasing the osmolarity of the medium. It can be observed from the data that cell growth, protein secretion, and, presumably, metabolism, are also depressed when the cultivation is conducted with a moderate addition of salt. The reduced metabolism and the resultant lowered CO_2 production may also contribute to the reduced bubble fractionation effect on the extracellular proteins. A moderate amount of cell fractionation did occur, however, at the 10 h point, despite the reduced volume of CO_2 bubbles.

Sparging additional gas into the bioreactor from the bottom is expected to lead to an increase in the degree of protein stratification, by increasing the amount of bubble surface area to which the proteins adhere. Displayed in Fig. 4 are the data for Experiment 3, in which 0.14 vvm of air was sparged into the bioreactor. The largest total protein separation ratio achieved in this experiment was 1.8 (at 10 h), which is larger than the highest separation ratio (1.7 at 4 h) attained in the non-sparged run (Experiment 1). The data depicted in Fig. 4 indicate that the overall total extracellular protein concentrations were significantly reduced at all time points under air-sparged conditions, while the cell growth rate increased, when compared to all other experiments.

In Experiment 4 (Fig. 5), 0.14 vvm of carbon dioxide was sparged into the bioreactor during the cultivation. These conditions produced the largest separation ratio of all four cultivation experiments: 2.2 at the 10 h time point. The separation ratios of 1.9 at 4 h and 1.8 at 7 h were also higher than those at the same time points in the other experiments. When compared to the standard non-sparged run (Experiment 1), conducted in the environment of only its naturally produced CO_2, the data for Experiment 4 indicate that during the first few hours of the cultivation process the cell mass and overall protein secretion values lagged their standard early rate of increase by about 3 h. Perhaps this was due to the presence of exogenous CO_2 within the medium.

In Table 1, the data for the additional medium samples taken during Experiment 4, taken both 5 min before and after each normal sample time point, are presented. In all cases, the differences between the three samples taken 5 min apart were within experimental error, indicating that these short time changes in concentration are either small or smoothly evolving. We would expect, therefore, that abrupt changes (such as those observed in our preliminary cultivation studies [1]), do not occur here.

The total extracellular protein separation ratios at the 7 h sample time point for the non-sparged experiments of this bioreactor and the previous bioreactor [2], are compared in Fig. 6 to elucidate the effect of an increased H/D ratio bioreactor on protein stratification. As would be expected for a bubble fractionation scheme,

Table 1. Cultivation medium extracellular protein concentrations for Experiment 4, additional samples were taken shortly before and after the normal sample times to check for significant cycling in the protein concentration data

Cultivation time point	Bioreactor position	Total extracellular protein concentration		
		5 min before sample time $\mu g\ ml^{-1}$	Normal sample time $\mu g\ ml^{-1}$	5 min after sample time $\mu g\ ml^{-1}$
4 h	Top	189	181	186
	Middle	100	105	108
	Bottom	98	98	100
7 h	Top	250	256	258
	Middle	145	140	144
	Bottom	140	142	149
10 h	Top	386	380	394
	Middle	174	180	181
	Bottom	172	175	180

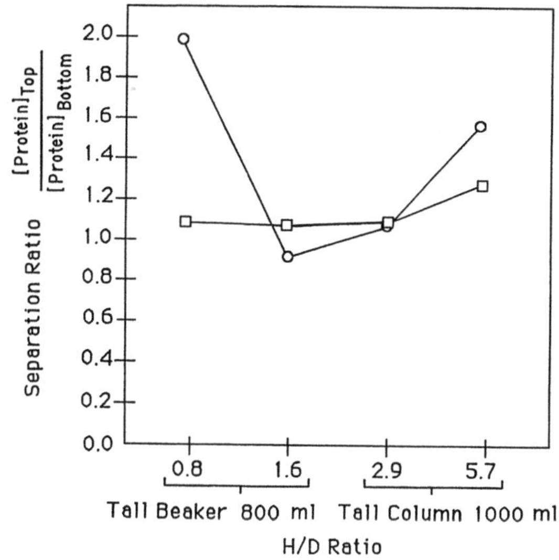

Fig. 6. Comparison of cultivation medium total protein separation ratios for two different size bioreactors. ○, Non-salted cultivation medium; □, 0.6 M NaCl cultivation medium

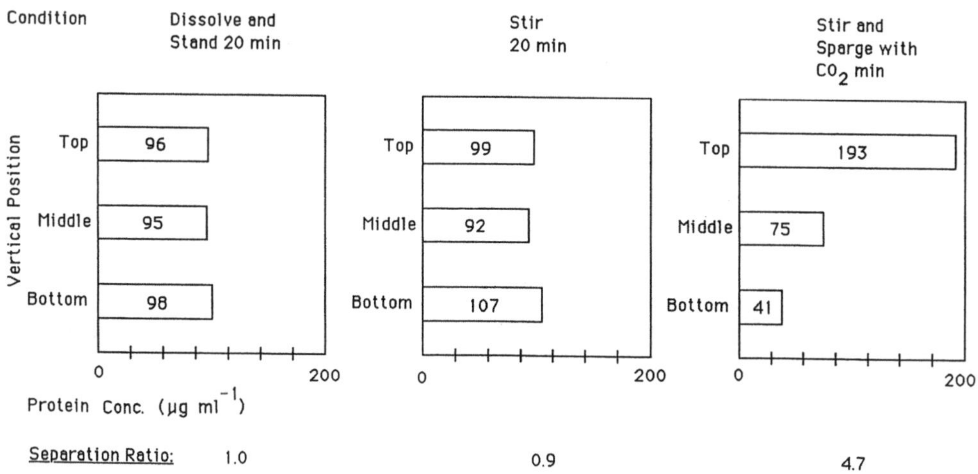

Fig. 7. Total protein assays for Experiment 5, non-cultivation, sparged with CO_2 (0.14 vvm)

increasing the H/D ratio enhances the separation effect for both the case with salt and the one without salt.

The results of the non-cultivation experiment (Experiment 5) are shown in Fig. 7. Sampling the partially-purified invertase protein mixture after it stood undisturbed for 20 min within the bioreactor, and then again after being stirred for 20 min at the same rate used during the cultivation experiments, indicated that no significant top to bottom total protein gradient had been established in either case. However, after sparging and stirring the mixture for 20 min, sampling the mixture revealed an almost 5 times higher total protein concentration at the top position of the bioreactor as compared to the bottom (separation ratio = 4.70). This observation tends to further corroborate the premise that bubble fractionation is the primary means of separating proteins in our bioreactor.

4.2 Influence of Cultivation Medium Conditions on the Concentrations of Individual Polypeptides

After separation of the individual polypeptides present within each medium sample by SDS-gel electrophoresis and quantification of the visualized protein bands on the gel, employing a laser densitometer, nearly all of the extracellular polypeptides were found to exhibit some degree of surface activity by demonstrating higher concentrations in the top position medium sample (and usually to a lesser degree, the middle) when compared to the bottom position medium sample. This protein surface activity results in vertical position differences for particular proteins if the conditions for bubble fractionation exist, e.g. the presence of sufficient CO_2 bubbles. In all of the cultivation experiments, quantification of individual polypeptides in the medium samples from the 0 h time point (before CO_2 bubbles were produced from the cultivation and, in the sparged experiments, before sparging had begun) indicated no fractionation effects were occurring for any of the proteins.

It is observed in the following data that polypeptides of different molecular weights in the medium mixture, though present in the same environment, can display a different degree of stratification within the bioreactor. This may be due to the nature of bubble fractionation and the amphoteric quality of proteins. All protein molecules possess both positively and negatively charged groups. Accordingly, the net charge on a protein will depend on the pH of its environment. The pH value of a water-protein solution at which there is a net charge of zero on a dissolved protein molecule is called the isoelectric point, pI, of that protein [14]. Since water is a polar solvent, the pI is at the pH where the polypeptide is least soluble in water. Therefore, if the cultivation medium pH is near the isoelectric point of one of its proteins, that protein will be highly insoluble in the medium and tend to come out of solution (by adhering to the rising bubbles within the medium) to a greater

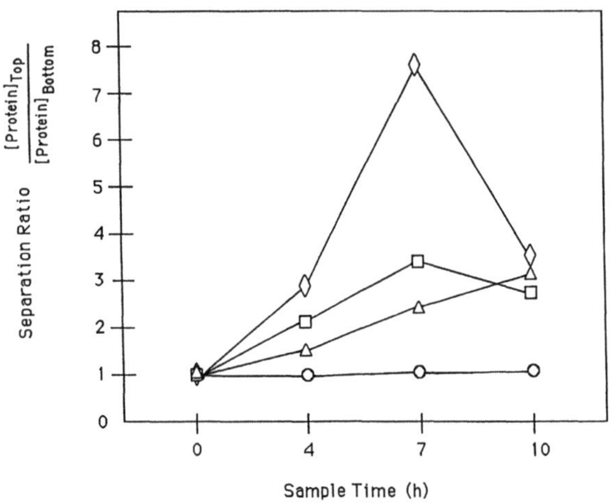

Fig. 8. Separation ratios of four proteins from the cultivation medium of Experiment 1, non-sparged.
Polypeptide molecular weight:
○ 39 kDa
△ 62 kDa
□ 78 kDa
◇ 95 kDa

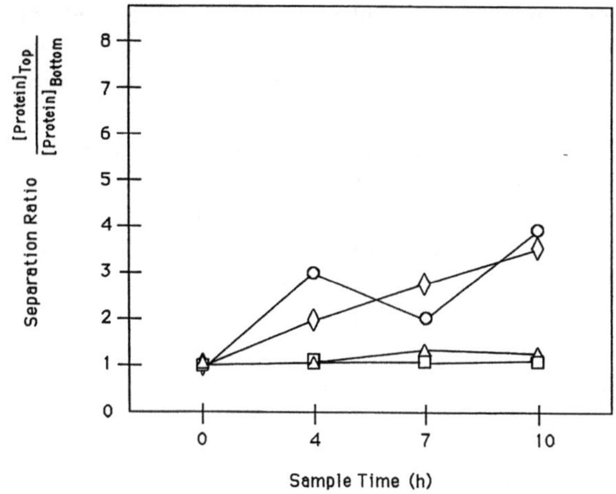

Fig. 9. Separation ratios of four proteins from the cultivation medium of Experiment 2, non-sparged, 0.6 M NaCl.
Polypeptide molecular weight:
○ 39 kDa
△ 62 kDa
□ 78 kDa
◇ 95 kDa

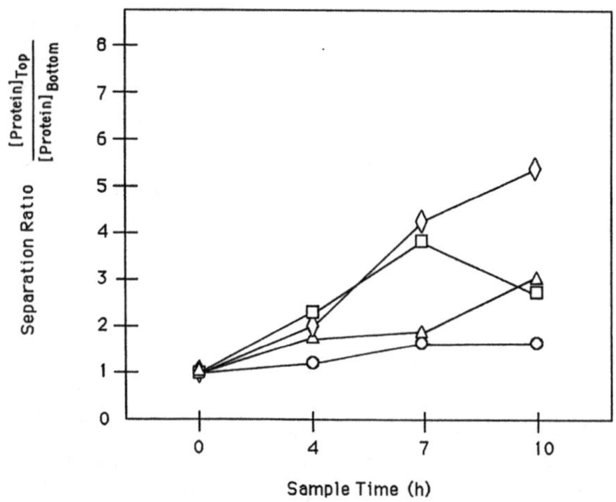

Fig. 10. Separation ratios of four proteins from the cultivation medium of Experiment 3, sparged with air (0.14 vvm).
Polypeptide molecular weight:
○ 39 kDa
△ 62 kDa
□ 78 kDa
◇ 95 kDa

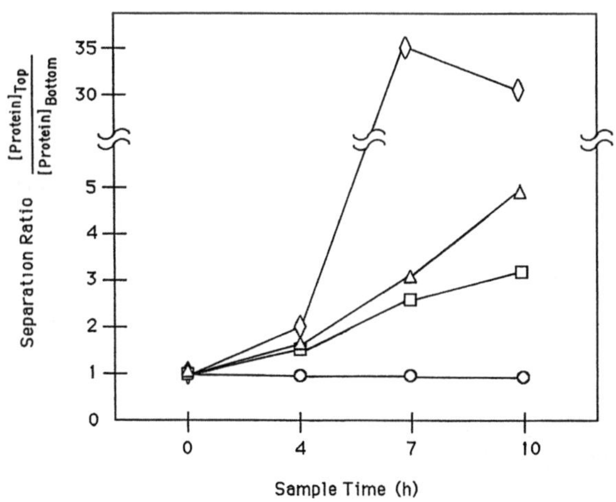

Fig. 11. Separation ratios of four proteins from the cultivation medium of Experiment 4, sparged with CO_2 (0.14 vvm).
Polypeptide molecular weight:
○ 39 kDa
△ 62 kDa
□ 78 kDa
◇ 95 kDa

extent than one whose isoelectric point is dissimiliar to that of the medium. Since each secreted polypeptide generally has a different isoelectric point from the others, we expect each one to respond with varying degrees of fractionation at a given medium pH.

Four significant polypeptides present in the medium were chosen for individual study, since they were all present in the four experiments. These polypeptides had approximate molecular weights of 39 kDa, 62 kDa, 78 kDa, and 95 kDa, determined by comparison of their Rf values with those of the standard molecular weight markers. It is interesting to note that except for the first two, these molecular weights differ by approximately 14 kDa, in keeping with Savageau's "14 kDa rule" proposed for significant bacterial proteins [15].

The 62 kDa and 78 kDa proteins represented a significant portion of the total extracellular protein present in the cultivation medium (~5 to 20% each). The 39 kDa and 95 kDa proteins were present in smaller concentrations (~ <1 to 5% each). Figs. 8, 9, 10, and 11 show the separation ratio for each polypeptide at each of the four sample time points in Experiments 1, 2, 3, and 4, respectively. The separation ratio here is defined as the concentration of a polypeptide at the top sample position divided by its concentration at the bottom sample position (concentrations derived from the densitometer analysis of the dried gel electrophoresis slabs). The separation ratios for each polypeptide at the 7 h point in the non-sparged experiment, using our current bioreactor, are compared with those of the bioreactor used in our previous study [2] in Fig. 12. The enhancement in the separation of the 95 kDa polypeptide is quite significant in the tall cylinder bioreactor when compared with the previous shorter tall beaker bioreactor.

It can be seen in Figs. 8 through 11 that the 95 kDa protein responds especially well to the effects of bubble fractionation during the cultivation process under all of our experimental conditions. The pI of this protein, therefore, may be close to the pH of our cultivation medium in these controlled pH experiments (pH = 5). The separation ratio for this polypeptide reached a high of 35 at the 7 h point in the carbon dioxide sparged experiment (Experiment 4, Fig. 11), and maintained high separation at the 10 h point, with a separation ratio of 30. The

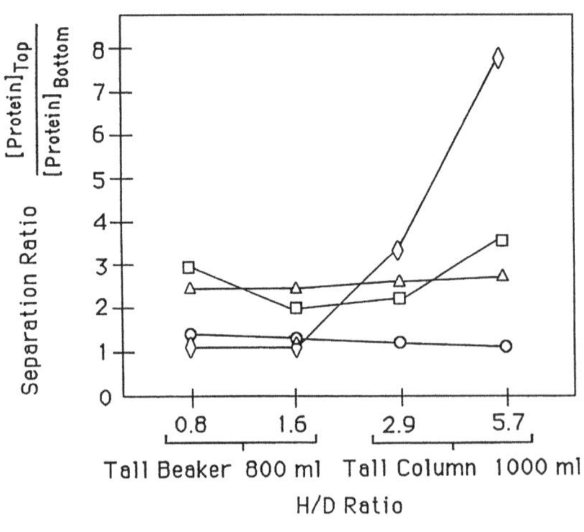

Fig. 12. Comparison of separation ratios of four polypeptides for two different size bioreactors. Polypeptide molecular weight:
○ 39 kDa
△ 62 kDa
□ 78 kDa
◇ 95 kDa

enhanced separation ratio of the 95 kDa polypeptide in Experiment 4, when compared to the other experiments, seems to be a result of the additional bubble surface area provided by the sparged CO_2. The highest separation ratio in the air sparged experiment (Experiment 3, Fig. 10), 5.4, was also achieved with this same protein, at the 10 h point.

The reduced stratification of the 95 kDa polypeptide in Experiment 3 was possibly due to the low concentration of this polypeptide under air sparged conditions (most of the extracellular polypeptides demonstrated reduced absolute concentrations in Experiment 3, as seen in the total protein levels for Experiment 3, shown in Fig. 4). Other factors contributing to the reduced stratification of proteins under air sparged conditions when compared to CO_2 sparged conditions might be: (a) sparged air bubbles may have a larger diameter (thus, a reduced surface area for the same volume of gas) than sparged CO_2 bubbles (possibly due to surface tension differences between these two gases and the medium), (b) reduced protein binding at the air-medium interface when compared to the CO_2-medium interface (due to differing electrostatic effects between these gases and the medium or different surface tensions at the interfaces), (c) in an air saturated medium, the extracellular polypeptides may assume a quaternary structure that is less hydrophobic or less likely to adhere to a gas-liquid interface than in a CO_2 saturated medium, or (d) the increased cell density in the cultivation medium under air sparged conditions may tend to dislodge the proteins from the rising bubbles, by impact or abrasion, in the well stirred medium.

The data from Experiments 1 and 3 indicate that the 78 kDa protein seems to be produced in the greatest concentration and shows the highest degree of fractionation at the 7 h time point, with reduced overall concentration and fractionation being exhibited at 10 h. In these experiments, the 7 h time point was within the period of most rapid growth. Experiment 4 exhibited the most rapid growth (and greatest separation for the 78 kDa polypeptide) at the 10 h point. Perhaps this polypeptide is significantly digested after the peak period of growth in the non-CO_2 sparged runs of Experiments 1 and 3, thus reducing its absolute concentration in the medium during the log phase. This digestion could then result in diminished separation ratios after that point of the cultivation (as demonstrated in Experiments 1 and 3). The 62 kDa protein also seemed most prevalent during the period of most rapid growth and demonstrated significant fractionation then also (except in the salted medium experiment, Experiment 2). Interestingly, the 62 kDa polypeptide generally demonstrated enhanced stratification within the medium past the period of most rapid growth by displaying increased fractionation at the 10 h point, despite being present in reduced amounts, when compared to the 7 h time point (except in the salted medium experiment). Both the 62 kDa and the 78 kDa polypeptides display little fractionation in the added salt experiment. Perhaps these polypeptides may become more hydrophilic in an ionic environment.

The 39 kDa protein, though not present in high concentrations in any of the experiments, displayed the unusual characteristic of achieving its best separation ratio during the added salt experiment. Perhaps, this indicates that its structure is made more surface active in an ionic environment.

5. Conclusions

Proteins present within the extracellular medium during a yeast cultivation process are spacially fractionated (concentrated at a particular vertical position) within the cultivation vessel due to the phenomenon of bubble fractionation, despite moderately vigorous mixing. The degree of fractionation is influenced by the conditions under which the cultivation takes place. Fractionation is enhanced under CO_2 sparged conditions, presumably because there is more bubble surface area available for the adsorption of hydrophobic proteins. Conducting the cultivation in a tall and narrow bioreactor enhances the protein fractionation effect. The addition of moderate levels of salt to the medium reduces the fractionation for most (but not all) of the proteins. Individual polypeptides respond to changes in medium conditions with varying degrees of fractionation. Perhaps a cultivation process can be conducted under a series of predetermined conditions to sequentially eliminate undesirable proteins and subsequently concentrate the desired protein within the medium.

Many questions must be answered before the bubble fractionation phenomena can be applied to industrial use. For example: (a) How important is the effect of the nature of the sparged gas on the degree of fractionation? Would sparging with an inert gas such as nitrogen increase stratification? (b) Is there an optimum bubble size for fractionation of a specific polypeptide, or is smaller always better? (c) Is stirring the medium a significant factor in the fractionation effect; Should stirring be reduced to a minimum to lessen back mixing within the medium? (d) Would specific salts added to the medium enhance the separation of particular polypeptides? (e) Could the medium pH be modified to tailor the cultivation for enhanced separation of a specific polypeptide? If so, how close to the pI of the desired polypeptide need the medium be for significant stratification? How quickly does the fractionation effect reach equilibrium?

Satisfactory answers to these questions may permit bubble fractionation to become a viable method for initial protein purification on an industrial scale.

6. Acknowledgements

The authors thank Elizabeth A. Shipp and the staff of the laboratories of Dr. T. R. Soderling (Howard Hughes Medical Center, Vanderbilt University), Dr. B. J. Danzo (Vanderbilt University Medical Center), Dr. D. J. Wilson (Chemistry Department, Vanderbilt University), Dr. N. K. Pandey (Research Centre, Hindustan Lever Limited, Bombay, India), and Don-Hee Park (Chonnam National University, Kwangju, Korea) for their guidance and assistance in this research.

7. References

1. Park, D. H., Malaney, G. W., Tanner, R. D.: Kinetics of Protein Secretion and Fractionation in Baker's Yeast Fermentations. Paper presented at the First Japan-U.S. Biotechnology Conference, Honolulu, December, 1984 and published in J. Chem. Tech. Biotechnol. *39*, 85 (1987)
2. Effler, W. T., Pandey, N. K., Tanner, R. D., Malaney, G. W.: Preconcentration (Within the Broth)

Secreted Extracellular Proteins During a Baker's Yeast Fermentation Process. Biot. and Bioeng. Symp. Series, No. 16, C. Scott, ed., p. 633 (1986)
3. Pandy, N. K., Effler, W. T., Park, D. H., Tanner, R. D., Malaney, G. W.: Study of Proteins Secreted by Growing *Saccharomyces cerevisiae*. Current Science, Vol. 55, No. 23, p. 1172 (1986)
4. Monmaney, T.: Science 85, Vol. 6, No. 6, p. 30 (1985)
5. Lewis, H. W.: Biotechnology in Japan. Office of Naval Research, Far East, Science Bulletin *10*, 3 (1985)
6. Mammalian Protein: Secretion Gained in Filamentous Fungi. Chem. and Eng. News, Vol. 63, No. 37, p. 4 (1985)
7. Maxon, W., Johnson, M. J.: Ind. End. Chem. *45*, 2554 (1953)
8. Lowry, O. H., Rosebrough, N. J., Farr, A. L., Randall, R. J.: J. Biol. Chem., *193*, 265 (1951)
9. Laemmli, U. K., Favre, M.: J. Mol. Biol. *80*, 575 (1973)
10. Wray, W., Boulikas, T., Wray, V., Hancock, R.: Anal. Biochem. *118*, 197 (1981)
11. Andrews, A. T.: Electrophoresis: Theory, Techniques, and Biochemical and Clinical Applications, p. 37, Oxford University Press, Oxford 1981
12. Novick, P.: The Molecular Biology of the Yeast *Saccharomyces*: Metabolism and Gene Expression, p. 394, Cold Spring Harbor Monograph Series 1982
13. Mahler, H. R., Cordes, E. H.: Biological Chemistry, p. 93, Harper & Row, New York 1971
14. Tanford, C.: Physical Chemistry of Macromolecules, p. 241, John Wiley & Sons, New York 1963
15. Blanchard, F.: Michigan Today, Vol. 18, No. 3 (1986)

4.5 Supercritical Fluid Bioreactor

Kozo Nakamura

Department of Agricultural Chemistry, University of Tokyo, Bunkyo-ku, Tokyo 113, Japan

Contents

The supercritical fluid has generally the density similar to that of liquid and the viscosity similar to that of gas. It has a solvent power like liquid and diffuses in the solid matter as easily as gas. The density of the supercritical fluid, which influences the solubility, is controllable by regulation of pressure or temperature, and this characteristic is available for product recovery. Therefore it is interesting to consider the use of the supercritical fluid as a medium for enzymatic reaction. The enzymatic acidolysis of triglyceride was studied using the supercritical carbon dioxide as a reaction medium in the batch reactor. The reaction could proceed like in n-hexane, and the stability of lipase was fairly good if it was adsorbed in the solid carrier with a small amount of water.

1. Introduction

The supercritical fluids generally have similar density to liquids and similar viscosity to gases. They have solvent power like liquids and diffuse in solids as easily as gases. Among the supercritical fluids the supercritical carbon dioxide (abbreviated as SCC hereafter) is safe for human beings, and its critical pressure and temperature, 7.38 MPa and 31.3 °C respectively, are suitable for its application in the processing of food. Therefore SCC has attracted many researchers' interest in its use for the extraction of natural food substances [1].

The supercritical fluids may be also an interesting medium for unit operations other than extraction, and recently a review on reactions in supercritical fluids has been written [2]. Several advantages are expected in use of SCC as a reaction medium:

(a) The low viscosity and the high diffusivity of a supercritical fluid can contribute to efficient mass transfer,

(b) The solubility of substrates and products can be controlled by regulation of pressure and/or temperature, so that their separation may become easy, and

(c) SCC is harmless to human beings.

Bioproducts and Bioprocesses
Editors: Fiechter, Okada, Tanner

The reaction in a supercritical fluid means automatically the high pressure reaction, and the favorable effect of high pressure on some enzymatic reactions can be expected.

The present authors and coworkers started the study on the supercritical fluid bioreactor in 1983, and applied for a patent in 1984 [3] after the enzyme, lipase, experimentally proved to work in SCC [4,5]. There are the other two groups of researchers who have been also interested in use of supercritical fluids as a medium for enzymatic reaction. Randolph et al. [6] studied on the enzymatic hydrolysis of disodium p-nitrophenyl phosphate to p-nitrophenol in SCC containing 0.1 vol% water. They reported that about 70% of the substrate could be hydrolyzed in 4 h under the condition of 35 °C and 10.1 MPa. The activity of alkaline phosphatase was reported not to change after 24 hours' contact of the enzyme with SCC. Hammond et al. [7] used the polyphenol oxidase adsorbed on the glass beads to oxidize p-cresol or p-chlorophenol to their corresponding o-benzoquinones in SCC (36 °C, 34 MPa) or supercritical fluoroform (34 °C, 34 MPa). The substrate conversions were reported to be 70 to 80% in either static or flow reaction, but the rapid polimerization of the products o-benzoquinones occured as similary as when the enzymatic oxidation was conducted in hexane.

The data of enzymatic reactions in the supercritical fluids are still limited, and the experimental results should be added further to evaluate SCC as a medium for enzymatic reaction.

2. Pressure Effects on Enzymes

The enzymatic reaction in supercritical fluids is conducted at a high pressure, and the effect of pressure on enzymes is briefly reviewed here.

According to the transition state theory the rate constant k is expressed below using Boltzmann's constant k_B, Planck's constant h, and gas constant R;

$$k = r\left(\frac{k_B T}{h}\right)\exp\left(-\frac{\Delta G^{\neq}}{RT}\right) = r\left(\frac{k_B T}{h}\right)\exp\left(\frac{\Delta S^{\neq}}{R}\right)\exp\left(-\frac{\Delta H^{\neq}}{RT}\right) \tag{1}$$

where $\Delta G^{\neq}$, $\Delta S^{\neq}$ and $\Delta H^{\neq}$ are transition state activation free energy, entropy and enthalpy respectively. r is transition coefficient.

The relationship derived from the thermodynamic equilibrium condition is applied to Eq. (1), and the rate constant based on the molar concentration depends on pressure as follows [8];

$$\left(\frac{\delta \ln k}{\delta p}\right) = -\frac{\Delta V^{\neq}}{RT} + \beta_0 \Sigma \upsilon_i \tag{2}$$

where υ_i are the stoichiometric coefficients, negative for reactants and positive for products. β_0 is the compressibility coefficient of the solvent, and $\Delta V^{\neq}$ is the activation volume defined as the difference between the volume of the activated state and the volumes of the initial reactants.

The effect of pressure on the enzymatic one-substrate reaction is also evaluated with the activation volume $\Delta V^{\neq}$ consisting of two parts, the volume change of the catalytic step, $\Delta V_c^{\neq}$, and the volume change of the binding step, $\Delta V_{ass}^{\neq}$, [3];

$$\Delta V^{\neq} = \Delta V_c^{\neq} + \left(\frac{K_m}{K_m + C}\right) \Delta V_{ass}^{\neq} \tag{3}$$

where C and K_m are substrate concentration and Michaelis constant respectively.

The volume changes $\Delta V_c^{\neq}$ and $\Delta V_{ass}^{\neq}$ were measured for several proteases [9], and both of them are negative in case of thermolysin, which means that the reaction can be promoted at higher pressure.

The loss of enzymatic activity is brought out with the denaturation of enzyme protein, and its kinetic can be considered thermodynamically. The free energy change of reaction or denaturation, ΔG, has been expressed as follows [8];

$$\Delta G = \Delta G_0 + \Delta V_0 (p - p_0) - \Delta S_0 (T - T_0) + \tfrac{1}{2} \Delta\beta (p - p_0)^2$$
$$+ 2\,\Delta\alpha (p - p_0)(T - T_0) - \Delta C_p \left\{T\left[\ln\left(\frac{T}{T_0}\right) - 1\right] + T_0\right\} \tag{4}$$

where ΔG_0, ΔV_0 and ΔS_0 are the changes of Gibbs free energy, volume and entropy found in the state (p_0, T_0). The coefficients $\Delta\alpha$, $\Delta\beta$ and ΔC_p are defined as follows;

$$\Delta\alpha = \left(\frac{\delta\,\Delta S}{\delta p}\right)_T = -\left(\frac{\delta\,\Delta V}{\delta T}\right)_p \tag{5}$$

$$\Delta\beta = \left(\frac{\delta\,\Delta V}{\delta p}\right)_T \tag{6}$$

$$\Delta C_p = T\left(\frac{\delta\,\Delta S}{\delta T}\right)_p . \tag{7}$$

The activation volume, $\Delta V^{\neq}$, can be expressed below if the relationship of Eq. (4) is applied to Eq. (1);

$$\Delta V^{\neq} = \Delta V_0^{\neq} + \Delta\beta(p - p_0) + \Delta\alpha(T - T_0) \tag{8}$$

The above short review of the pressure effects on enzymes can be useful when the experimental results of the enzymatic interesterification in SCC are discussed. There are, however, the other pressure effects to influence the activity of enzyme as well as its stability through the pH change because the dissociation of buffer ions and the solubility of CO_2 can vary with pressure. CO_2 may react with the residues of special amino acids possibly to induce the denaturation of enzyme protein. Those are the extra factors to be taken into account when the real mechanism of the pressure effect on enzymes is elucidated.

3. Enzymatic Interesterification of Triglyceride in SCC

The melting characteristic of triglyceride depends not only on the kinds of the component fatty acids but also on their distribution in the positions of triacylglycerol. Cocoa butter, for example, contains almost equal amount of oleic acid, stearic acid and palmitic acid, but the unsaturated fatty acid, oleic acid, is located mostly in 2-position while the saturated ones, stearic acid and palmitic acid, occupy selectively 1- and 3-position. Several food companies in the world are interested in the enzymatic interesterification to produce the cocoa butter substitute from the cheaper oils [10]. This reaction was selected to test SCC as a medium of enzymatic reaction.

3.1 Experimental Method

(1) Materials

Triolein and stearic acid were purchased from Nakarai Chemicals, Ltd., Japan. The purity of triolein was measured by analyzing the content of oleic acid changed from 64% to 73%. The purity of stearic acid was also measured to be 95%. The buffer used was 0.3 M TES (i.e. N-Tris(hydroxymethyl)methyl-2-aminoethane sulfonic acid, pH 6.5) prepared by dissolving the powdery reagent of TES (Wakoo Junyaku Co., Japan) into the deionized and distilled water. The lipase purchased from Seikagaku Kogyo Co. Ltd. was of *Rhizopus delemar* origin and its activity was 600 unit mg^{-1} solid. One unit corresponded to liberating 50 μmol equivalent of fatty acid from olive oil emulsion in the reaction period of 150 h at 30 °C. The diatomaceous earth used as a disperser was the filter aid (Manville, Celite No. 503).

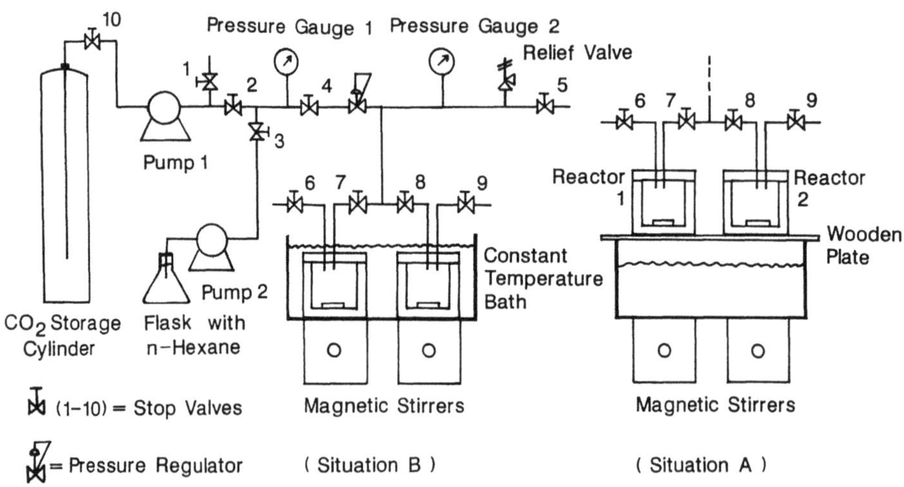

Fig. 1. Experimental apparatus of batch reaction

(2) Reaction Experiments

The experimental apparatus is schematically shown in Fig. 1. The reactor was a 40 ml high pressure vessel in which 10 or 50 mg lipase (powder), 10 to 400 μl TES buffer, 0.2 g triolein, 0.2 g stearic acid and 0.8 g Celite were loaded. The two reactors were put into the bath of constant temperature and connected in parallel through the pressure regulator to the pump which increased the pressure of liquid carbon dioxide flowing out of storage cylinder. When the pressure gauge indicated the arrival at the desired value, the stop valve was closed to shut off the feed of carbon dioxide and simultaneously the pump was switched off. The reactor was magnetically stirred during the reaction period. After the reaction the reactor was cooled with dry ice/acetone until the pressure fell down to the ambient pressure. The reactor was then opened and the contents were transferred into a beaker which was left in the room for sublimation of CO_2.

About 50 ml of diethyl ether was added in the reactor to recover the residual reaction mixture of substrate and product, and the solution was put into the beaker to dissolve substrate and product remaining after sublimation of CO_2. The solution was transferred into a test tube and the solvent was evaporated at 45 °C. The residue was redissolved in about 8 ml of n-hexane and the half amount of the solution was used in the analysis.

When the contact effect of SCC on the enzymatic activity was examined, a weighed amount (10–40 mg) of lipase was loaded in the reactor with 0.2 ml of 0.3 M Tris buffer (pH 5.6) and magnetically stirred for a certain length of time during contact with SCC. The lipase thus treated was recovered in the reactor with release of pressure and diluted with the buffer to be the concentration of 0.1 mg ml^{-1} for assay of its enzymatic activity.

(3) Assay

Triglyceride was separated from the reaction mixture and fatty acids were assayed with a minor modification of the method reported by Yokozeki et al. [11], and the details of the analysis had been explained elsewhere [4].

The residual triglyceride and the content of stearic acid were calculated as follows;

$$\text{Residual triglyceride} = \frac{(w_0/M_0) + (w_s/M_s)}{(w/M_0)} \tag{9}$$

$$\text{Content of stearic acid in residual triglyceride} = \frac{(w_s/M_s)}{(w_0/M_0) + (w_s/M_s)} \tag{10}$$

where w is the weight of oleic acid contained in the substrate triolein; w_0 and w_s are the weight of oleic acid or stearic acid respectively contained in the remaining triglycerides; M_0 and M_s are the molecular weight of oleic or stearic acid respectively.

The activity of the lipase exposed to SCC was assayed in the reaction where 1 ml of the diluted enzyme solution was added to the substrate solution consisting of 15 ml NaCl aqueous solution (0.01 M), 3 ml $CaCl_2$ aqueous solution (0.1 M) and 2 ml tributyrin. The reaction was conducted in a 100 ml beaker at 30 °C for 30 min, and stopped with addition of 20 ml mixture of acetone and ethanol (1:1). The amount of liberated free fatty acids was measured by titration using 0.05 N NaOH aqueous solution, and the end point of titration corresponded to pH 10.

3.2 Results and Discussion

(1) Interesterification

Figure 2 shows the effect of amount of buffer on the residual triglyceride and the content of stearic acid incorporated. The reactions were done for 6 h at 50 °C and 13.7 MPa.

The residual triglyceride decreased a little with increasing amount of buffer initially and then increased maximally to about 80%. This reverse may reflect the incorporation of fatty acids into the hydrolyzed triglyceride, diolein and monoolein. The residual triglyceride decreased again when the amount of buffer became excessive. The content of stearic acid changed with increase in the amount of buffer as similarly as the residual triglyceride. The decrease in the solubility of stearic acid may be the reason why the poorer interesterification was observed in the larger amount of buffer. This effect of amount of buffer was different from what was observed using n-hexane as a reaction medium.

The maxima of residual triglyceride and content of stearic acid shown in Fig. 2 are almost as same as those obtained in the interesterification using n-hexane [12]. The diesperser, Celite, was effective to produce good results as shown in Fig. 2 probably because the enzyme and the buffer both adsorbed were dispersed uniformly in the reactor.

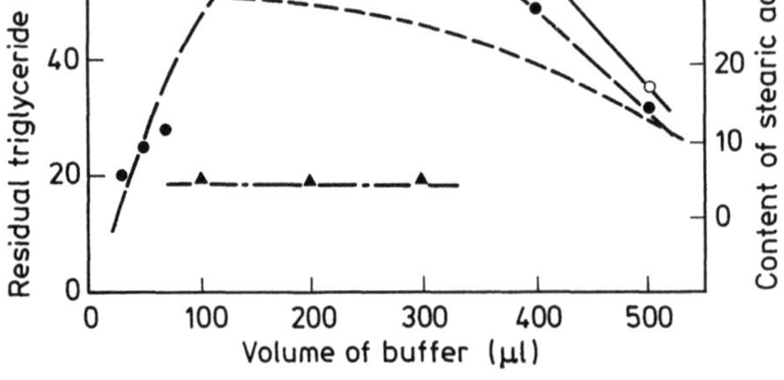

Fig. 2. Effect of buffer volume on residual triglyceride and content of stearic acid. Amount of lipase used was 50 mg and the enzymatic reaction was carried out for 6 h at 50 °C and 140 kg cm^{-2} (○, ●) with or (△, ▲) without Celite. The dotted line indicates the amount of oleic acid which remained in the unhydrolyzed triglyceride

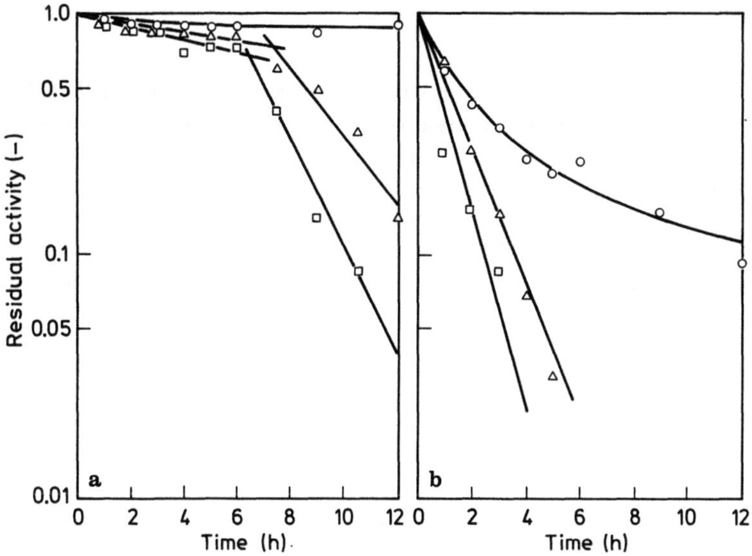

Fig. 3. Effect of pressure on stability of free lipase in SCC at **(a)** 35 °C, **(b)** 50 °C. Pressure: ○, 1 bar; △, 150 kg cm^{-2}; □, 300 kg cm^{-2}

(2) Stability of Lipase

Figure 3 shows the residual activity of the free lipase exposed to SCC for 1 to 12 h. The residual activity was calculated as the ratio of the activity of exposed lipase to the initial activity.

The lipase exposed to SCC lost its activity in two steps at 35 °C, that is, the residual activity decreased gradually during the first several hours and then declined with a larger rate constant of inactivation. At the higher temperature, 50 °C, the activity of the lipase began to decrease sharply soon after it contacted with SCC. The higher pressure could increase the concentration of carbon dioxide solved in the enzyme solution which might change the pH and react with some amino acid residues of the enzyme protein. Therefore conclusion should be drawn carefully that the activation volume of denaturation was negative, although the inactivation of the lipase was accelerated with pressure increase of SCC.

Figure 3 gives the results for the free lipase containing water more than five times as much as weight of enzyme. However, more important is the fact that the adsorbed lipase is more stable than its free form.

The activity of enzyme is related to the higher structure of protein, of which change can not be detected with gel electrophoresis. It is, however, appropriate here to refer to the study reporting that the molecular weight did not change in case of the lysozyme contacted at room temperature for 6 h with 30 MPa carbon dioxide or nitrogen gas [13].

4. Conclusions

The lipase reaction using SCC as a medium was studied, and it was found that the reaction of hydrolysis and esterification could occur in supercritical carbon dioxide as similarly as in organic solvents. Thermal effect seemed to be the main reason for inactivation of lipase in SCC, and some ill effect of carbon dioxide became less significant when the limited amount of water was used to bind the enzyme to the disperser. The experimental results so far obtained are encouraging me to accumulate the data of reaction rate and equilibrium for development of the continuous supercritical fluid bioreactor.

5. Nomenclature

C	$mol\ l^{-1}$	concentration
ΔC_p	$J\ K^{-1}$	quantity defined by Eq. (7)
ΔG	J	Gibbs free energy change
$\Delta G^{\neq}$	J	transition state activation free energy
h	J s	Planck's constant
$\Delta H^{\neq}$	J	transition state activation enthalpy
k		rate constant
k_B	$J\ K^{-1}$	Boltzmann's constant
K_m	$mol\ l^{-1}$	Michaelis constant
p	Pa	pressure
r	—	transition coefficient
R	$J\ K^{-1}$	gas constant
ΔS	$J\ K^{-1}$	entropy change
$\Delta S^{\neq}$	$J\ K^{-1}$	transition state activation entropy
T	K	temperature
ΔV	l	volume change
$\Delta V^{\neq}$	l	transition state activation volume
$\Delta V_c^{\neq}$	l	volume change of catalytic step
$\Delta V_{ass}^{\neq}$	l	volume change of binding step
0	—	standard state (suffix)
$\Delta\alpha$	$l\ Pa^{-1}$	constant defined by Eq. (5)
β_0	$l\ J^{-1}$	compressibility coefficient of solvent
$\Delta\beta$	$l\ K^{-1}$	constant defined by Eq. (6)
υ_i	—	stoichiometric coefficients

6. References

1. Randall, L. F.: Separation Sci. and Technol. *17*, 1 (1982)
2. Subramanian, B., McHugh, A.: Ind. Eng. Chem. Process Des. Dev. *25*, 1 (1986)
3. Nakamura, K., Yano, T.: Jpn. Pat. Appl. (1984)
4. Nakamura, K., Chi, Y. M., Yamada, Y., Yano, T.: Chem. Eng. Commun. *45*, 207 (1986)

5. Nakamura, K., Chi, Y. M., Yamada, Y., Yano, T.: Proc. World Congress III of chemical Engineering Vol. I 945 (1986)
6. Randolph, T. W., Blanch, W. H., Prausnitz, J. M., Wilke, C. R.: Biotech. Letters *7*, 325 (1985)
7. Hammond, D. A., Karel, M., Klibanov, A. M.: Applied Biochem. Biotech. *11*, 393 (1985)
8. Morild, E.: Adv. Protein Chem. *34*, 93 (1981)
9. Kunugi, S., Fukuda, M.: Seibutsu Butsuri (Biological Physics) *25*, 192 (1985)
10. Macrae, A. R.: in Biocatalysts in Organic Syntheses (Tramper, J., van der Plas, H. C., Linko, P. eds.), p. 195, Elsevier 1985
11. Yokozeki, K., Yamanaka, S., Takinami, K., Hirose, Y., Tanaka, K., Sonomoto, K., Fukui, S.: Eur. J. Appl. Microbiol. Biotechnol. *14*, 1 (1982)
12. Tanaka, T., Ono, E., Ishihara, M., Yamanaka, S., Takinami, K.: Agric. Biol. Chem. *45*, 2387 (1981)
13. Weder, J. K. P.: Food Chemistry *15*, 175 (1984)

4.6 Application of Affinity Adsorption in Antibiotic Production

Henry Y. Wang, Patrick Evans, Kevin Dykstra, Somesh Nigam, and Gregory S. Hyatt

Department of Chemical Engineering, The University of Michigan, Ann Arbor, Michigan 48109-2136, USA

Contents

Feedback regulation plays a significant role in controlling both primary and secondary metabolite biosyntheses. Product inhibition, repression, and product degradation are commonly observed in many bioprocesses. They can be avoided if the bioproducts can be separated during their biosynthesis. We have been investigating various engineering means to separate and concentrate bioproducts in-situ during growth without disrupting their synthesis. In this paper we also discuss the possibility of affinity adsorption using various biospecific ligands for in-situ product removal. Cycloheximide and thienamycin synthesis were used as model processes.

1. Introduction

Most industrial bioprocesses are limited by the end-products that are formed by the producing microorganisms. This limitation may come about in various ways. The product may elicit a general toxicity to the microorganism such as inhibition of growth. The product may also have a more specific action upon the cellular metabolism. In some cases the product may inhibit the activity of one or several enzymes that are responsible for product synthesis. In others the product may repress the synthesis of these enzymes. In any case, product limitation and product instability restrict the final product concentration and the volumetric productivity of a process and thus its economic viability.

Optimization of culture media and environmental conditions may yield only nominal increases in volumetric productivity if product limitation and instability are

Bioproducts and Bioprocesses
Editors: Fiechter, Okada, Tanner

significant. Two options are plausible for circumventing product limitations. One, the cells may be genetically engineered in order to cricumvent the limitations that are imposed upon product synthesis. Two, the product may be simultaneously removed from the culture broth as it is formed. Circumventing of product limitation via genetic means may be feasible if the limitation is by inhibition or repression mechanism. On the other hand, if a general toxicity is elicited by the product, genetic techniques will be limieted in their potential for success. The reason for this is that general toxicity is often directed towards different, nonspecific sites in the cell that are coded by different genes [4]. Furthermore, in the case where a hydrophobic product directs its action toward the plasma membrane, genetic manipulation and subsequent alteration of the membrane structure may only be practical to a limited extent and then may have other adverse side-effects. Additionally, a product-tolerant organism will not necessarily over-produce the product. For affections by product inhibition or repression, the biochemical pathways for the product and the genes that are responsible for it must be known a priori in order to circumvent the inhibition or repression mechanism. This task is especially noteworthy since the pathways for most industrial bioproducts, such as antibiotics, are not completely characterized.

The other option for circumventing product limitation and product instability is to simultaneously remove the product from the medium as it is formed. This strategy will not only alleviate end-product limitations but may also potentially decrease downstream product recovery costs. A number of methods of in-situ product removal have been proposed. For volatile products such as ethanol vacuum process has been proposed [9,20,21]. This technique is rather limited since most bioproducts are relatively nonvolatile. Membrane processes such as dialysis and ultrafiltration have been proposed but have disadvantages including membrane fouling and inability to concentrate the product [1,22,24]. The addition of solid adsorbents [3,8,10,25] or liquid extractants [13] directly to the medium have also been proposed; however, various problems are also associated with these techniques [19,26].

Another major problem of using non-specific adsorption or extraction methods has been that the desired products are usually concentrated with other contaminants of similar physico-chemical characteristics. Thus, subsequent isolation and purification become more difficult to achieve. One way to minimize these problems is to use affinity adsorption during bioprocesses. The term affinity adsorption or affinity chromatography is generally ascribed to a process of purification utilizing an immobilized molecule (the ligand) which has a specific affinity or attraction for the product to be purified. In the case of purification of a specific bioproduct, a variety of ligands (covalently bound to a support matrix) that exhibit specific and reversible binding affinity to this bioproduct can be used. The crude mixture is then passed through the matrix with the desired product being retained as it binds to the ligand. After the contaminants have been washed away, the bioproduct can be specifically eluted by appropriately altering properties of the wash solution such as pH, ionic strength, or temperature, or by adding a molecule recognized by the protein. This may result in a highly purified bioproduct of the order of several thousand-fold.

The basic concept is not new. Campbell and his co-workers [6] described the use of solid-phase immuno-adsorbents for the purification of antibodies in 1951. It was not until 1968, however, that widespread interest in the technique was generated. It was then that Cuatrecasas et al. [7] reported outstanding successes in enzyme purification with the process they called affinity chromatography. Many purification successes, theoretical studies, and review articles [28,29] dealing with the subject have been reported. Occasionally, affinity chromatographic procedures have been reported capable of purifying a protein from cell-free extract to apparent homogeneity in a single step. The technique has also been applied to larger scale enzyme purification. Robinson et al. [23] have reported the use of an affinity column which is operated automatically and contains 1.8 liters of gel. Industrial applications of affinity adsorption is steadily increasing due to the availability of numerous monoclonal antibodies and other affinity ligands.

It must also be emphasized that the technique as normally practiced and as described above is not a chromatographic process in the traditional sense. The column is instead operated as an adsorption column similar to an ion-exchange or activated carbon column. The term "biospecific adsorption" as suggested by Barry and O'Carra [5] is thus much more appropriate for the process. Also, most affinity adsorption processes are used for purification purposes rather than for the primary isolation step.

Not many articles have addressed the use of affinity adsorbents for primary separation purposes. Solid adsorbents are easily fouled and, because of their small size, they are difficult to recover from culture broths, therefore, expensive adsorbents cannot be used. Work is currently being done in our laboratory on circumventing these problems by immobilizing or encapsulating adsorbents in hydrogel matrices [19,26].

2. Model Systems

Cycloheximide and thienamycin, two structurally unrelated antibiotics produced by different strains of *Streptomyces* have been used in our studies. Even though little is known about the biosynthesis of cycloheximide, its production has been demonstrated to be feedback regulated, and thus, continuous removal of the antibiotic from the culture medium leads to tremendous increases in production [17,27]. Thienamycin is a new β-lactam antibiotic produced by *Streptomyces cattleya* [16]. It was found to be the most potent broad spectrum antibiotic discovered in the recent years. Thienamycin is also found to be a β-lactamase inhibitor and therefore resistant to enzymatic hydrolysis [15]. Unfortunately, thienamycin is found to be extremely sensitive to pH, metallic ions, nucleophilic reagents, and even high concentration of thienamycin itself [15]. Only chemically synthesized thienamycin derivative has been commercialized. Innovative approach to increase concentration of this process is needed to overcome the inherent low concentration and instability problem plaguing the biosynthesis of this antibiotic. We have been investigating the possibility of using affinity extraction during thienamycin synthesis to enhance productivity and stability.

3. Materials and Methods

3.1 Microorganism and Culture Media

The process and on-line extraction conditions of cycloheximide process have been described previously [27].

Streptomyces cattleya NRRL-8057 was also used in this study. Process inocula were prepared in a seed medium consisting of glucose, 10 g; yeast extract 10 g; KH_2PO_4, 182 mg; Na_2HPO_4, 190 mg; $MgSO_4 \cdot 7\,H_2O$, 50 mg; distilled water, 1 l. The pH of the medium was adjusted to 6.5. The medium consists of: corn meal, 20 g; yeast extract, 10 g; soy flour, 15 g; sodium citrate, 4 g; $CaCl_2 \cdot 2\,H_2O$, 0.5 g; $MgSO_4 \cdot 7\,H_2O$, 0.1 g; $CaCl_2 \cdot 16\,H_2O$, 0.01 g; $FeSO_4 \cdot 7\,H_2O$, 0.01 g; distilled water, 1 l.

3.2 Thienamycin Process

A 250 ml Erlenmeyer flask containing 50 ml of the seed medium was inoculated with *Streptomyces cattleya* from an agar slant and incubated for 48 h on a shaker (250 rpm) at 25 °C. After this incubation, 5 ml of this seed culture were used to inoculate shake flasks containing 50 ml of the medium. All media were sterilized at 121 °C for 20 min. The shake flasks were also incubated at 25 °C on a shaker until harvested. Aliquots were removed from the shake flasks for pH and cell measurements. 10 ml of well-mixed spent medium were placed into graduated conical centrifuge tubes and spun at 2400 rpm for 20 min. The packed cell volume (PCV) thus obtained would be used as index of growth. The supernatant was saved for antibiotic assay.

3.3 Thienamycin Assay Procedure

Thienamycin content of samples were determined by a microbiological assay procedure using *Staphylococcus aureus* NRRL B313 as the indicator organism [15]. For each assay, 100 l of supernatant sample after proper dilution, was pipetted onto

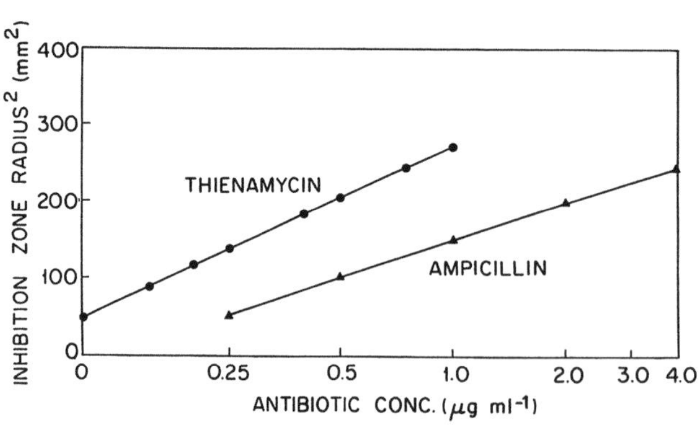

Fig. 1. Calibration curves of ampicillin and thienamycin using microbiological assay

a 0.5 inch diameter sterile paper disc, which was then placed onto the surface of the seeded agar. These plates were then inverted and incubated for 24 h at 37 °C. Any resulting inhibition zone size was measured with a Fischer-Lilly zone reader. The zone sizes were then correlated with a standard curve produced with various ampicillin standards. The thienamycin concentration was then expressed as equivalent units of ampicillin and could be converted to thienamycin concentration using a thienamycin standard curve (Fig. 1).

3.4 Thin Layer Chromatography

Thin layer chromatography was used to isolate and detect thienamycin from other organic contaminants in the supernatant. The silica gel plates were developed in a solvent chamber saturated with 70:30 (ethanol:water) mixture according to Kahan et al. [15]. Thienamycin was detected as a yellow spot using ethanolic solution of ninhydrin at an Rf value between 6.2 and 6.7.

4. Results and Discussion

4.1 On-Line Extraction of Cycloheximide

Feedback regulation of cycloheximide biosynthesis has been well documented by Kominek [17]. Our work has involved alleviation of the feedback regulatory effects of this antibiotic using in-situ product removal by adsorption onto immobilized neutral polymeric resin such as XAD-4 (Rohm and Haas, Philadelphia, Pa.) (Table 1). This work has included design and construction of a computer controlled bioreactor with adsorber system for on-line processing [20], computer modelling studies of the regulatory mechanisms controlling cycloheximide biosynthesis [11], characterization of the adsorption process, and engineering studies using mathematical models of the complete process system [12].

In designing a process system for in-situ product recovery, there are several important considerations. First, when free adsorbents are added directly to whole reaction

Table 1. Comparison of the performance of various adsorbents on cycloheximide biosynthesis

Type of resin	Addition time (log)	Total bioactivity[a] ($\mu g\ ml^{-1}$)	Residual medium[a] activity ($\mu g\ ml^{-1}$)
None	—	(675)	710
XAD-2	48	1090	195
XAD-4	48	1190	~10
XAD-7	48	1350	370
XE-348	48	1050	390
252	48	1210	640
XAD-4	140	880	250
XAD-7	140	650	360

[a] Activity at harvest time (log 144)

mixture, they are subject to fouling, resulting in reduced adsorption rates and resin capacity. Second, the small resin particles (~0.5 mm in diameter) are difficult to recover from the viscous culture medium. These problems were overcome in our laboratories by immobilizing the resin particles in a porous matrix [26]. In this manner, the capacity of the resin particles for cycloheximide was unaffected by cell debris and other components, such as antifoaming agents, and the rate of adsorption was comparable to rates observed in aqueous solution. In addition, the immobilized resin particles could be easily recovered from the culture medium and could be more completely regenerated for reuse than freely suspended adsorbents. Most of the polymeric resins are semi-specific to cycloheximide at best. Therefore, the purity of the adsorbed cycloheximide is not good and usually they are heavily contaminated with other molecules and similar physico chemical characteristics. Affinity extraction seems to be more appropriate for on-line extraction. We decided to investigate the use of affinity extraction in our thienamycin model process.

4.2 On-Line Extraction of Thienamycin

The time profile of a thienamycin process using *Streptomyces cattleya* NRRL-8057 is shown in Fig. 2. Cell growth occurred within the first 24 h. Antibiotics started to appear at 24 h and it reached a peak around 72–96 h. Degradation started to occur quickly afterwards, even though the pH was maintained around 6.5. This rapid degradation, so far, cannot be minimized through pH adjustment or other environmental manipulation.

We are interested in using affinity adsorption technique on-line with the thienamycin synthesis so that product inhibition and degradation can be minimized. The initial

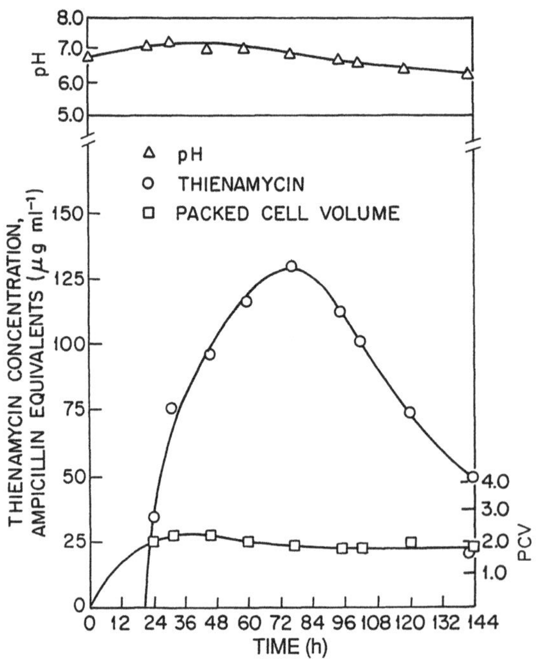

Fig. 2. Time profile of a thienamycin process. The peak is reached at 80 h. Degradation occurs rapidly afterwards

step here is to select a suitable ligand for affinity adsorption of thienamycin. The ligand must exhibit specific and reversible binding affinity for thienamycin and secondly it must be in a form that can be recovered easily later on.

In the cytoplasmic membrane of various bacterial cells, there are a number of proteins called Penicillin Binding Proteins (PBP) that bind specifically with the β-lactam antibiotics [2]. Different β-lactam antibiotics have different affinities for these PBP's. Some of these PBP's have been identified and purified. Therefore, it is possible for us to use these PBP's, especially the ones that exhibit high affinities for thienamycin as our specific ligands. Initially, we can use bacterial cells that contain these PBP's as ligands to demonstrate its feasibility. As a general rule, *Bacillus* supp. for example, has been shown to bind up to 3–10 g of penicillin per g of dry cells [18]. This loading capacity also depends on the physiological state of the cells [18]. Several researchers have reported reversible binding of penicillin antibiotics to various binding proteins on many other bacterial cells such as *Escherichia coli*, *Staphylococcus aureus* and *Streptococcus faecalis* cells. By immobilizing suitable bacterial cells in some form of gel matrix, it is conceivable to use these immobilized bacterial cells as a means to concentrate and purify thienamycin from the culture medium.

Adsorption of thienamycin to *Bacillus stearothermophilus*, a thermophilic bacterium is investigated. The reason we choose thermophilic bacteria is that because they generally do not grow at the process temperature and therefore, do not compete well for nutrient consumption if we want to add these bio-extractants directly to the reaction mixture.

Thienamycin is also a β-lactamase inhibitor [14]. Therefore, β-lactamase can also be used as the specific ligand for affinity adsorption of thienamycin. β-lactamase (Sigma, Mo) is first immobilized onto cyanogen bromide activated sepharose 4B beads (Pharmacia, Uppsala, Sweden). Estimated immobilized β-lactamase activity is about 0.5 mole penicillin per min per ml of beads. The immobilized β-lactamase can be used to test the "binding" ability of thienamycin molecules from the reaction mixture.

The adsorption of thienamycin from reaction mixtures onto thermophilic bacterial cells and immobilized β-lactamase is shown in Figs. 3 and 4. It shows that all these

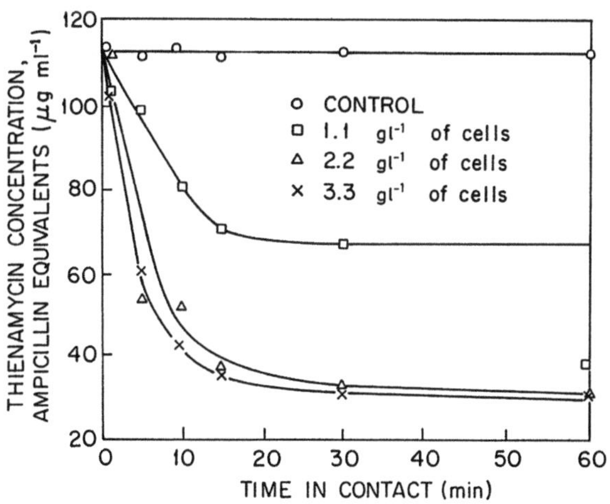

Fig. 3. Adsorption of thienamycin by *B. stearothermophilus*

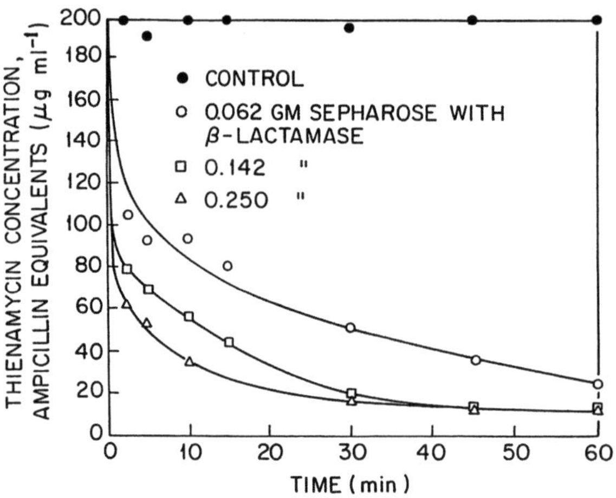

Fig. 4. Adsorption of thienamycin by immobilized β-lactamase

adsorbents can be utilized to adsorb thienamycin from the medium in 1–2 h. The adsorption was determined by measuring the decrease in bioactivity in medium using the disc-diffusion assay. It appears that bacterial cells, as used, are much more effective as an adsorption agent than other non-specific adsorbents such as XAD-2 resin. As shown in Table 2, the bacterial cells have a loading capacity 100 times higher than the resin on a weight basis. Immobilized β-lactamase beads are also more effective than polymeric resins (Table 2).

The disappearance of the thienamycin from the medium is not due to destruction or deactivation. By adding pyridine into the supernatant (up to 2 vol % of final concentration), thienamycin can be fully recovered to 90–100% of the normal concentration. In the case of using immobilized β-lactamase as an adsorbent, aqueous pyridine is not effective in re-extracting thienamycin out indicating very strong binding or destruction.

We can also use other β-lactam antibiotics such as ampicillin to compete for the active sites and therefore displacing the immobilized thienamycin. A TLC plate is used to demonstrate this possibility. After addition of bacterial cells or immobilized β-lactamase, the thienamycin spot (Rf = 0.65) generally disappeared. After addition of ampicillin to the solution, we see an increase in thienamycin spot which indicates

Table 2. Loading capacities of various adsorbents of thienamycin

Types of adsorbent	Eq. conc. C_{eq} [a] μg ml^{-1}	Loading capacity mg[a] g^{-1} adsorbent
XAD-2 resin	40–100	0.1–0.8
Immobilized β-lactamase	10–50	2–8
B. stearothermophilus cells	40–90	40–95

[a] Ampicillin equivalent

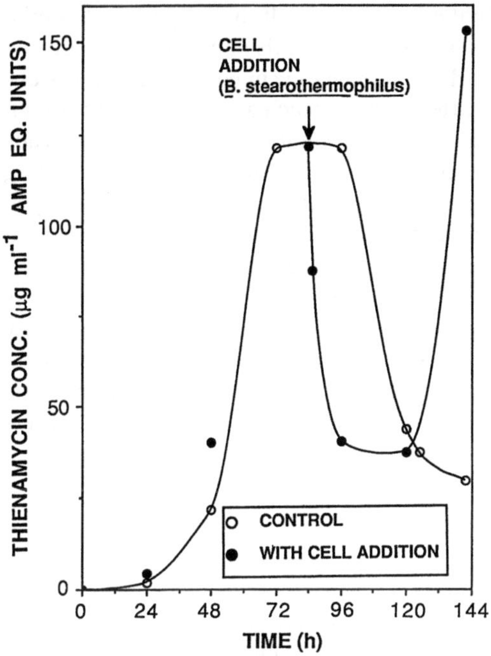

Fig. 5. In-situ addition of *B. stearothermophilus* into a thienamycin process

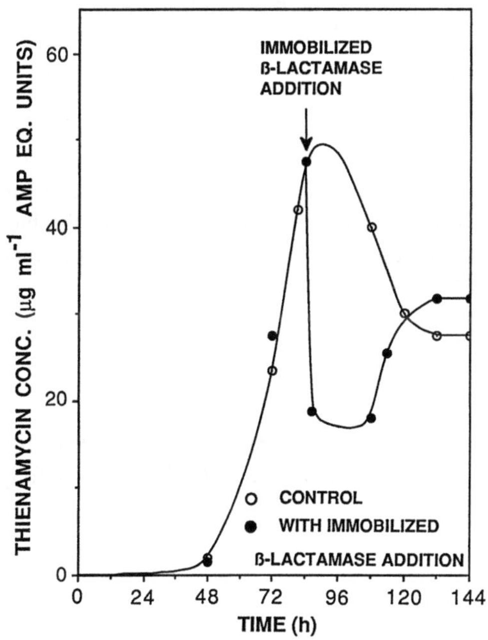

Fig. 6. In-situ addition of immobilized β-lactamase into a thienamycin process

that ampicillin in high concentration may replace the thienamycin from the active sites. We also used other β-lactam antibiotics in this experiment as control.

Since thienamycin can be extracted out of the culture medium using bioadsorbents such as bacterial cells and immobilized β-lactamase, we are interested to find out if this can be achieved during antibiotic biosynthesis. In one experiment, sterile preparation of *B. stearothermophilus* cells was added to a thienamycin process during its peak synthesis (Fig. 5). An immediate drop in thienamycin concentration was observed. The antibiotic synthesis resumed afterwards to reach a maximum higher than the control run. A similar experiment was done using immobilized β-lactamase beads (Fig. 6). An immediate drop in thienamycin concentration was observed indicating adsorption with antibiotic synthesis still occurred afterwards. Again, the final antibiotic concentration using in-situ extraction is higher than the control run. In both cases, we were unsuccessful in increasing final thienamycin concentration by adding pyridine solution afterwards. This is probably due to the presence of excess proteinous materials in the spent medium interfering with the extraction. Currently we are collaborating with Prof. Tadayuki Imanaka at Osaka University (Osaka, Japan) to do protein engineering on these PBP and β-lactamases for the purpose of affinity adsorption to optimize the adsorption and release kinetics and direct recovery of β-lactam antibiotics.

5. Conclusion

The dilute nature of the products produced in various bioprocesses is primarily due to the product regulation and toxicity to the producing cells. Both genetic manipulation and engineering improvement can be used to help alleviate these problems. We have been investigating the possibility to remove the inhibitory bioactive products, such as antibiotics, continuously during the production phase. In the case of cycloheximide biosynthesis, on-line extraction using both nonspecific resin adsorption and solvent extraction can adequately be used to achieve higher product yields. The problem associated with the non-specific nature of the bioseparation can be minimized by immobilizing the adsorbents in reversible hydrogel matrices such as calcium alginate. Another viable alternative is to use biospecific adsorbents during the bioprocess. For thienamycin synthesis, our preliminary results indicate that this approach seems feasible and should be investigated further.

6. Acknowledgement

This work was supported in part by NSF Research Grants CBT-84-06686 and ECE-86-12969. Typing and editorial assistance by Cynthia Miller is gratefully acknowledged.

7. References

1. Abbott, B. S. and Gerhardt, P.: Biotech. Bioeng. *12*; 577 (1970)
2. Abraham, E. P.: Scientific American *6*, 76, (1981)
3. Bahadur, K. and Saroj, K. K.: Japan J. Microb. *4*; 43 (1960)
4. Baronofsky, J. J., Schreuis, W. J. A., and Kashket, E. R.: Appl. Environ. Microbiol. *48*; 1134, (1984)

5. Barry, S. and O'Carra, P.: Biochem. Journ. *135*, 595, (1973)
6. Campbell, D. H., Leuscher, E., and Lerman, L. S.: Proc. Natl. Acad. Sci. USA *37*, 575, (1951)
7. Cuatrecasas, P., Wilchek, M., and Anfinsen, C. B.: Proc. Natl. Acad. Sci. USA *61*, 636–643, (1968)
8. Cho, D. Y. and Jahnke, H. K.: U.S. Patent 3, 515, 717, (1970)
9. Cysewski, G. R. and Wilke, C. R.: Biotech.Bioeng. *19*; 1125, (1977)
10. Denkewalter, R. G. and Gillin, J.: Anslegeshrift 1062891, (1959)
11. Dykstra, K. and Wang, H. Y.: Annals of the New York Academy of Sciences, *506*: 511–522, (1987)
12. Dykstra, K., Li, X-M. and Wang, H. Y.: Biotech. Bioeng. *32*, 356 (1988)
13. Finn, R. K.: J. Ferment. Technol. *44*: 305, (1966)
14. Georgopapadakou, N. H. and Liu, F. Y.: Antimicrob. Chemotherap. *18*: 1, 148—157, (1980)
15. Kahan, J. S., Kahan, F. M., Goegelman, R., Currie, S. A., Jackson, M., Stapley, E. O., Miller, A. K., Hendlin, D., Mochales, S., Hernandez, S., Woodruff, H. B., and Birnbaum, J.: J. Antibiotics, *32*, 1, (1979)
16. Kahan, J. S., Kahan, F. M., Goegelman, R., Currie, S. A., Jackson, M., Stapley, E. O., Miller, T. W., Miller, A. K., Hendlin, D., Mochales, S., Hernandez, S., Woodruff, H. B., and Birnbaum, J.: U.S. Patent 4, 081, 548, (1978)
17. Kominek, L. A.: Antimicrob. Chemotherap. *7*: 856–860, (1975)
18. Lawrence, P.: J. Bacteriol., *108*, 662, (1971)
19. Nigam, S. C. and Wang, H. Y.: ACS Symp. Ser. 314, 151 (1986)
20. Payne, G. F.: Ph. D. Thesis; The University of Michigan, 1984
21. Ramalingham, A. and Finn, R. K.: Biotech. Bioeng. *19*: 583, (1977)
22. Raymond, R. L., Jamison, V. W., and Hudson, J. O.: Appl. Microbiol. *17*: 512, (1969)
23. Robinson, P. J., Wheatley, M. A., Janson, J. C., Dunnhill, P., and Lilly, M. D.: Biotechnol. Bioeng. *16*, 1103, (1974)
24. Schultz, J. S. and Gerhardt, P.: Bacteriol. Review, *33*: 1, (1969)
25. Tone, H., Kitai, A., and Ozaki, A.: Biotech. Bioeng. *10*; 689, (1968)
26. Wang, H. Y. and Sobnosky, K.: ACS Symp. Ser. *271*, 123, (1985)
27. Wang, H. Y.: Annals of the New York Academy of Sciences, *413*, 313, (1984)
28. Wankat, P. C.: Analytical Chem., *46*, 11, 1400, (1974)
29. Whitesides, G. M. and Nishikawa, A. H.: Affinity Chromatography in Applications of Biochemical Systems in Organic Chemistry. Edited by J. B. Jones, C. J. Sih, and D. Perlman. pp. 929–968, Wiley Interscience, 1976

4.7 Application of Cross-Flow Filtration for High Density Cultivation and Regulation of *trp* Promoter

Takeshi Kobayashi

Nagoya University, Nagoya 464-01, Japan

Abstract

A new fermentation system with continuous separation by cross-flow filtration using a ceramic filter was developed. By applying this system, high concentration cultivation of lactic acid bacteria and regulation of *trp* promoter for induction of a foreign gene by adjusting tryptophan concentration in a medium were carried out.

Streptococcus cremoris, *Lactobacillus casei* and *Bifidobacterium longum* were grown in the new fermentor up to high concentrations of 81, 49 and 54 g dry cell weight per liter, respectively. The cell productivities achieved by the fermentation system were 5 to 15 times as high as that of each corresponding conventional batch cultivation.

Escherichia coli harboring the recombinant plasmid containing β-galactosidase gene fused to *trp* promoter was cultivated in the fermentor with cross-flow filtration. In the initial stage of the cultivation, tryptophan concentration in a medium was kept at 170 μg ml^{-1}. The cross-flow filtration was started when optical density of cells at 570 nm became 10. Tryptophan concentration dropped to low level within 30 min and an efficient expression of β-galactosidase gene was started. By this two stage cultivation, very high cell mass was achieved (60 g dry cell weight per liter) and amount of produced β-galactosidase was about 10% of total cellular protein.

Bioproducts and Bioprocesses
Editors: Fiechter, Okada, Tanner

4.8 Extraction of Volatile Fatty Acids from Spent Medium with a Supported Liquid Membrane

Takuo Yano, Pakorn Nuchnoi, Naomichi Nishio, and Shiro Nagai

Department of Fermentation Technology, Faculty of Engineering, Hiroshima University, Shitami, Saijo-cho, Higashi-Hiroshima 724, Japan

Contents

Experimental and analytical studies on extraction of volatile fatty acids (VFAs) from aqueous solution using a supported liquid membrane (SLM) were carried out. Teflon soaked with 20% (w/w) of tri-n-octyl phosphine oxide (TOPO) in kerosene was used as a supporting membrane. The extraction rate of VFAs transferred from acid source across the SLM to the sink was measured. It was observed that only undissociated form of VFAs can penetrate into the SLM and the complex formation between TOPO and VFA (1/1 molar ratio) inside the membrane enhanced a flux of VFA across the membrane. This phenomenon was analyzed by the mathematical models for transfer of permeate across the SLM by carrier-facilitating mechanism. In this extraction system, diffusion of the VFA-TOPO complex inside the membrane was most probably the rate-limiting step for the transfer of VFAs.

1. Introduction

1.1 Objectives

In anaerobic wastewater treatments, volatile fatty acids (VFAs) are intermediately produced by acidogens, finally converted to methane by methanogens. These acids severely inhibit the growth of acidogens together with other microorganisms concerned.

Here, continuous extraction of VFAs from spent medium by means of a supported liquid membrane (SLM) system was attempted to develop a bioconversion process for the production of VFAs from biomass resources.

Bioproducts and Bioprocesses
Editors: Fiechter, Okada, Tanner

1.2 Supported Liquid Membrane (SLM)

There have been many attempts to select carriers capable of facilitating the transport of objective substance through a SLM [1]. However, the application of membrane in this area has not been well distributed compared to electrolysis, electrodialysis, ultrafiltration and reverse osmosis. These transport phenomena through SLM were probably first observed and studied in physiological processes of living systems involving diffusional transport of oxygen in hemoglobin and myoglobin solution [2,3]. This remarkable process has attracted a number of researchers to utilize it for industrial applications. This led to the introductory use of synthetic reactive membranes soaked in suitable solution for gas separation [4-7]. The transport phenomena of metals through SLM have been also studied [8-12].

Although the utilization of SLM for removing desired chemicals was generally possible by selecting a suitable carrier, solvent and supporting membrane, almost all of the studies done so far on SLM were metal ions extraction or gas separation. The application of this process for VFAs has not been progressed [13], probably due to the absence of effective carriers (extractants). However, several efforts for the extraction of VFAs have been done to recover VFAs from dilute concentration of industrial aqueous waste and from spent medium for economic point of view [14-19]. In a recent review on extraction chemistry of the acids, the liquid-liquid extraction of organic acids employing different solvents and effective extractants has been exhaustively discussed [19].

Thus, it was aimed in this work to study whether or not SLM becomes another strategic means to extract VFAs. The characterizations of the transport of VFAs across SLM containing TOPO (tri-n-octyl phosphine oxide) in kerosene were investigated and mathematical models to predict the process behaviour were discussed.

2. Extraction of VFAs by Liquid-Liquid Extraction System

The liquid-liquid extraction experiments were carried out to elucidate the important role of TOPO-kerosene solution for the extraction of VFAs and to investigate the reaction mechanism between VFAs and TOPO. Several concentrations of VFA in water were vigorously shaken with equal volume of organic solvent which consists of 20% w/w tri-n-octyl phosphine oxide (TOPO) in kerosene or TOPO-free kerosene in separatory funnels. After separating the aqueous phase from the funnel, the VFA concentration in aqueous phase was measured, whereas the VFA in organic phase was re-extracted with 1 M NaOH solution in separatory funnel to measure the extracted VFA in alkaline solution. From the VFA content both in organic and aqueous phases, the equilibrium curves were depicted by usual method [20].

Formic acid was determined colorimetrically by Lang's method [21,22] while the other fatty acids were determined by a gas chromatography under the described conditions [23].

Figure 1 shows the equilibrium distribution curves of VFAs between aqueous and kerosene phases (without and with 20% w/w TOPO). The reaction schemes can be expressed as follows.

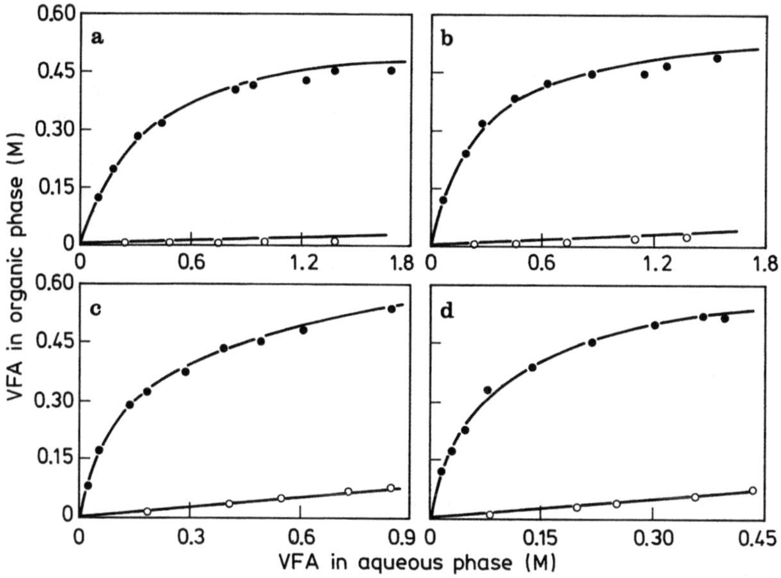

Fig. 1a–d. Equilibrium distribution curves of the tested volatile fatty acids (VFAs) between aqueous and organic phases. Organic phases: ●, 20% w/w TOPO in kerosene; ○, TOPO-free kerosene. Tested VFAs: **a** formic acids; **b** acetic acid; **c** propionic acid; **d** *n*-butyric acid

$$\underset{C_{1aq}}{HA_{aq}} \rightleftarrows \underset{C_4}{H^+} + \underset{C_5}{A^-} \qquad K_a = C_4C_5/C_{1aq} \tag{1}$$

$$\underset{C_{1aq}}{HA_{aq}} \rightleftarrows \underset{C_{1or}}{HA_{or}} \qquad K_p = C_{1or}/C_{1aq} \tag{2}$$

$$\underset{C_{1or}}{HA_{or}} + \underset{C_2}{TOPO} \rightleftarrows \underset{C_3}{Ha \cdot TOPO} \qquad K_c = C_3/C_2C_{1or} \tag{3}$$

where K_a, K_p and K_c are dissociation constant, partition coefficient and complex formation constant respectively. C_4 and C_5 are molar concentration of H^+ and A^- in aqueous phase. C_2 and C_3 are concentration of TOPO and HA · TOPO complex in organic phase, and C_{1aq} and C_{1or} are concentration of undissociated acid in aqueous and organic phases respectively.

From Eqs. (1) to (3) the relationship describing the distribution of acid between immiscible phases could be written as:

$$\log (C_3/C_2) = \log (C_{1aq}) + \log (K_cK_p) . \tag{4}$$

Based on Eq. (4) log-log plotting of C_3/C_2 against $C_{1\,aq}$ gave straight lines with slope equal to 1 (see Fig. 2) indicating that acid-TOPO complex consists of equi-molar basis. From the data of TOPO-free kerosene in Fig. 1, K_p values were estimated from each slope of straight line. K_c values were estimated from each value of intercept [K_cK_p in Eq. (4)] substituting K_p value into K_cK_p.

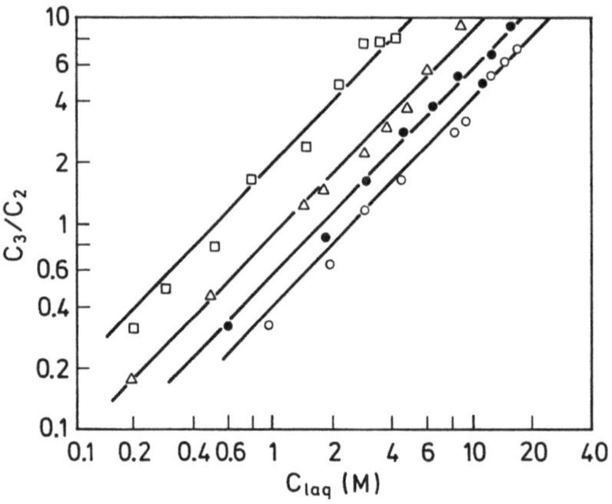

Fig. 2. A log-log plotting of concentration ratio of acid-TOPO complex to TOPO vs concentration of undissociated acid in aqueous phase. Symbols: ○, formic acid; ●, acetic acid; △, propionic acid; □, *n*-butyric acid. Data from Fig. 2. *Solid lines* were obtained by least square regression method based on Eq. (4) adjusting values of K_c and K_p to fit the experimental data

3. Extraction of VFAs by SLM System

3.1 Preparation of SLM

The preparation of SLM was as follows: a hydrophobic polytetrafluoroethylene membrane (PTFE, Teflon NFT-5200, Nitto Electric Industrial Co., Ltd. Osaka, Japan) (see Table 1) was soaked in organic solution (kerosene with or without 20%

Table 1. Physical properties of NFT-5200 Teflon membrane

Properties	Value	Unit
Pore size	0.2	μm
Porosity	0.7	—
Average thickness	62.5	μm
True density	2.25	$g\,l^{-1}$
Appearance density	0.675	$g\,l^{-1}$
Dry weight/area	42.5	$g\,m^{-2}$
Weight fraction of adsorbed kerosene	0.46	—

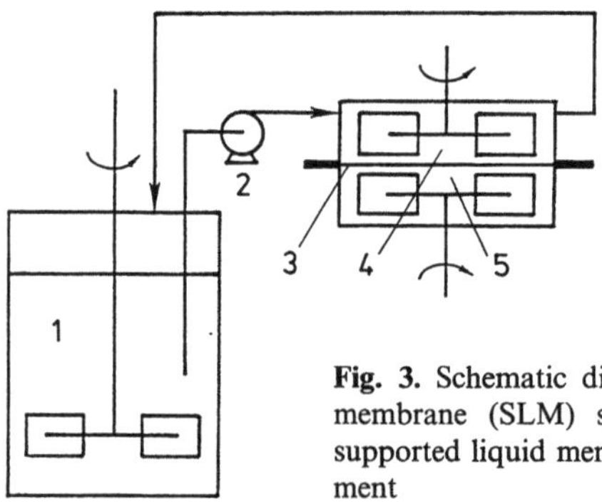

Fig. 3. Schematic diagram of experimental set up of supported liquid membrane (SLM) system: *1*, main vessel; *2*, circulating pump; *3*, supported liquid membrane; *4*, acidic compartment; *5*, alkaline compartment

w/w TOPO) at 37 °C for 1 h after draining and weighing the membrane, this was clamped between two compartments of glass vessels with stirrer on each side (see Fig. 3).

3.2 Extraction of VFAs by SLM System

The experiments on transport of VFA across the SLM was carried out as follows: as shown in Fig. 3, 1 M NaOH solution was filled up in one side, whereas VFA solution was circulated in the other side from the main vessel. The pH and temperature were controlled in the main vessel where samples were taken out to determine the remaining VFA at certain intervals.

The time courses of extraction of VFAs by SLM are shown in Fig. 4. The straight line obtained in semi-log plotting implied that the extraction rate followed a first order reaction, i.e., proportional to concentration of each acid in the bulk fluid of acid compartment. As expected, the decrease in pH increased the extraction rate, suggesting the importance of undissociated form of acid in this transfer process. To envisage extraction mechanism as a function of pH, the slopes of each line in Fig. 4 were reassessed by the relative mass flux of the acid across the membrane and these values were plotted against (pH — pK_a) based on the modified Henderson-Hessel-

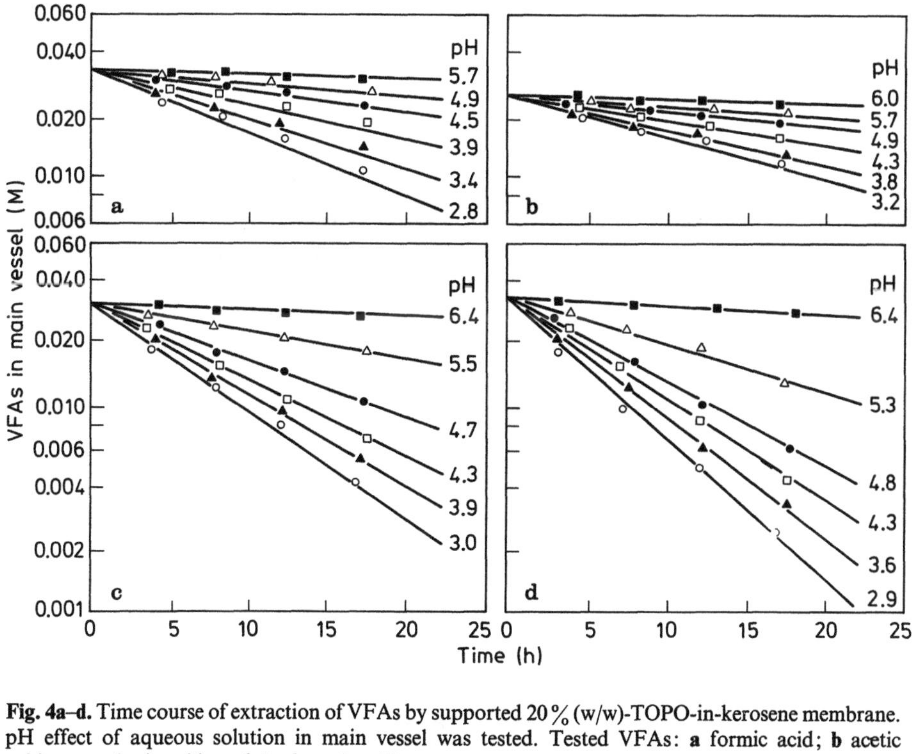

Fig. 4a–d. Time course of extraction of VFAs by supported 20% (w/w)-TOPO-in-kerosene membrane. pH effect of aqueous solution in main vessel was tested. Tested VFAs: **a** formic acid; **b** acetic acid; **c** propionic acid; and **d** *n*-butyric acid

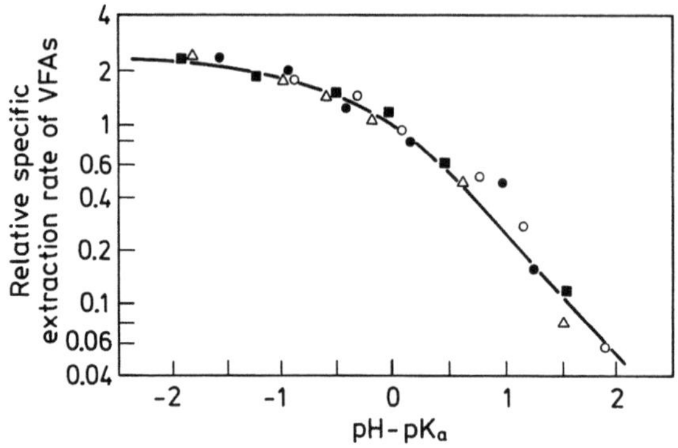

Fig. 5. Effect of (pH − pK_a) on relativc specific extraction rate of VFAs across the membrane. Relative specific extraction rates were derived from Fig. 4 setting as unity when pH = pK_a, based on Henderson-Hesselbalch equation [24], pH = pK_a + log [A^-]/[HA]. Tested VFAs: ○, formic acid; ●, acetic acid; △, propionic acid; and ■, *n*-butyric acid

balch function [24] (see Fig. 5). It was clear that effective extraction of VFAs was possible when the pH was equal to or lower than the pK_a value of VFAs.

The values of the various dimensionless parameters in the governing equations for this experiment are displayed in Table 2. Although all of them have physical meaning and affect on permeation rate of solute across the membrane (in Table 2, expressed as facilitation factor), the most important parameter, Damkohler number (γ), is the ratio of two characteristic terms, i.e., diffusion of complex molecule (L^2/D_{3mem}) to decomplexation reaction ($1/k_b$). It is pointed out [25] that the facilitated transport phenomenon requires diffusion across the liquid film while the reaction is taking place. The phenomenon of diffusion and reaction can be viewed as parallel events, for example, when a rapid reaction occurs, diffusion may be rate-limiting step, or vice versa. The value of Damkohler number from this experiment implies that characteristic time of diffusion (L^2/D_{3mem}) is about 2–10 folds of the reaction time ($1/k_b$). Therefore, diffusion is the rate-limiting step of this transport phenomenon.

Table 2. The physical parameters for the extraction of VFAs through a TOPO-kerosene supported liquid membrane (detail to obtain the value of parameters, see Fig. 7)

VFAs to be transported	Fourier number of mass transfer τ	Damkohler number γ	Mobility ratio λ	Dimensionless equilibrium constant $\varkappa$	Facilitation factor F
Formic acid	340–4400	5.9	290–380	0.40–0.30	26.7–27.6
Acetic acid	500–4600	4.8	120–160	0.37–0.28	5.8– 5.9
Propionic acid	340–3800	4.0	40– 80	0.41–0.23	3.2– 3.4
n-Butyric acid	220–1800	2.1	40– 50	0.73–0.51	2.8– 3.0

3.3 Extraction Mechanism of SLM

Figure 6 shows the extraction mechanism of SLM. This membrane was permeable to the undissociated acid (HA), but not permeable to anion (A^-). Undissociated acid made the complex with TOPO at acidic side and released TOPO at alkaline side. Because undissociated acid was dissociated to A^- and H^+ in alkaline compartment, undissociated acid concentration in acidic side became to be larger than one in alkaline side. So undissociated acid extraction was promoted.

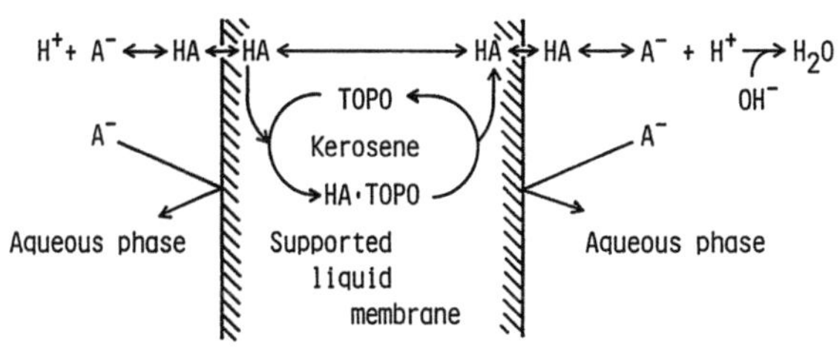

Fig. 6. Schematic diagram of extraction mechanism of SLM

3.4 Problems of SLM System

This SLM system has two problems. One is a low pH of acid phase, the other is a low diffusion rate of membrane. About the former problem, in general, as few microorganisms can anaerobically grow at low pH, this has to be isolated from abnormal environments.

About the later problem, a membrane with thinner thickness should be used for the test operation. However, the thinner thickness brought weaker resistance against mechanical strength leaving behind the problems to be solved.

4. Mathematical Model

Analytical models for transfer of permeate across the SLM by carrier-facilitation have been developed as shown in many articles. Among these, Goddard [26, 27] provided an extensive review of the theory for this type of transport. Based on the reviews [26, 27], to derive the important variables that influence upon the transfer rate of VFAs across the hydrophobic SLM, the equations below were derived.

Assuming that the undissociated form of monocarboxylic VFA (HA) combines chemically with TOPO ($mol^{-1}\,mol^{-1}$) in the membrane phase, the reaction scheme and kinetic function of net complex formation, R are given by

$$HA + TOPO \underset{k_b}{\overset{k_f}{\rightleftharpoons}} HA \cdot TOPO \tag{5a}$$

and

$$R = k_f C_{1\,mem} C_{2\,mem} - k_b C_{3\,mem} \tag{5b}$$

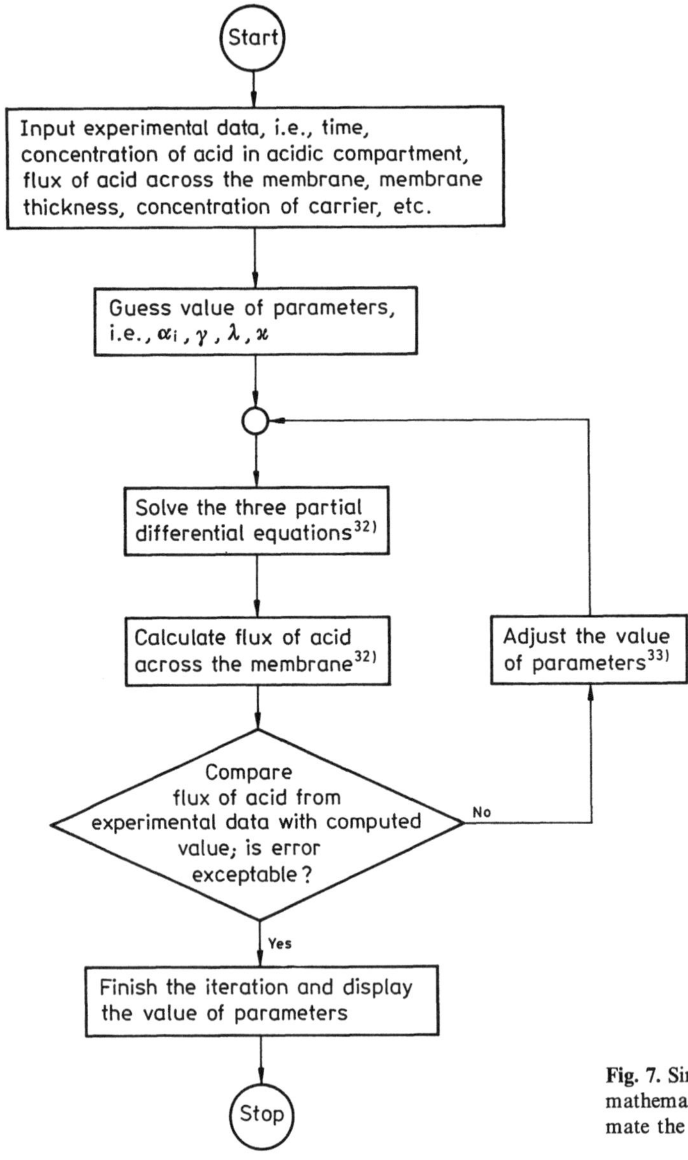

Fig. 7. Simplified algorithm using mathematical model to estimate the value of parameters

where, HA · TOPO is the acid-TOPO complex; C_1, C_2 and C_3 denote the molar concentration of HA, TOPO, and HA · TOPO respectively; k_f and k_b are the forward and backward reaction rate constants respectively; subscript mem refers to the organic phase inside the membrane.

The coupling of chemical kinetics with transient one-dimensional mass-transfer across the reactive membrane gives the three simultaneous partial differential equations which serve as the governing equations.

$$\frac{\partial C_{i\,mem}}{\partial t} = D_i \frac{\partial^2 C_{i\,mem}}{\partial x^2} + \beta_i R\,, \qquad i = 1, 2, 3 \tag{6a}$$

where D_i is diffusion coefficients of various species; t and x are time and coordinate in direction of acid transport respectively; and β_i is generation-depletion index for reaction. By nondimensionalizing, Eq. (6a) becomes

$$\frac{\partial \theta_i}{\partial \tau} - \alpha_i \frac{\partial^2 \theta_i}{\partial \zeta^2} - \beta_i \gamma \xi_i \Phi = 0\,. \tag{6b}$$

The subscripts 1, 2, and 3 mark HA, TOPO, and HA · TOPO respectively and the dimensionless variables are

$$\begin{aligned}
&\zeta = x/L && \tau = D_1 t/L^2 \\
&\varkappa = k_f C^{\delta}_{1\,mem}/k_b && \Phi = \varkappa \theta_1 \theta_2 - \theta_3 \\
&\gamma = k_b L^2/D_3 && \lambda = D_3 C^{\circ}_{7\,mem}/D_1 C^{\delta}_{1\,mem} \\
&\alpha_i = D_i/D_1 && \beta_i = \begin{array}{ll} -1 & \text{for } i = 1 \\ +1 & \text{for } i = 3 \end{array} \\
&\xi_i = \begin{cases} \lambda & \text{for } i = 1 \\ \alpha_3 & \text{for } i = 2, 3 \end{cases} && \theta_i = \begin{cases} C_{i\,mem}/C^{\delta}_{1\,mem} & \text{for } i = 1 \\ C_{i\,mem}/C^{0}_{2\,mem} & \text{for } i = 2, 3 \end{cases}
\end{aligned} \tag{6c}$$

where L is membrane thickness. $C^{\circ}_{2\,mem}$ and $C^{\delta}_{1\,mem}$ denote TOPO concentration in membrane at initial condition and free permeate (HA) concentration on the surface of the membrane respectively. In obtaining the solution of Eq. (6b), nine constraints, i.e., three from initial conditions and six from boundary conditions, are required. The initial conditions are

$$\Theta_1(\zeta, 0) = \Theta_3(\zeta, 0) = 0\,, \qquad \Theta_2(\zeta, 0) = 1 \tag{7}$$

and the boundary conditions are

$$\theta_1(0, \tau) = 1\,, \qquad \theta_1(1, \tau) = K_{p\,mem} C_{1\,si}/C_{1\,so} = 0 \tag{8a}$$

$$\frac{\partial \theta_i(0, \tau)}{\partial \zeta} = \frac{\partial \theta_i(1, \tau)}{\partial \zeta} = 0 \quad \text{for} \quad i = 2, 3 \tag{8b}$$

where subscripts si and so indicate the bulk state of permeate-sink and permeate-source aqueous solution respectively; K_{pmem} is the partition coefficient of acids between membrane and aqueous solution.

Equation (8a) is based on the assumption that the undissociated acid concentrations on both membrane surfaces are equilibrated to those in the bulk aqueous phases. Eq. (8b) implies that TOPO and HA · TOPO can only mobilize inside the membrane, but cannot penetrate out to aqueous phases.

When the free permeate molecule is much smaller than the carrier molecule, the carrier and the carrier-permeate complex molecules can be assumed to be roughly equivalent size. Hence, three of the four flux conditions in Eq. (8b) are almost independent at pseudo-steady state. It is necessary to specify further restriction which can be obtained from stoichiometric constraints. The restriction is given in Eq. (9)

$$\int_0^1 (\theta_2 + \theta_3)\, d\zeta = 1 . \tag{9}$$

The total diffusion flux of HA across the membrane can be calculated from

$$J = D_1 \left(\frac{\partial C_{1\,mem}}{\partial x}\right)_{x=0} . \tag{10}$$

To obtain this value, it is necessary to know the concentration gradients by solving Eq. (6a). In the case of simple passive diffusion transport (no carrier in the membrane), the model can be simplified as single partial differential equation. The analytical solution of this equation together with boundary and initial conditions is easily obtained by Eq. (11)

$$\theta_1 = 1 - \zeta + \sum_{n=1}^{\infty} (-2/n) \sin(n\pi\zeta) \exp(-n^2\pi^2\tau) . \tag{11}$$

On the other hand, for carrier facilitated transport (carrier exists in the membrane), the model is composed of sets of partial differential equations and can not be solved exactly [28]. Instead of this, the solutions are obtainable by either approximate analytical solutions or numerical calculations [28, 29]. The model equations similar to the above are popularly solved by finite difference approximation technique described elsewhere [30–32].

Since the reversible reaction between carrier and the solute to be transported allows for selectively increase flux of the solute across the membrane, it is fashionable to express the performance of this process in terms of the facilitation factor. This enhancement index is defined as the carrier-mediated flux divided by the flux in the absence of the carrier

$$F = \frac{(J)_{\lambda>0}}{(J)_{\lambda=0}} \tag{12}$$

where J is calculated by Eq. (10).

5. Conclusions

(1) The SLM designed for transport of VFA was prepared with Teflon membrane by soaking TOPO in kerosene.
(2) This SLM was permeable for the undissociated VFAs, but not to anion forms.
(3) The pH of VFA's compartment was a significant operating condition for transport of VPAs.
(4) The SLM used was stable after continuous operation for 2 weeks.

6. Acknowledgements

Most part of this article is a reproduction from Journal of Fermentation Technology, *65*, 30 (1987). The authors wish to thank the Society of Fermentation Technology, Japan for permission of the reproduction.

7. Nomenclature

A	transfer area of membrane, m^2
A^-	dissociated acid
C_1	concentration of HA, kmol m^{-3}
C_2	concentration of TOPO, kmol m^{-3}
C_3	concentration of HA · TOPO complex, kmol m^{-3}
C_4	concentration of H^+, kmol m^{-3}
C_5	concentration of A^-, kmol m^{-3}
D	diffusion coefficient, m^2 s^{-1}
F	facilitation factor, dimensionless
HA	undissociated acid
HA · TOPO	acid-TOPO complex
H^+	proton
J	diffusion flux across the membrane, kmol s^{-1} m^{-2}
K_a	dissociation constant, kmol m^{-3}
K_c	complex formation constant, m^3 $kmol^{-1}$
K_p	partition coefficient, dimensionless
k	reaction rate constant, m^3 $kmol^{-1}$ s^{-1} or l s^{-1}
L	membrane thickness, m
M	molecular weight, kg $kmol^{-1}$
n	integer number
R	kinetic function for genetic generation of acid-TOPO complex, kmol m^{-3} s^{-1}
SLM	supported liquid membrane
T	absolute temperature, K
TOPO	tri-*n*-octyl phosphine oxide
t	time, s
V	molar volume, m^3 $kmol^{-1}$
VFA	volatile fatty acids
x	co-ordinate in direction of acid transport, m

Greek letters

α	dimensionless diffusivity coefficient
β	generation-depletion index, i.e. +1 for generation and −1 for depletion
γ	Damkohler number, ratio of characteristic diffusion time of complex, L^2/D_3, and backward reaction time, $1/k_b$
ε	porosity of membrane, dimensionless

ζ	dimensionless coordinate along membrane thickness
Θ	dimensionless concentration
$\varkappa$	dimensionless equilibrium constant
λ	mobility ratio, complex to free-permeate
μ	viscosity of supporting liquid, N s m^{-1}
ξ	dimensionless parameter
ψ	correction factor, dimensionless
φ	dimensionless kinetic function
τ	dimensionless time or Fourier number of mass transfer
ω	diffusion tortuosity of membrane, dimensionless

Subscripts

aq	aqueous phase
b	backward reaction
f	forward reaction
i	species index
mem	membrane phase
or	organic phase
si	sink-acid solution
so	source-acid solution

Superscripts

$^{\circ}$	at initial condition
δ	at upstream face of membrane where acid solutions are contacted

8. References

1. Schultz, J. S., Goddard, J. D., and Suchdeo, S. R.: AIChE J. *20*, 417 (1974)
2. Wittenburg, J. B.: Physiol. Rev. *50*, 559 (1970)
3. Kreuzer, F. and Hoofd, L. J. C.: Respir. Physiol. *15*, 104 (1972)
4. Otto, N. C. and Quinn, J. A.: Chem. Eng. Sci. *26*, 949 (1971)
5. Otto, N. C. and Quinn, J. A.: Biotechnol. Bioeng. Symp. *3*, 311 (1972)
6. Suchdeo, S. R. and Schultz, J. S.: Chem. Eng. Sci. *29*, 13 (1974)
7. Donaldson, T. L. and Quinn, J. A.: Chem. Eng. Sci. *30*, 103 (1975)
8. Lee, K. H., Evans, D. F., and Cussler, E. L.:AIChE J. *24*, 860 (1978)
9. Takahashi, K. and Takeuchi, H.: J. Chem. Eng. Jpn. *18*, 205 (1985)
10. Harada, M., Yamazaki, F., Adachi, M., and Eguchi, W.: J. Chem. Eng. Jpn. *17*, 527 (1984)
11. Caracciolo, F., Cussler, E. L., and Evans, D. F.: AIChE J. *21*, 160 (1975)
12. Danesi, P. D.: Sep. Sci. Tech. *19*, 857 (1984–1985)
13. Ishikawa, H., Murakami, T., Hata, M., and Hikita, H.: Chem. Eng. Sci. *34*, 123 (1985)
14. Khon, P. M.: Chem. Eng. *83*, 58 (March 15, 1976)
15. Helsel, R. W.: Chem. Eng. Prog. *73*, 55 (May 1977)
16. Levy, P. F., Sanderson, J. E., Kispert, R. G., and Wise, D. L.: Enzyme Microb. Tech. *3*, 207 (1981)
17. Levy, P. F., Sanderson, J. E., and Wise, D. L.: Biotechnol., Bioeng. Symp. *11*, 239 (1981)
18. Busche, R. M., Shimshick, E. J., and Yates, R. A.: Biotechnol. Bioeng. Symp. *12*, 249 (1982)
19. Kertest, A. S. and King, C. J.: Biotech. Bioeng. *28*, 269 (1986)
20. Wardell, J. M. and King, J. C.: J. of Chem. and Eng. Data *23*, 144 (1978)

21. Lang, E. and Lang, H.: Z. Anal. Chem. *20*, 8 (1972)
22. Sleat, R. and Mah, R. A.: Appl. Environ. Microbiol. *47*, 884 (1984)
23. Nishio, N., Kitaura, S. and Nagai, S.: J. Ferment. Technol. *60*, 423 (1982)
24. Lehninger, A. L.: Principles of Biochemistry, Worth Publishers, Inc., New York 1982
25. Noble, R. D., Sep. Sci. Technol. *20*, 577 (1985)
26. Goddard, J. D., Schultz, J. S. and Bassett, R. J.: Chem. Eng. Sci. *25*, 665 (1970)
27. Goddard, J. D., Schultz, J. S. and Sucheo, S. R.: AIChE J. *20*, 625 (1974)
28. Joshi S. V., Astarita, A. and Savage, D. W.: AIChE Symp. Ser. *202*, 63 (1981)
29. Hoofd, L., and Kreuzer, F.: AIChE Symp. Ser. *202*, 123 (1981)
30. Carnahan, B., Luther, H. A. and Wilkes, J. O.: Applied Numerical Methods, John Wiley & Sons, New York 1969
31. Cherchhouse, R. F.: in: Handbook of Applicable Mathematics, (Laderman W., ed.), John Wiley & Sons, New York 1981
32. Folkner, C. A. and Noble, R. D.: J. Membrane Sci. *12*, 289 (1983)
33. Schmidt, R.: Adv. in Nonlinear Parameter Optimization, Springer-Verlag, Berlin 1980

5. Sensoring and Control of Bioprocesses

5.1 Microbiosensors for Clinical and Process Analysis

Isao Karube and Eiichi Tamiya

Research Center for Advanced Science and Technology, University of Tokyo,
4-6-1 Komaba, Meguro-ku, Tokyo 153, Japan

Contents

Microbiosensors for urea and adenosine triphosphate (ATP) are prepared with ion sensitive field effect transistor (ISFET) modified by enzymes such as urease and H^+-ATPase respectively. The Micro-urea sensor gives a linear relationship between the initial rate of the output gate voltage change and the logarithmic value of urea concentration between 16.7 and 167 mM. Determination of urea is possible within 30 s. The optimum pH is 6.0–6.5, at 37 °C. The system can be used for 20 d with only slight loss of enzymatic activity. The micro-ATP sensor can be used for the ATP determination. A linear relationship was obtained between the initial rate of change of the differential gate output voltage and the logarithm of the ATP concentration over the range 0.2–1.0 mM ATP-sensing system exhibited a response to 1 mM ATP for at least 18 d.

On the other hand, amperometric microbiosensors for glucose and glutamate are developed using an oxygen electrode and a micro-hydrogen peroxide electrode prepared by semiconductor fabrication technology. A glucose sensor consisted of a hydrogen peroxide sensor and immobilized glucose oxidase (GOD) membrane. A linear relationship was observed between glutamate concentration in the range from 5 to 50 mM and the current decrease.

1. Introduction

The determination of glucose, urea, glutamate and ATP is very important in clinical and industry process analyses. Conventional methods for their determinations are mostly based either on a direct color reaction [1,2] or on the spectrophotometric measurement of reaction products. These methods, however, involve complicated and delicate procedures and the assay times are rather long because of the several reactions involved. The spectrophotometric methods cannot be applied directly to colored samples or biological fluids. Alternative systems based on electrochemical sensors have been developed. Electrochemical sensors employing immobilized biocatalysts have definite advantages. Namely, an enzyme sensor possesses excellent selectivity for biological substrates and directly determine single compounds in a complicated mixture without need for a prior separation step. The development of

Bioproducts and Bioprocesses
Editors: Fiechter, Okada, Tanner

a simple, inexpensive assay is therefore of interest and the miniaturization of enzyme sensors is especially important for clinical analysis. Semiconductor fabrication technology has permitted the development of ISFET and micro electrodes which have been utilized as pH and enzyme-based sensors [3].

2. Micro Urea Sensor

The determination of urea in serum and other biological fluids is important for diagnostic clinical purposes. The development of simple inexpensive assays is of importance for the determination of urea in clinical work and miniaturization of the enzyme sensors is particularly important. In this study, ISFET was modified by urease-immobilized membrane and it was applied to urea determination. The ISFET device was prepared by the similar method described by Matsuo and Esashi [4]. The gate insulator of the ISFET is composed of two layers; the lower is thermally grown silicon dioxide, and the upper is silicon nitride, which is sensitive to hydrogen ions and also has a barrier effect on ion penetration. The thickness of the silicon dioxide or nitride layer is ca. 0.1 μm. Before enzyme immobilization, the organic membrane was formed over the surface of the silicon nitride by vapor deposition or by dipping. The ISFET was fixed inside a belljar and 100 μl of γ-aminopropyl triethoxysilane was vaporized at 80 °C, 0.5 Torr for 30 min, followed by 100 μl of aqueous 50% (w/v) glutaraldehyde solution vaporized under the same conditions. The ISFET thus chemically modified was stored at room temperature for ca. 24 h to complete the reaction. In the dipping method, 250 mg of cellulose triacetate was dissolved in 10 ml of dichloromethane; 200 μl of aqueous 50% (w/v) glutaraldehyde solution and 1 ml of 1,8-diamino-4-aminomethyloctane were added, and the mixture was stirred for 30 min. The tip of the ISFET, where the organic membrane had been formed over the gate insulator by vapor deposition or dipping, was immersed in 20 mg ml^{-1} urease solution at 4 °C for 24 h. In this way, urease was immobilized on the surface of the organic membrane, producing an enzyme FET.

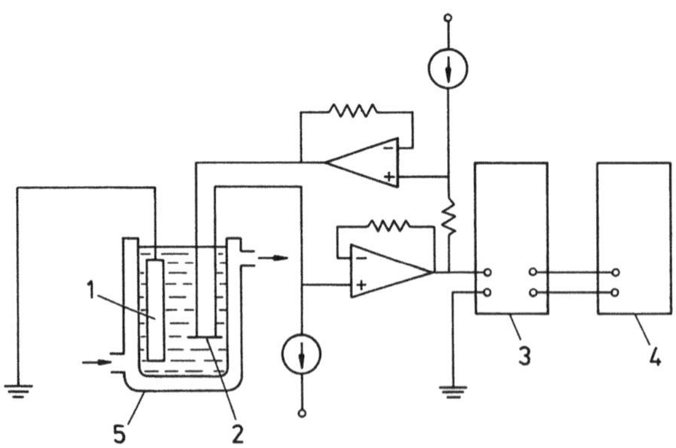

Fig. 1. Circuit diagram for measuring the gate voltage, V_g: *1*, Ag/AgCl reference electrode; *2*, treated FET; *3* electrometer; *4* recorder; *5* electrochemical cell

The circuit diagram for measuring the gate voltage (V_g) is shown in Fig. 1. The Ag/AgCl reference electrode was placed in the same solution as the enzyme FET. A gate voltage, V_g, was applied across the reference electrode and the source of the enzyme FET. The surface potential on the silicon nitride insulator of the FET is affected by solution pH, with a concomitant change in the gate voltage proportional to the change in surface potential. Therefore, the surface potential change on the silicon nitride insulator of the enzyme FET, caused by variation in pH, can be measured as a change in the gate voltage. In this case, the initial voltage applied across the source and drain was constant at 1.0 V, with a constant current of 150 μA. The gate voltage obtained was displayed on a recorder. The total volume of buffer in the cell was 4.4 ml of 5 mM Tris-HCl. The enzyme FET and reference electrode were allowed to stand in this solution at 37 °C for 10–20 min. After the gate voltage had reached a steady value, 100 μl aliquots of urea solution were injected into the cell and the change in V_g was recorded. All solutions were thermostated at 37 ± 1 °C.

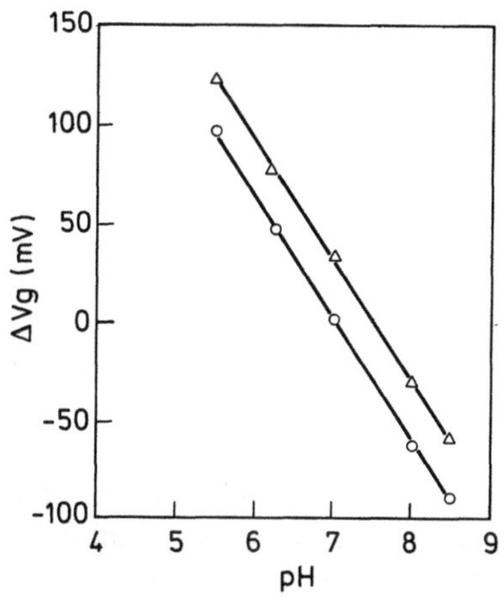

Fig. 2. Effect of the thin organic membrane on the pH response of the ISFET at 37 °C; ○, ISFET without membrane; △, ISFET with organic membrane. V_g is the difference between the gate voltage of the ISFET without the membrane at pH 7.0 and the gate voltage measured at other pH values

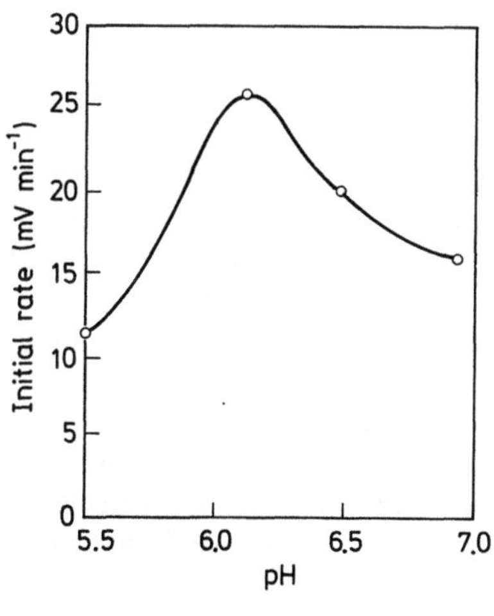

Fig. 3. Effect of pH on the initial rate of change of the gate voltage of the enzyme FET in 167 mM urea solution. The initial rates were measured 30 s after injection of the urea solution, at 37 °C

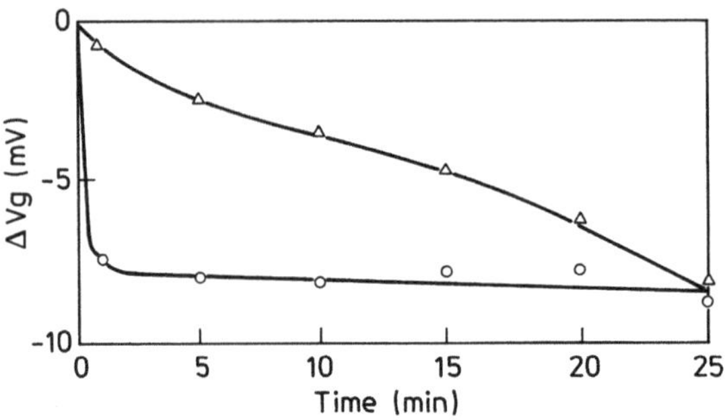

Fig. 4. Comparison of response curves for thin and thick organic membranes, for urea injected into pH 6.5 buffer solution at 37 °C, to give a final concentration of 167 mM: ○, thin membrane; △, thick membrane

Figure 2 shows the effect of the organic membrane, formed by vapor deposition over the gate insulator, on the pH response of the ISFET. The linear V_g-pH characteristics of the ISFET with and without the membrane are essentially the same over the pH range 5.5–8.5. The pH sensitivity was about 1 mV pH^{-1} at 37 °C, which is substantially in agreement with Nernstian behavior. Figure 3 shows the effect of pH on the initial rate of change of the gate voltage, 30 s after injection of urea to give a final concentration of 17 mM. The optimum pH for free urease is 6.5–7.5, but for immobilized urease the greatest activity was at pH 6. This may be attributed to an alkaline microenvironment around the immobilized urease resulting from localization of ammonium ions. Subsequent experiments were done at 37 °C and pH 5.

Figure 4 shows a comparison of the response curves to 167 mM urea for sensors with organic membranes formed by vapor deposition (giving a thin membrane) and by dipping (giving a thick membrane). The response time with the thin membrane was very rapid; that with the thick membrane was very long (many minutes). It was expected that the thickness of the membrane would affect the response time of the enzyme FET. The thin membrane was estimated to be ca. 0.1 μm thick and the thick membrane 100 μm. The time responses of the thin-membrane FET to urea addition were 30 s to achieve 85% of the steady-state value, and 1 min to achieve 95%. The initial rate of the gate voltage change 30 s after injection was plotted against the logarithmic value of urea concentration. The calibration graph for urea was linear over the range 16.7–167 mM, the initial rate being 2 mV min^{-1} at the lower extreme and 13 mV min^{-1} at the higher limit. At <16.7 mM urea, the response could not be distinguished from that for the buffer alone. Likewise, the responses to >167 mM urea were almost the same as that for 167 mM urea. The results indicate the possibility of determination of urea within 30 s. When the urea sensor was tested over a period of three weeks for 167 mM urea, the reproducibility of measurements was within 5 mV min^{-1}. Between measurements, the enzyme FET was stored at 4 °C. Over this period, the sensitivity decreased by about 10%.

3. Micro ATP Sensor

The determination of adenosine-5-triphosphate (ATP) is important in industrial process and clinical analysis and is usually based on spectrophotometry [5] or bioluminescence [6,7] these methods are rather lengthy because of the multistep reactions involved. The enzyme H^+-ATPase catalyzes the hydrolysis of ATP and orthophosphate. Several studies of its function in biological membranes have been reported [8-11]. In this communication, the preparation and characteristics of a bioelectrochemical system for ATP determination based on ISFET and H^+-ATPase, are described. The H^+-ATPase (E.C. 3.6.1.3 from thermophilic bacterium PS 3) was prepared as described by Yoshida et al. [8]. This enzyme is classified as thermophilic F 1 or TF 1. The sensor system consists of two ISFETs; one having a crosslinked polyvinylbutyral resin /H^+-ATPase membrane (ATP-sensitive FET), and the other having only the crosslinked polyvinylbutyral resin and 1 ml of 1,8-diamino-4-aminomethyloctane were dissolved in 10 ml dichloromethane; after stirring for ca. 30 min, this polymer solution was added dropwise over the gate insulator of each FET. The FETs were then immersed in 5% (v/v) glutaraldehyde solution at room temperature for 24 h to promote the crosslinking reaction of the amino group of 1,8-diamino-4-aminomethyloctane with glutaraldehyde. The H^+-ATPase was immobilized of the FET by immersing the tip into a 5 mg ml^{-1} H^+-ATPase solution at 4 °C for 16–24 h. The response of the FET to ATP was measured in differential voltage mode, by measuring the difference in gate output between the ATP-sensitive gate and the pH-sensitive gate. An Ag/AgCl reference electrode was placed directly in the solution with the two FETs. The gate voltage was measured between the Ag/AgCl reference electrode and the source of the enzyme-coated FET and the pH-sensitive FET. Variation in solution pH change the surface potential on the silicon nitride insulators of the two FET, giving rise to a proportional change in the gate output voltage. In this case, the voltage between source and drain is held constant at 4.0 V, the current between source and drain also being constant at 500 μA. The differential gate output voltage between the FETs was displayed on a recorder. The gate output voltage/pH characteristics of the FETs with and without the polyvinylbutyral was 54–61 mV pH^{-1} at 36–41 °C.

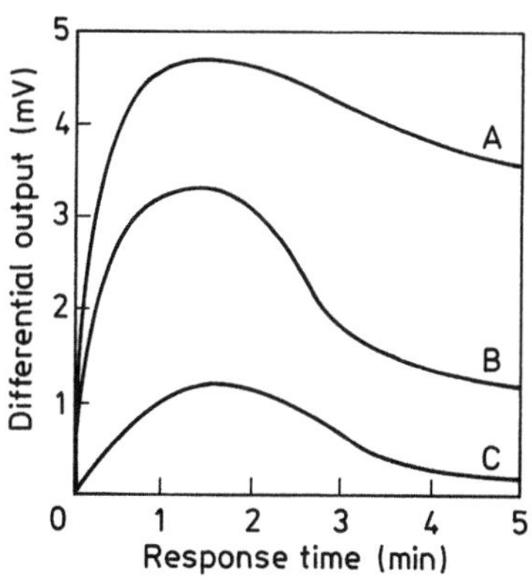

Fig. 5. Responses to ATP: *A*, 1.0 mM; *B*, 0.6 mM; *C*, 0.2 mM. ATP was injected into the buffer solution at pH 7.0, 40 °C, at time zero

Both FETs showed a linear relationships between the gate output voltage and pH over the pH range 5–10.

The temperature was held constant at 40 °C because this is the usual temperature for the H^+-ATPase assay [8,9]. The pH of greatest in biological systems, and pH 7.0 was used in subsequent experiment; the activity was then 27% of that of pH 9.0. The response of the system for different ATP concentrations are shown in Fig. 5. There is a rapid initial increase in the differential output, followed by a gradual decay. A similar response has been reported for a urea sensitive enzyme-FET [12]. It is considered that the initial rapid increase in the differential output is due to the diffusion of the hydrogen ions generated by the enzyme reaction through the membrane on the enzyme-FET. The gradual drop in the output can be attributed to immobilized enzyme causes a decrease in enzyme activity and the substrate concentration surrounding the immobilized enzyme gradually decreases because of diffusion limitations in unstirred solution. The initial rate of change in the differential gate output voltage after injection was plotted against the logarithm of the ATP concentration. A linear relationship was obtained in the range 0.2 to 1.0 mM ATP at pH 7.0. No detectable response was observed below 0.2 mM ATP compared to the buffer solution, and above 1.0 mM, the response did not increase appreciably. The selectivity of the system is indicated in Fig. 6.

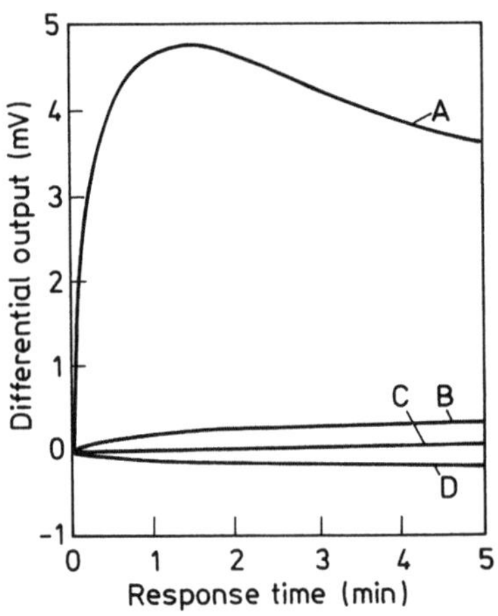

Fig. 6. Responses of the ATP-sensing system to various compounds *A*, ATP; *B*, glucose; *C*, creatinine; *D*, urea. Solutions (1 mM) were injected into the buffer pH 7.0, 40 °C

Only slight responses were obtained for 1 mM glucose, urea or creatinine when the measurements were made in the differential mode. The responses found are attributed to slight unknown differences in the samples affecting the differential voltage between the enzyme-FET and the pH-sensitive FET, and not to lack of H^+-ATPase selectivity. Between measurements, the coated FETs were stored at 4 °C. The ATP-sensor system showed a 90% decreased response to 1 mM ATP after 18 d, probably because of enzyme denaturation during storage and operation.

4. Micro-Glucose Sensor

The determination of glucose in blood samples is important in the clinical field, and the development of bioelectrochemical devices considerably help routine laboratory work. Many enzyme sensors have been developed and some sensors are used for clinical analysis.

The development of miniaturized and implantable enzyme sensors employing micro-transducers are required in medical field. Therefore, a micro-hydrogen peroxide (H_2O_2) sensor has been developed utilizing the currently available silicone

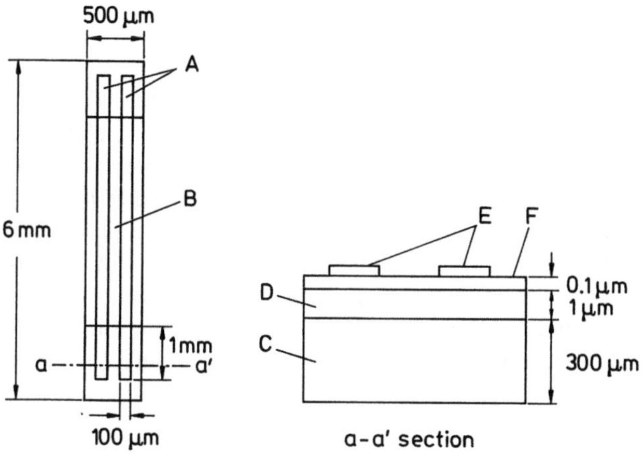

Fig. 7. Schematic diagram of a microelectrode A, E; Au, B:Ta_2O_5, C:Si, D:SiO_2, F:Si_3N_4

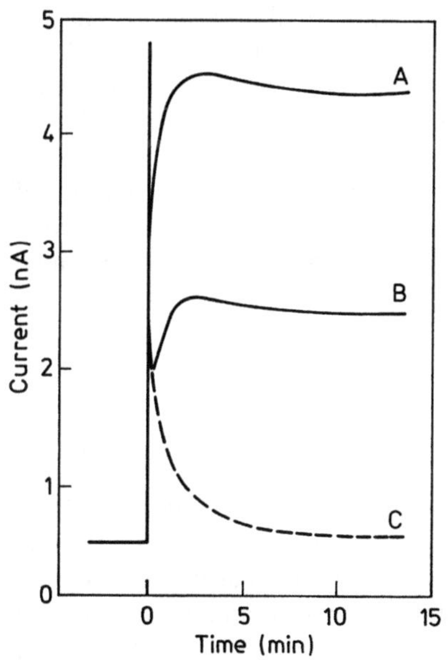

Fig. 8. Glucose response curve. Experiments were performed at 37 °C, pH 7.0. *A*, 10 mg dl^{-1} glucose; *B*, 5 mg dl^{-1} glucose; *C*, no glucose

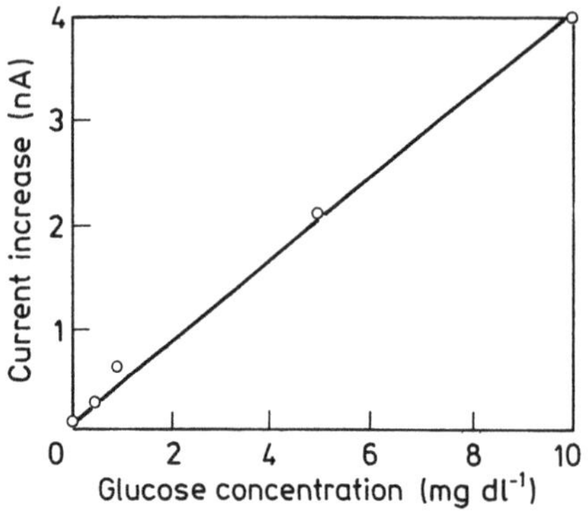

Fig. 9. Glucose calibration curve. Experiments were performed at 37 °C, pH 7.0

fabrication technology. The structure of micro-H_2O_2 sensor is shown in Fig. 7. Micro-Au electrodes were created on the silicon nitride surface using the vapor deposition method and partially insulated by coating with Ta_2O_5. The H_2O_2 electrode was placed in the sample solution containing H_2O_2 and the overpotential fixed at 1.1 V. The output current of the sensor immediately increased and reached a steady state value within 1 min. A linear relationship was observed between the H_2O_2 concentration and the steady state current in the range 1 μM to 1 mM H_2O_2. This electrode was then employed as the transducer in a micro-glucose sensor. The modified microelectrodes were then immersed in GOD solution containing BSA and glutaraldehyde, the GOD becoming chemically bound to the surface of the micro-electrode by a shiff linkage. Figure 8 shows a typical response curve for the micro-glucose sensor. The output current increased after injection of a sample solution, steady state being reached within 5 min. Figure 9 shows a calibration curve for the micro-glucose sensor.

A linear relationship was obtained between the current increase (the current difference between the initial and steady state currents) of the sensor system and concentration of glucose (0.1–10 mg dl^{-1}). Examination of the selectivity of the micro-glucose sensor indicated no response to other compounds such as galactose, mannose, fructose and maltose. Therefore, the selectivity of this sensor for glucose is highly satisfactory. Furthermore, the long-term stability of the sensor was examined with a sample solution containing 10 mg dl^{-1} glucose. The output current of the sensor system was almost constant for more 15 d and 150 assays. Therefore, this micro-glucose sensor possesses both selectivity and good stability, its potential use as a micro-glucose sensor being very good.

5. Micro-Glutamate Sensor

The determination of L-glutamate is very important in the clinical analysis, because glutamate was product for glutamicoxaloacetic transaminase (GOT) and glutamic-pyruvic transaminase (GPT) whose activities were used as medical indexes.

Various glutamate sensors, containing immobilized enzyme and electrochemical device, have been developed for the fermentation and food industries. Glutamate

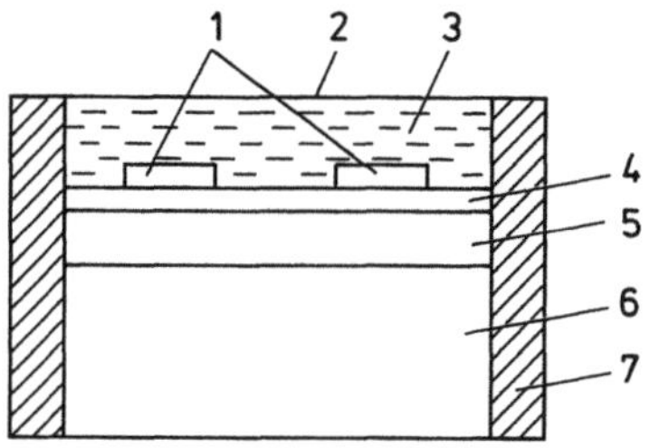

Fig. 10. Schematic diagram of a oxygen electrode. *1*, Au; *2*, Teflon membrane; *3*, 0.1 M KOH; *4*, Si_3N_4; *5*, SiO_2; *6*, Si; *7*, Silicon rubber

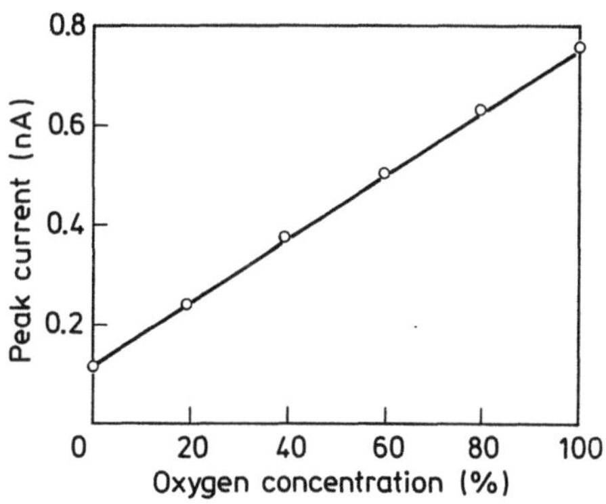

Fig. 11. Calibration curve of the oxygen electrode. Potential range: $-1.25 \sim 1$ V, scan rate: 100 mV s^{-1}

oxidase catalyses the oxidation of glutamate, oxygen being consumed by the reaction. Therefore, an oxygen sensor can be employed as the transducer for a glutamate sensor. A micro-oxygen sensor was developed by modifying the micro-Au electrodes, prepared as previously described.

Figure 10 shows the structure of a micro-oxygen sensor. The micro-oxygen sensor consisted of a gas permeable teflon membrane, two micro-Au electrodes and 0.1 M KOH solution for an electrolyte. The characteristics of the micro-oxygen electrodes were examined with a sample solution containing various concentrations of dissolved oxygen (oxygen and nitrogen mixed gas was introduced to a sample solution). A cyclic voltammogram of the micro-oxygen electrode was measured by changing the volume ratio of oxygen. A peak current was observed due to reduction of oxygen, when the terminal voltage of about 1.1 V was applied to two Au electrodes.

A linear relationship was observed between the oxygen concentration and the peak current from the electrode (Fig. 11). This result indicated that micro-oxygen electrode can be used for the oxygen determination. Therefore, the microoxygen electrode was employed for the transducer of microglutamate sensor. Glutamate oxidase was immobilized on a cellulose triacetate membrane containing glutaraldehyde and triamine (1,8-diamino-4-aminomethyloctane).

The glutamate oxidase membrane was placed on the teflon membrane of the micro-oxygen sensor and covered with a nylon net. Application of a glutamate sample solution to the sensor system produced a rapid drop in the current output to a steady state value, resulting from glutamate oxidation. Figure 12 shows relationship between current decrease and the glutamate concentration. When the

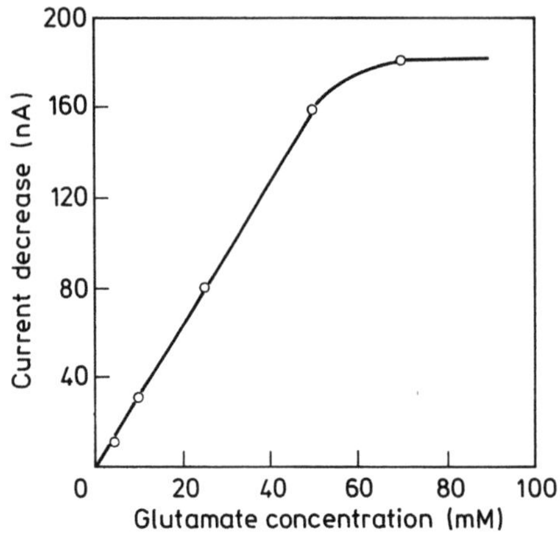

Fig. 12. Calibration curve of the micro-glutamate sensor. Experiments were performed at 40 °C, pH 7.5

current decrease at 5 min was used as a measure of the micro-glutamate sensor, a linear relationship was observed between the glutamate concentration in the range from 5 to 50 mM and the current decrease. The selectivity of the sensor for glutamate was found to be satisfactory and hence its application to clinical analysis is very important.

6. Conclusions

The employment of semiconductor fabrication technology enables the miniaturization of biosensor, and also allow measurement using only minute amounts of the sample solution and enzymes. Moreover, the use of semiconductor devices allows integration and the consequent preparation of multi-functional biosensors that can be implanted within the human body. Mass production is also possible for ICs and transistors reducing their prices and making them suitable as a disposable type of biosensor transducer.

7. References

1. H. L. Resenthal, Anal. Chem. *27*, 1980 (1955)
2. R. L. Lemar and D. Bootzin, Anal. Chem. *29*, 1233 (1957)
3. R. A. Lienado and G. A. Rechnitz, Anal. Chem. *4*, 1109 (1974)
4. T. Matsuo and M. Esashi, Sens. Actuators *1*, 77 (1981)
5. T. Bucher, Biochim. Biophys. Acta 292 (1947)
6. M. Deluca, in (A. Meister, Ed.), Advances in Enzymology, Wiley, New York *44*, 37 (1976)
7. L. J. Blum, P. R. Coulet and D. C. Gautheron, Biotech. Bioeng. *27*, 232 (1985)
8. M. Yoshida, N. Sone, H. Hirata and Y. Kagawa, J. Biol. Chem. *250*, 7910 (1975)
9. M. Yoshida, N. Sone, H. Hirata and Y. Kagawa, J. Biol. Chem. *252*, 3480 (1977)
10. T. Hamamoto, K. Ohno and Y. Kagawa, J. Biochem. *91*, 1759 (1982)
11. Y. Kagawa, J. Biochem. *95*, 295 (1984)
12. Y. Miyahara, F. Matsu, T. Moriizumi, H. Matsuoka, I. Karube and S. Suzuki, in T. Seiyama, K. Fueki, J. Shiokawa and S. Suzuki (Eds.), Chemical Sensory, Kodansha, Tokyo, Elsevier, Amsterdam, p. 513, 1983

5.2 Use of Infrared Spectroscopy for the On-Line Multicomponent Analysis and Control of Bioprocesses

Janice A. Phillips

Bioprocessing Institute, Lehigh University, Mountaintop Campus, 111 Research Drive, Bethlehem, PA 18015, USA

Abstract

Current strategies for the control of bioprocesses are based on the use of material balances to indirectly estimate the status of the culture. Until recently, direct measurements of substrate and product concentrations, other than that of gaseous components, has been hampered by the lack of sensors suitable for use in a sterile environment. However, infrared spectroscopy has been shown to be useful in the quantitation of β-lactam production in the *Penicillium* fermentation, pyruvate utilization in the *E. coli* process, and simultaneous glucose uptake and ethanol formation in the *Saccharomyces cerevisiae* process. This technique potentially offers the unique advantage of allowing for resolution of multiple components in the reaction mixture. Its general utility as a "bioreactor sensor" will be determined by the sensitivity and stability of the instrumental analysis, the speed and accuracy of the mathematical techniques for resolution of the spectra, and the implementation of a satisfactory sampling system. This paper will address these problems and their resolution for the on-line analysis with and use of the IR spectrophotometer in the closed-loop control of the *Saccharomyces cerevisiae* process.

Bioproducts and Bioprocesses
Editors: Fiechter, Okada, Tanner

5.3 The Use of Fluorometry in Monitoring and Control of Cell Cultures

Arthur E. Humphrey, Kendall Brown, John Horvath[1], and Hratch Semerjian[1]

Bioprocessing Institute, Lehigh University, Bethlehem, PA 18015, USA

Fluorometry is an attractive means of monitoring bioprocesses. It needs only a light pipe or quartz port for on-line attachment. It can be done in an non-invasive manner. Not only can one measure the cellular activity by fluorometrically monitoring reduced pyridine nucleotide concentrations of the system but one can also monitor a number of other parameters including T, pH, cell, substrate, and product concentrations and it can be done with a single sterilizeable probe.

Table 1. Historical evolution of the use of fluorescence for monitoring bioprocesses

Year	Researchers	Subject
1954	Chance	Spectrophotometry of intracellular respiratory products
1957	Duysens & Amesz	Fluorescence spectrophotometry of NADH in intact cells
1959	Chance & Thorell	Fluorescence measurement of mitochondrial NADH
1964	Chance, Estabrook & Ghosh	Damped oscillations of NADH in yeast cells
1970	Harrison & Chance	Fluorimetric monitoring of NADH in continuous yeast culture
1970	Harrison, Harms & Humphrey	On-line monitoring of bioprocesses by culture fluorescence
1977	Ristroph, Watteeuw, Armiger & Humphrey	Effect of glucose and pO_2 on culture fluorescence
1978	Zabriskie & Humphrey	Estimation of cell mass and growth rate from culture fluorescence
1983	Einsele, Ristroph & Humphrey	Estimation of mixing times and glucose uptake by culture fluorescence
1984	Scheper, et al.	Use of culture fluorescence for bioprocess control
1986	Armiger, et al.	Control based upon culture fluorescence
1986	Armiger, et al.	Extensive correlation of culture fluorescence with cell concentration
1986	Horvath, Semerjian, Brown & Humphrey	Use of fluorescence for monitoring cell, substrate and product concentrations in bioprocesses

[1] Located at the Chemical Metrology Division of the National Institute of Standards and Technology in Gaithersburg, MD, USA.

Bioproducts and Bioprocesses
Editors: Fiechter, Okada, Tanner

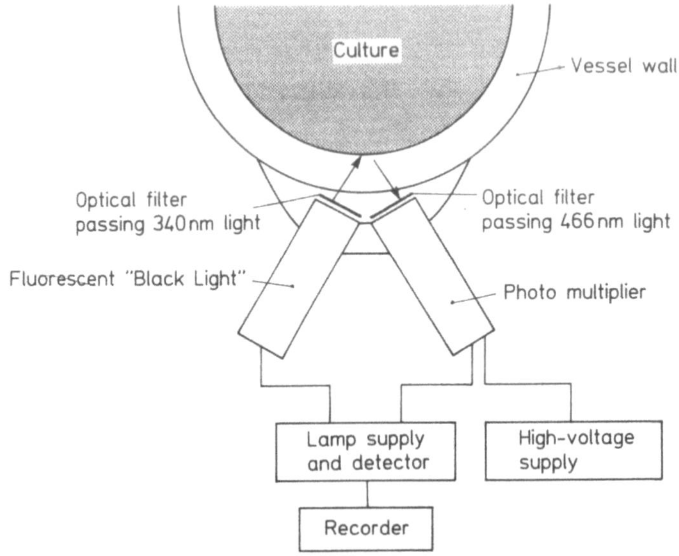

Fig. 1. Basic device for monitoring reduced pyridine nucleotide concentration of media

The application of fluorometry to monitoring a bioprocess system (Table 1) was first introduced by Duysens and Amesz [1] in 1957. They noted that the fluorescence spectrum for aerobic suspensions of Baker's yeast was similar to that for NADH and the signal could be enhanced by additions of glucose. This technique was derived from the work of Chance [2] who studied the spectrophotometry of intracellular pigments in respiring yeast. Later Chance and Thorell [3] demonstrated this fluorescence was located in the yeast mitochondria where NADH is concentrated. Then, in 1964, Chance, et al. [4] used fluorometry to investigate oscillations in starved yeast. This work established fluorometry as a laboratory technique to follow intracellular metabolism.

In 1970 Harrison and Chance [5] built a device to specifically monitor changes in the reduced pyridine nucleotide of a yeast culture. This device used an excitation light of 340 nm and monitored the back scattered fluorescence at 466 nm (Fig. 1). It was adapted by Harrison, Harms, and Humphrey in 1970 [6] to monitor continuously and on-line the cellular activity of pilot scale processes. It is the basis of the two fluorometric instruments (BioChem Technology and Ingold) presently marketed to monitor culture fluorescence.

The first extensive study to relate the fluorescence not only to cell activity but also to cell biomass was that of Zabriskie and Humphrey in 1978 [7]. (See Fig. 2.) They showed that the log of the fluorescent signal could be linearly related to the log of the cellular concentration for a number of bioprocesses (Fig. 3). Also, that culture fluorescence at 466 nm was not only an indication of cellular concentration but also concentration levels of energy substrate and dissolved oxygen in the culture system. (See Figs. 4a and b.) In 1977 Ristroph, et al. [8] used a fluorometric signal to control the alcohol fed concentration to a *Candida utilis* SCP process. This led Einsele, Ristroph and Humphrey [9] in 1978 to utilize culture

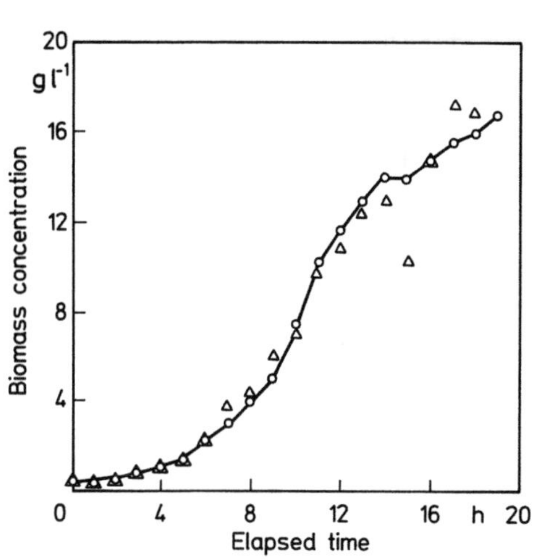

Fig. 2. Typical fluorescence patterns (after Zabriskie and Humphrey). *Saccharomyces cerevisiae* under non-limiting carbon and oxygen source and constant pH and temperature conditions. △, biomass by fluorescence; ○, biomass by dry weight

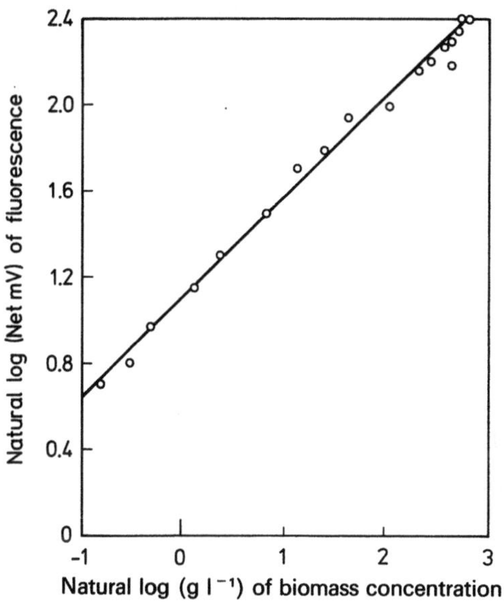

Fig. 3. Logarithmic relationship of fluorometric signal (F) and cell concentration (X) for Baker's yeast

fluorescence to estimate mixing times and glucose uptake by comparing the dynamic response of oxygen sensitive fluorophors with that of intracellular fluorescence. This work also encouraged first BioChem Technology and later Ingold to market fluorometric devices as a means of measuring on-line cellular concentrations and activity.

The Zabriskie and Einsele researches also encouraged a number of investigators, including Armiger, et al. [10)] and Scheper, et al. [11)] to study fluorometry as a means of bioreactor control. It has recently been established by Armiger that it is possible to monitor a variety of cell concentrations accurately in assorted media both in anaerobic and aerobic systems when the dissolved oxygen and energy substrate concentrations were maintained above their limiting concentrations [12)].

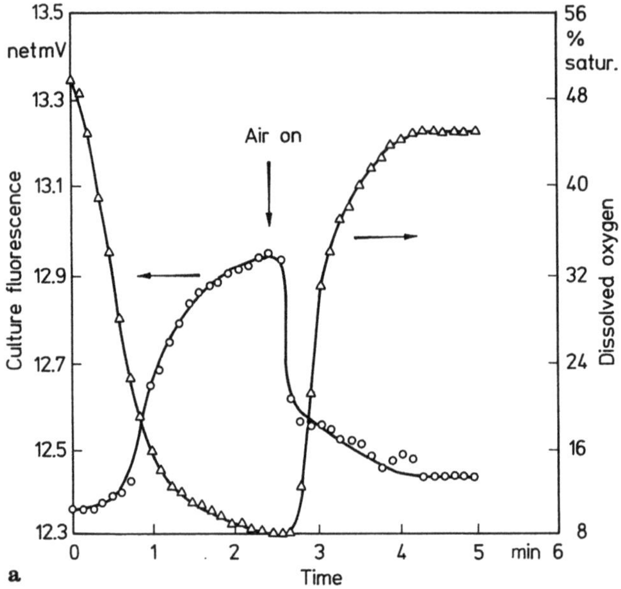

Fig. 4a.
Effect of O_2 starvation on culture fluorescence

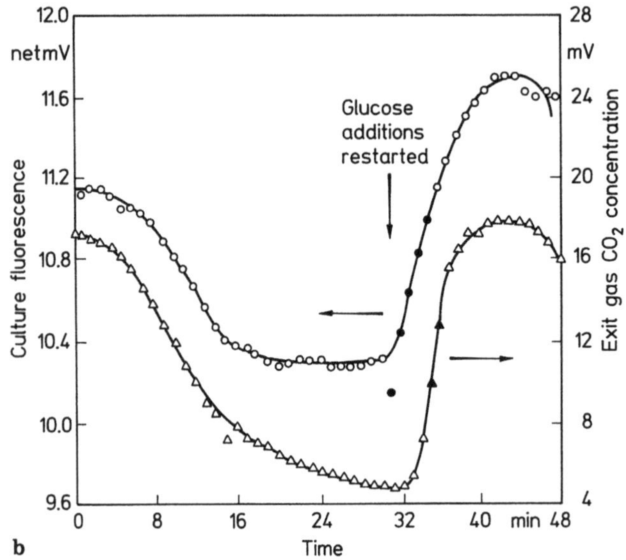

Fig. 4b.
Effect of glucose starvation on culture fluorescence

It is necessary to establish first by experimentation the value of a culture fluorescence factor that relates the fluorescence signal to cell concentration for each microbial species in the different media (Table 2). Indeed, Armiger, et al. [12] have established that the whole medium fluorometri signals when properly characterized are as accurate, if not more so, than off-line dry weight and turbidimetric measurements of cellular concentrations. Unfortunately, this need to characterize

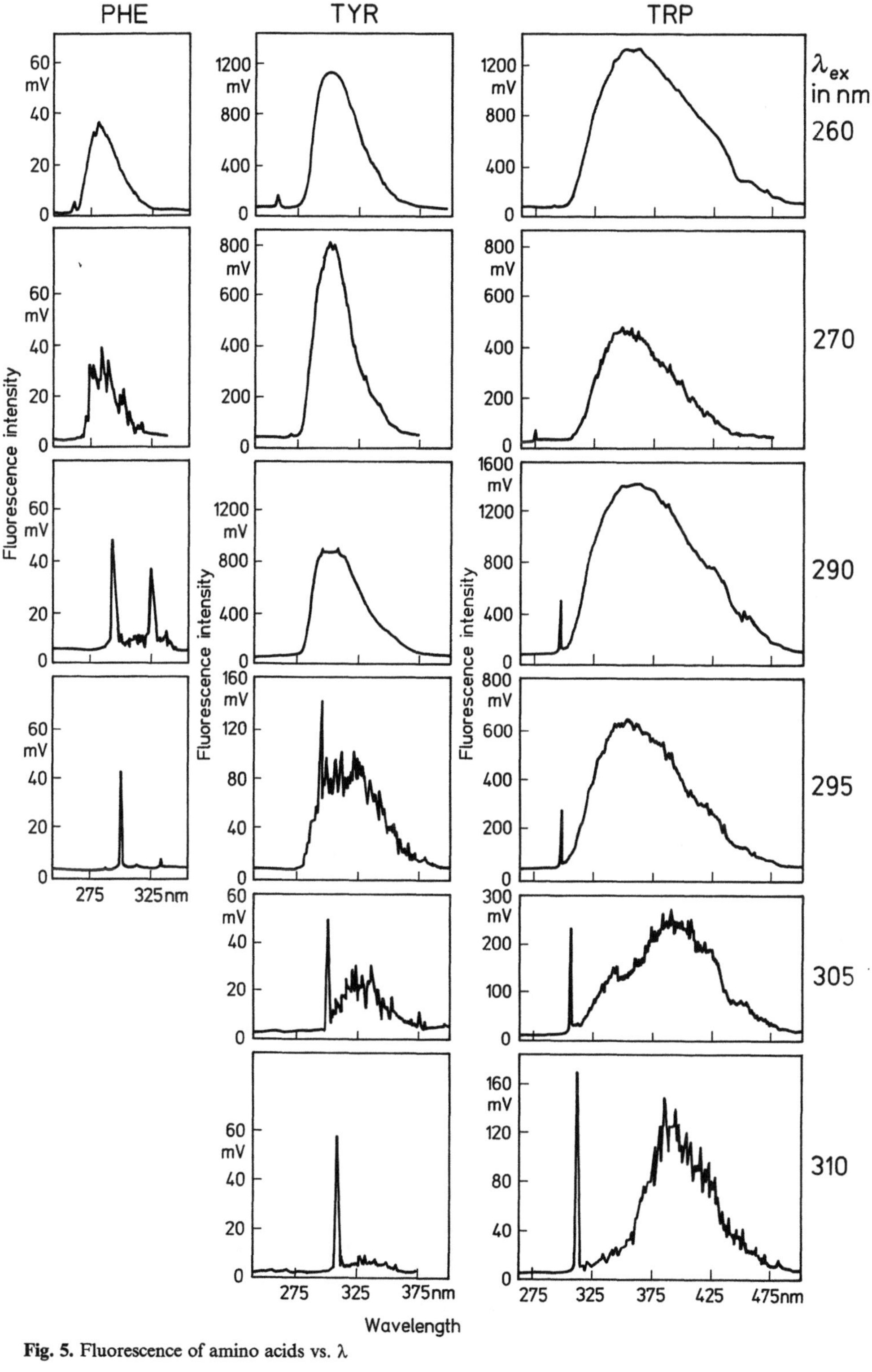

Fig. 5. Fluorescence of amino acids vs. λ

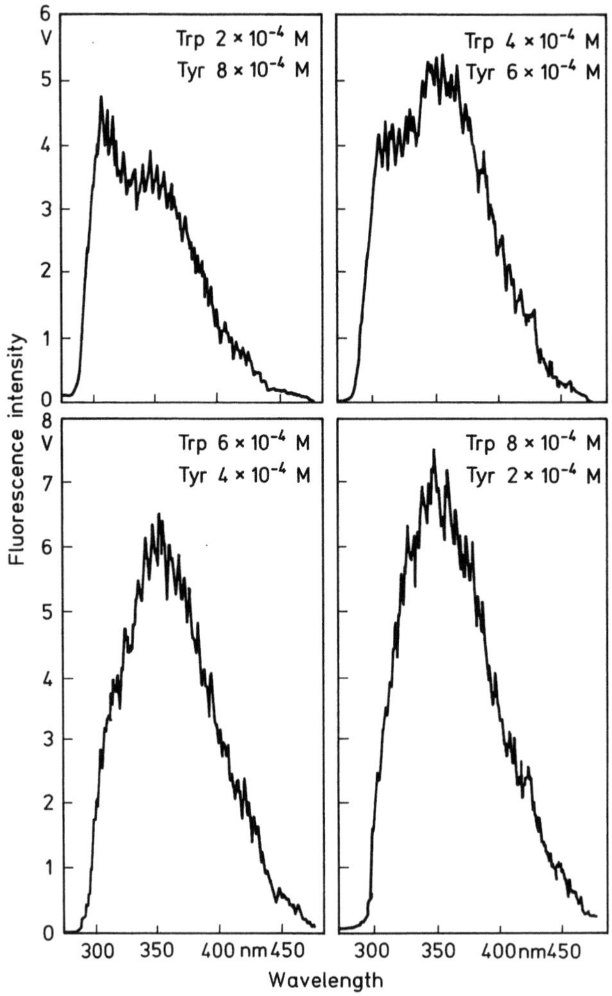

Fig. 6. Fluorescence spectra of amino acid mixtures excited at 280 nm

each process, medium and culture conditions, plus maintain the energy substrate and pO_2 well above limiting conditions has made many biotechnologists reluctant to adopt fluorometric techniques for monitoring their processes.

During the year 1986 some biotechnologists [13] came to appreciate that laser induced scanning fluorometry might provide one of the most attractive possibilities to monitoring instantaneously in an on-line, non-invasive manner many parameters of a biosystem. This includes not only cell concentrations but also that of amino acid containing substrates and products. Semerjian and Horvath [14] have shown that the aromatic amino acids — phenylalanine, tryosine and tryptophan — have unique fluorescent spectra when excited by laser beams in the 260–310 nm wavelength regions (see Fig. 5). This excitation/emission matrix can be used to identify and measure individual aromatic amino acid species in

Table 2. Culture fluorescence factor (CFF) for various bioprocesses (from Armiger et al. [12])

Organism	Media	Metabolic state	CFF ($g\,l^{-1}$ NFU^{-1})
E. coli K 12	semi-synthetic/glucose	exponential phase	0.0073
Recombinant *E. coli* K 12	semi-synthetic/glucose	exponential phase	0.0073
Alcaligenes eutrophus	basal media/fructose	exponential phase	0.0052
Bacillus amyloliquefaciens	glucose media	exponential phase	0.0100
	maltose media		0.0160
M. mucosa	semi-synthetic/methanol	$\mu = 0.14\,h^{-1}$	0.019
		$\mu = 0.19\,h^{-1}$	0.012
Clostridium acetobutylicum	semi-synthetic/glucose	acidogenic phase	0.0470
		solventogenic phase	0.023
Pediococcus	corn steep liquor	anaerobic/exponential	1.1300
	APT	anaerobic/exponential	0.1100
S. cerevisiae	glucose	anaerobic	0.0170
	glucose	anaerobic/EtOH inhibition	0.0190
	glucose	aerobic	0.0023
	molasses	anaerobic	0.1500
Penicillium chrysogenum	semi-synthetic/CSL	exponential phase	0.0800

a multicomponent mixture (Fig. 6) and specific polypeptides (Fig. 7). Furthermore, amino acid concentrations of from 10 μg to 1 g l^{-1} can be detected (see Fig. 8). Since the aromatic amino acid concentrations are usually less than 10% of the total amino acid concentration, this means that laser induced scanning fluorometry has the potential to detect individual proteins (enzymes, etc.) in reaction mixture in the range of 0.1 mg to 10 g l^{-1}.

More recently it has been shown by Junker and Wang [15] that certain fluorophors when added to media can be used to monitor culture temperature, pH, dissolved oxygen, and phosphate concentrations. Indeed, for small size (1 to 5 liter) processes the use of fluorophors in monitoring such culture parameters can be done at a cost of only a few dollars and appears to be a very attractive means to monitoring such reactors. Unfortunately, this technique is cost prohibitive in large scale processes unless very cheap fluorophors become available.

In the biotechnology laboratories at Lehigh University we have taken two approaches to the utilization of culture fluorescence. In the first instance we are convinced that monitoring the reduced pyridine nucleotide fluorescence of whole media and correcting the fluorescent signal for effects of dissolved oxygen and energy substrate concentrations will allow one to monitor directly and in a noninvasive way cell concentrations in most cases. While experimentally quantifying the relationships for each system takes time, nonetheless once this is done the fluorometric signal can be the most accurate as well as instantaneous and continuous means for monitoring cell concentration on-line and then time differentiation of that signal to obtain cell growth rates. Figure 9 is typical of the results we

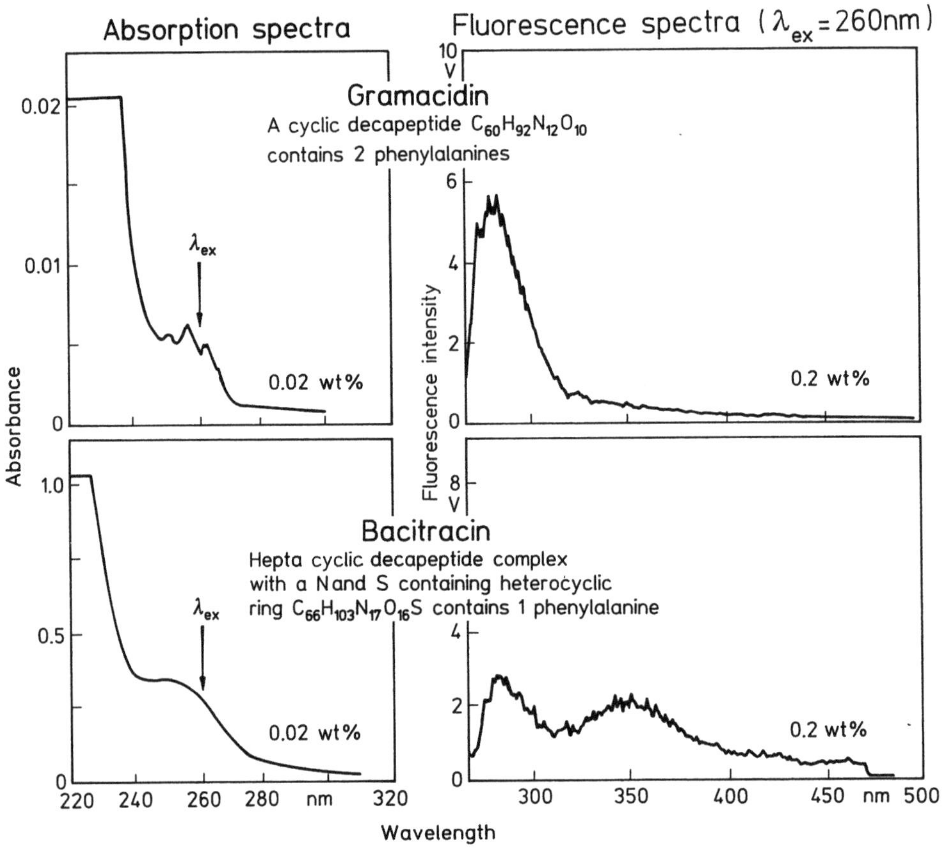

Fig. 7. Fluorescence spectra of polypeptide antibiotics

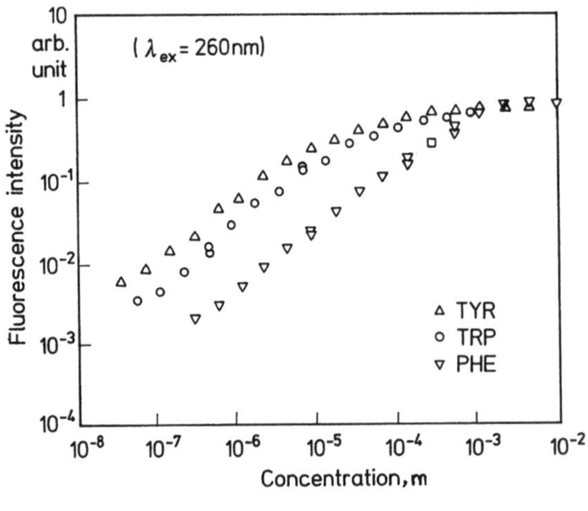

Fig. 8. Intensity vs. concentration.
Limits of detection:
PHE 1×10^{-7}
TYR 5×10^{-9}
TRP 3×10^{-8}

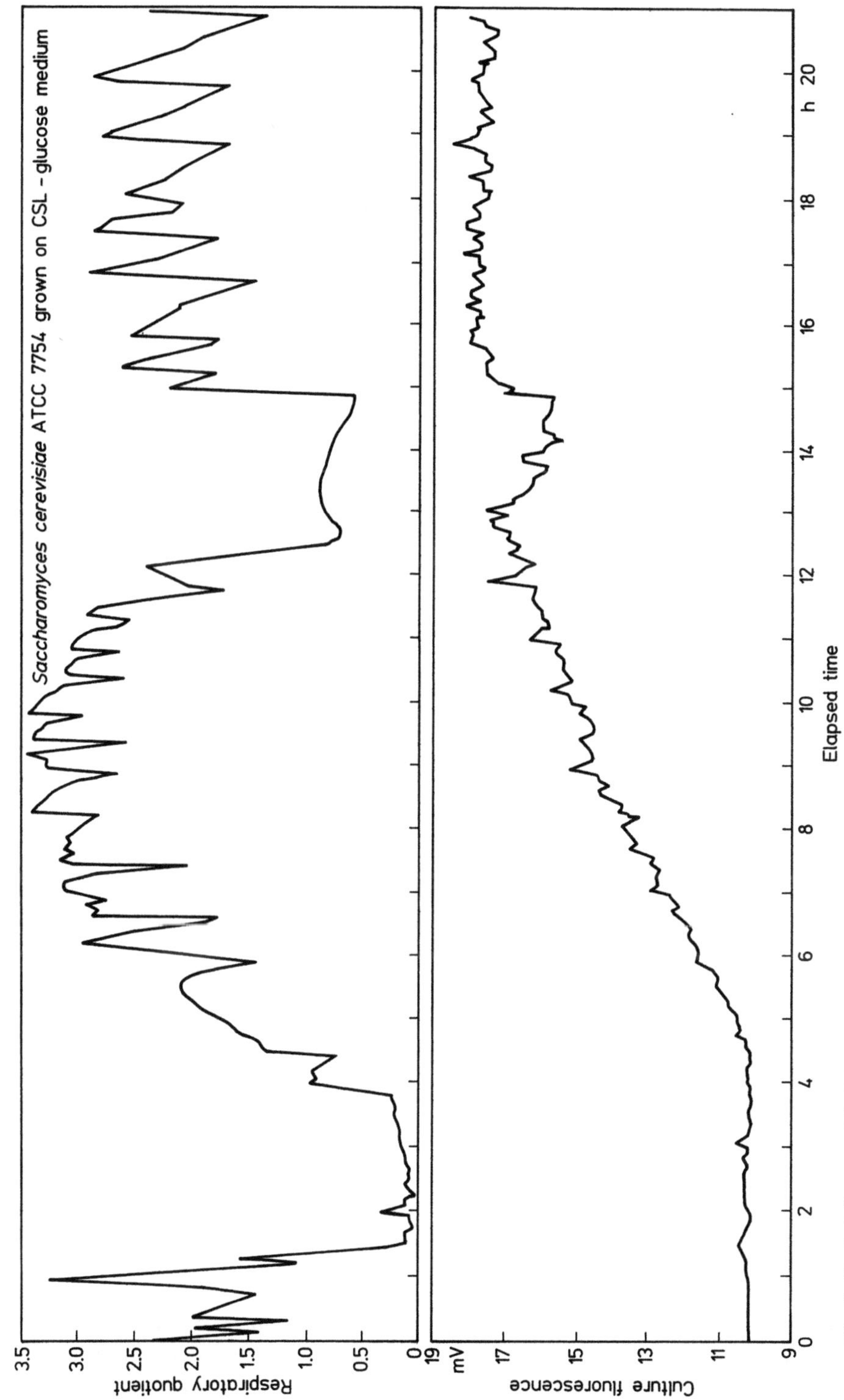

Fig. 9. Typical fluorometric data

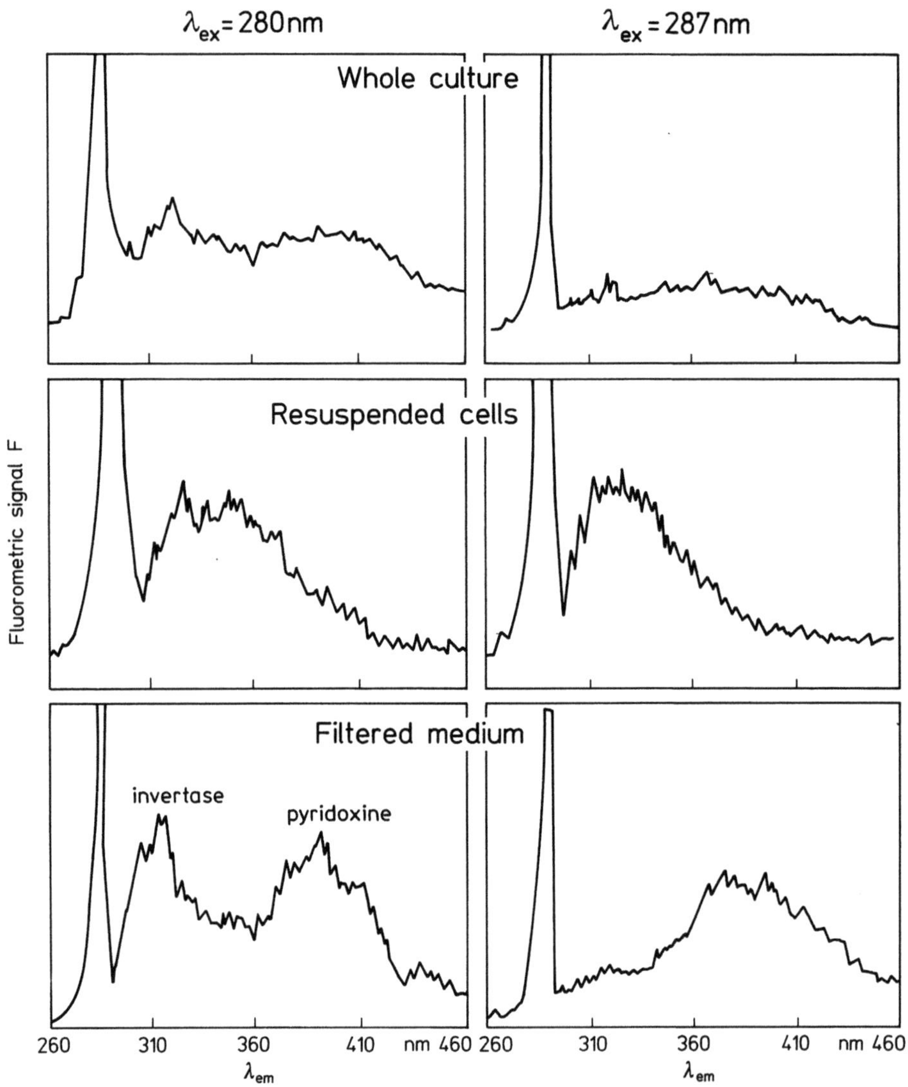

Fig. 10. Fluorometric spectra for *S. cerevisiae* RTY 110 pRB58 grown on GYP medium

obtained in on-line fluorometric monitoring. In the second approach we feel that if one is to perform ultimately meaningful control of bioprocess systems it will be necessary to also monitor substrate and product concentrations. As a consequence, we have been investigating laser scanning fluorescent spectra of several aromatic amino acid containing proteins such as peptone, invertase, and bacitracin in various media in an attempt to discern whether certain substrate and product concentrations can be continuously monitored in addition to cell concentrations by fluorometry. We believe that with the use of a light pipe system that utilizes several unique excitation wavelengths and then monitors perhaps a half dozen unique

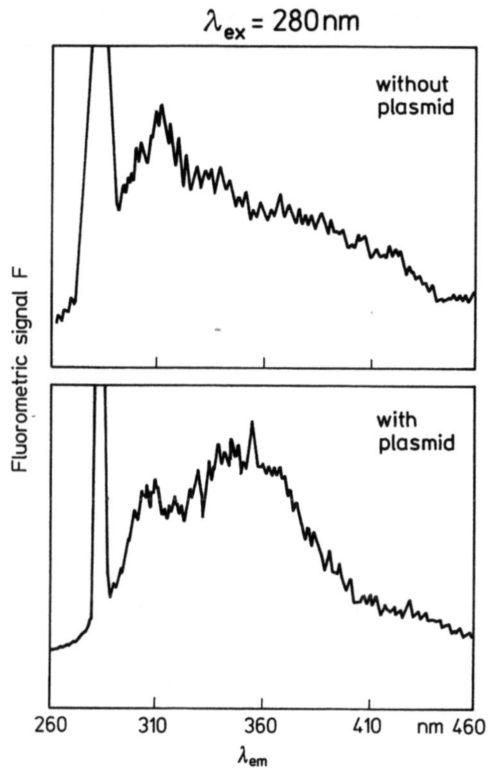

Fig. 11. *S. cerevisiae* RTY 110 grown on GYP medium — comparison of plasmid (pRB58) containing and plasmid cured yeast whole medium system

fluorescent wavelengths, it will be possible to monitor amino acid (or protein) uptake and enzyme product formation as well as cell concentration and growth rate in certain yeast processes. For example, it is possible to differentiate between and monitor bacitracin and gramicidin concentrations in media.

At the present time we are using a yeast (*Saccharomyces cerevisiae*) auxotroph (RTY 110, leu-, uracil-, suc-) that has a plasmid (pRB58) containing the leucine (leu +) and invertase (suc +) genes plus several other appropriate markers as a model system for studying the potential of fluorometric monitoring since the clone expresses the suc + gene in the form of a glycosylated invertase. To date we have excited various plasmid containing and plasmid cured yeast cultures at 280 and 287 nm grown on both rich and defined media. There clearly is a difference between the plasmid containing and plasmid cured cultures (see Fig. 11). Also, yeast and invertase separately contribute to the overall fluorescent spectra (Fig. 10).

We are now extending the excitation spectrum to higher (300 nm) and lower (260 nm) wavelengths in hopes that the excitation/emission spectra can be deconvoluted using on-line computer techniques with sufficient accuracy to measure cell and invertase concentrations for control purposes. We also are attempting to develop a tryptophan auxotroph of RTY110 in order to see if we can use fluorometric techniques for monitoring and controlling tryptophan when used as a growth limiting substrate. We are very optimistic on the basis of preliminary results. We believe that we can follow tryptophan uptake and invertase release in addition to cell concentrations by scanning fluorometry. What needs to be determined is how many excitation wavelengths must to be scanned to get sufficient data to achieve the desired accuracy for control purposes.

Acknowledgements

We wish to acknowledge the support of the National Institute of Standards and Technology, Division of Chemical Metrology in this work.

References

1. Duysens, L. N. M. and Amesz, J. "Fluorescence Spectrometry of Reduced Phosphopyridine Nulceotide in Intact Cells in the Near-Ultraviolet and Visible Region," Biochim. Biophys. Acta *24*, 19 (1957)
2. Chance, B., "Spectrophotometry of Intracellular Respiratory Pigments," Science *120*, 767 (1954)
3. Chance, B. and B. Thorell, "Fluorescence Measurements of Mitochondrial Pyridine Nucleotide in Aerobiosis and Anaerobiosis," Nature *184*, 931 (1959)
4. Chance, B., Estabrook, R. W. and Ghosh, A. "Damped Sinusoidal Oscillations of Cytoplasmic Reduced Pyridine Nucleotide in Yeast Cells, Proc. Natl. Acad. Sci. *51*, 1244 (1964)
5. Harrison, D. and Chance, B. "Fluorimetric Technique for Monitoring Changes in the Level of Reduced Nicotinamide Nucleotides in Continuous Culture of Microorganisms," Appl. Microbiol. *19*, 446 (1970)
6. Harrison, D., Harms, C. and Humphrey, A. "Design, Instrumentation, and Control of Fermentation Systems". Paper presented at the Tenth International Congress of Microbiology, Mexico City, August 9–15 (1970)
7. Zabriskie, D. W. and Humphrey, A. "Estimation of Fermentation Biomass Concentration by Measuring Culture Fluorescence," Appl. Environ. Microbiol. *35*, 337 (1978)
8. Ristroph, D. et al., "Experience in the Use of Culture Fluorescence for Monitoring Fermentations," J. Ferment. Technol. *55*, 599 (1977)
9. Einsele, A., Ristroph, D. L. and Humphrey, A. "Mixing Times and Glucose Uptake Measured with a Fluorometer," Biotech. Bioeng. *20*, 1487 (1978)
10. Armiger, W. B., Lee, J. F., Montalvo, L. M. and Forro, J. R. "Fed-Batch Control Based Upon the Measurement of Intracellular NADH," Proc. Bio-Expo '86, pp. 353–369, April 29–May 1, 1986
11. Scheper, J. et al., "Measurement of Culture Fluorescence for Fermentation Control," Dechema-Monographien *95*, 83 (1984)
12. Armiger, W. A. et al., "On-Line Measurement of Cell Concentration — The FluoroMeasure® System Users Manual," Chapter 2, BioChem Technology, Inc. (1986)
13. Semerjian, H., Horvath, J., Brown, K. and Humphrey, A. "Problems and Opportunities for Monitoring Fermentations by Scanning Fluorometry," poster presented at Biochemical Engineering V, Engineering Foundation Conference, Henniker, NH, July 28–August 1, 1986
14. Semerjian, H. and Horvath, J. "LIF for Sensing in Bioreactors," paper presented at the Intl. Symp. on Biosensors, 192nd ACS National Meeting, Anaheim, CA, Sept. 7–12, 1986
15. Junker, B. and Wang, D. I. C. "Use of Fluorometric Techniques in Monitoring Fermentation Systems," paper presented at Biochemical Engineering V, Engineering Foundation Conference, Henniker, NH, July 28–August 1, 1986
16. Alberti, J. C., Brown, K., Humphrey, A. E., and Phillips, J. A. "Use of Spectroscopy for the Monitoring and Control of Biological Unit Processes," paper presented at the Intl. Symp. on Biosensors, 192nd ACS National Meeting, Anaheim, CA Sept. 7–12, 1986

5.4 Mass Production of Intracellular Metabolite by Fully Automatic Fed-batch Culture of Microorganism

Tsuneo Yamane, Takahiro Suzuki and Shoichi Shimizu

Laboratory of Bioreaction Engineering, Department of Food Science and Technology, School of Agriculture, Nagoya University, Chikusa-ku, Nagoya 464-01, Japan

Contents

Attempts were made to produce two kinds of intracellular metabolites, poly-β-hydroxybutyric acid (PHB) and thiostrepton (TS), by automatic fed-batch cultures up to high cell mass concentrations. At 170 h of cultivation of a methylotroph, 150 g l^{-1} of PHB (its cellular content was about 64%) was obtained from methanol with 20% of yield. To maintain PHB synthetic activity at a high level, the ratio of methanol and ammonia feedings (C/N ratio of feeding) was gradually raised according to the increase in PHB content with a computer-aided automatic feeding system. At 220 h of cultivation of *Streptomyces laurentii*, 10.5 g l^{-1} of TS (its cellular content was about 7%) was obtained from glucose, corn steep liquor and defatted soy bean meal. To keep high TS production rate and to avoid the degradation of TS formed, a solution of these nutrients whose composition had carefully been determined experimentally was automatically supplied with a pH-stat mode. A general equation of direct cost for intracellular metabolite production composed of both yield and productivity was proposed. Based on the cost equation, advantages of the fed-batch culture up to high cell mass concentration over conventional batch culture were discussed concerning intracellular metabolite production.

1. Introduction

Metabolites produced by microbial reactions accumulate in two broadly classified localities, i.e., intracellularly or extracellularly. At the present state of art, one cannot alter the nature of the metabolite excretion (accumulation of the metabolite inside the cells which is originally excreted into the culture medium, or inversely secretion of the metabolite which accumulates inside the cells originally). Although some plasmid vectors permit the transformed microorganisms to secrete gene products from the cell into the culture medium [1-3], it is still not easy to control at

Bioproducts and Bioprocesses
Editors: Fiechter, Okada, Tanner

one's own will the locality of the metabolite produced. Accordingly, the classification of the metabolite formation with respect to its locality in the microbial culture is still an effective first step to develop kinetics and process optimization.

In case of a metabolite which accumulates intracellularly, its final concentration [g l^{-1}] in batch or fed-batch culture, P_f, is simply given by

$$P_f = \gamma_f X_f \tag{1}$$

where γ_f is the final content of the metabolite in the whole dry cell [g g^{-1}] and X_f is the final dry cell concentration [g l^{-1}]. On the other hand, when all other items of cost such as fixed cost, labor cost and cost of downstream pocessing are put aside, the direct, proportional cost of production for one run of batch or fed-batch comprises mainly direct raw materials cost and running cost (power cost such as electricity, gas etc.). The direct raw material cost of the microbial reaction arises from medium preparation. If the substrates amounting to S_i (i = 1, 2, ... , n) are used in the operating time, θ_f, of the single run, the direct cost of metabolite produced [yeng^{-1}], ¥ , is given by

$$¥ = \frac{\sum_{i=1}^{n} k_i S_i + k_\theta \theta_f}{\gamma_f X_f V_f}$$

$$= \frac{1}{V_f} \sum_{i=1}^{n} k_i \frac{S_i}{\gamma_f X_f} + \frac{k_\theta}{V_f} \frac{\theta_f}{\gamma_f X_f} , \tag{2}$$

where k_i is the cost of each substrate [yeng^{-1}], k_θ is the operating cost per h [yen h^{-1}] and V_f is the volume of final medium in a bioreactor which is assumed constant. When the metabolite is produced by a fed-batch culture, the amount of substrate used is the sum of initial substrate prepared for the culture medium and the amount of the substrate supplemented during the operation as

$$S_i = s_i(0)\ V(0) + \int_0^{\theta_f} F_i(\theta)\ d\theta \tag{3}$$

where $s_i(0)$ is the initial concentration of the substrate in the medium, V(0) is the initial volume of the culture and $\int_0^{\theta_f} F_i(\theta)\ d(\theta)$ is the total amount of the substrate fed during the cultivation. In a microbial reaction, metabolite yield [g g^{-1}], Y_{P/S_i}, and volumetric productivity of the metabolite [g l^{-1} h^{-1}], P_r, are defined as

$$Y_{P/s_i} = \frac{\gamma_f X_f V_f - \gamma_0 X_0 V_0}{S_i} \fallingdotseq \frac{\gamma_f X_f V_f}{S_i} \tag{4}$$

$$P_r = \frac{r_f X_f}{\theta_f} . \tag{5}$$

Substitution of Eqs. (4) and (5) into Eq. (2) gives

$$¥ = \sum_{i=1}^{n} \frac{k_i}{Y_{P/s_i}} + \frac{k_\theta}{P_r} \frac{1}{V_f} . \tag{6}$$

When pure oxygen gas or oxygen-enriched air is supplied to achieve high cell mass concentration, its cost must be included in the cost of raw materials. Thus, the first term on the right side of Eq. (2) is related to the metabolite yield from the substrate consumed [$g\ g^{-1}$] and the second term to the productivity of the metabolite formed [$g\ l^{-1}\ h^{-1}$].

From Eq. (2) it is straightforward that the direct cost is minimized by obtaining from the used amount of substrate

1) the greatest amount of cells
2) having the highest content of the metabolite concerned
3) in the shortest operating time.

Since the initial substrate concentration in a batch is not high, X_f attainable from the batch operation is rather small. When one executes a semi-batch operation by supplying carefully all nutrients necessary for cell growth, one can get very high concentration of cells over 150 [$g\ l^{-1}$], as we have shown with an aid of computer [4]. Semi-batch (or fed-batch) culture can be classified broadly into

(1) those without feedback control and
(2) those with automatic feedback control [5].

Initially, rather simple feeding modes without feedback control were usually used. Recent technical innovations for measurement and control in biotechnology have made possible more advanced feeding modes with automatic feedback control. The basic advantage of fed-batch microbial process is that the concentrations of nutrients fed into the culture liquid of the bioreactor can be controlled voluntarily by changing the feed rate. Because the content of intracellular metabolite changes significantly with variations of nutrients' concentrations in the medium, the fed-batch technique is quite effective to make culture condition optimal for the metabolite biosynthesis. Combination of the fed-batch culture technique to high cell mass concentration with the environmental control optimal for maximizing content of the intracellular metabolite concerned makes it possible to produce the metabolite with a considerably reduced production cost. Here we report mass productions of two useful intracellular metabolites, poly-β-hydroxybutyric acid (PHB) and thiostrepton (TS) by fully automatic fed-batch cultures of microorganisms.

PHB is a carbon-energy storage material accumulated inside a variety of microorganisms under appropriate condition such as limitation of nitrogen. Solid PHB is a biodegradable thermoplastic polyester with very interesting characteristics and is utilizable in various ways similar to many conventional synthetic plastics now in use. For instance, it can be moulded, reinforced with inorganic fillers, spun into a fiber or formed into a film with excellent gas barrier properties. TS is an antibiotic substance accumulated mainly inside of several microorganisms such as *Streptomyces* species and possesses marked antibiotic properties with a wide spectrum of activity against Gram-positive bacteria compared with that of penicillin. It has the effect to make animal domestics fleshy by administering in the feed.

2. Mass Production of PHB by Fed-batch Culture [6-8]

2.1 Materials and Methods

A methylotroph, *Protomonas extorquens* sp. K, was screened from 51 strains of methylotroph. Its optimal pH, temperature and methanol concentration were 7.0, 30 °C and 0.5 g l^{-1}, respectively. Its maximum specific growth rate was 0.27 h^{-1} on a minimum synthetic medium. The strain was cultivated in a small agitated-aerated vessel. The initial volume of the culture was 0.75 l. The concentration of methanol was maintained automatically at 0.5 g l^{-1} by use of a control system composed of a porous Teflon tubing sensor, a flame ionization detector and a micro-computer. A preset PID and exit CO_2-dependent stepwise program was employed [9]. A peristaltic pump connected with a stepping motor was used to supply methanol, and the rotation speed of the motor was controlled by the output signal from the microcomputer.

To evaluate the effect of the proportion of methanol to ammonia, C/N ratio, in the feeding solution on the PHB production, fed-batch cultures were carried out, in each of which ammonia supply was coupled with methanol feeding from the start. Ammonia water (33%) was mixed into the methanol and this mixture of methanol and ammonia (CN solution) was supplied. Six fed-batch cultures were carried out with different values of C/N ratios (mol methanol per mol ammonia).

Finally to get a high concentration of PHB effectively the following fed-batch culture was carried out. The early phase of the cultivation was a cellular growth phase in which ammonia was fed by pH-stat mode. After the cell concentration had reached 70 g l^{-1}, the C/N ratio of feeding was gradually changed from the lowest value to higher one according to the increase in PHB content. In this culture two

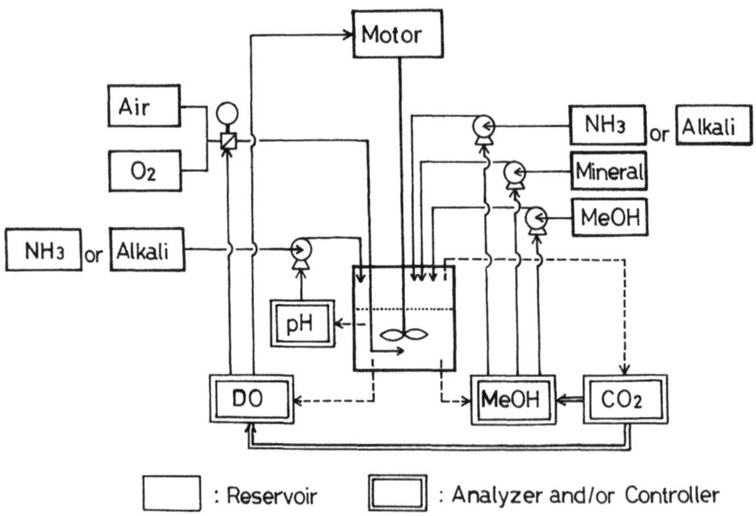

Fig. 1. System for fed-batch culture with controlled C/N ratio of feeding

pumps were used, one to supply methanol and the other ammonia water. Each pump was operated according to the step pulses from the microcomputer. During the cultivation PHB content was indirectly determined on the basis of the rate of CO_2 evolution, while C/N ratio of feeding was calculated by a program stored in the microcomputer. The fed-batch culture system of this method is shown schematically in Fig. 1.

2.2 Results and Discussion

2.2.1 Specific PHB Production Rate

When the microorganism produces and accumulates PHB intracellularly, the whole dry cell mass (X_{total}) [$g\,l^{-1}$] consists of two parts, namely PHB (X_{PHB}) [$g\,l^{-1}$] and residual biomass ($X_{residual}$) [$g\,l^{-1}$], where $X_{residual}$ is calculated by subtracting the PHB concentration from the total cell concentration ($X_{residual} = X_{total} - X_{PHB}$). As a kinetic model of microbial PHB formation, a linear relationship between $dX_{PHB}/(X_{PHB}\,dt)$ and the quotient, X_{total}/X_{PHB}, was proposed for the batch culture of *Alcaligenes eutrophus* strain H16 under chemolithotrophic growth condition (O_2, H_2 and CO_2)[10]. Later, another model correlating linearly the volumetric PHB formation rate, dX_{PHB}/dt, with elapsed time was proposed[11] from the same data obtained by Heinzle and Lafferty. We proposed a more realistic model for PHB formation from methanol as follows[7].

Since the volume of medium, V_{med}, increases with elapsed time in a fed-batch culture to high cell mass concentration, the whole dry cell mass in the medium is given by

$$X_{total}V_{med} = X_{PHB}V_{med} + X_{residual}V_{med}\,. \tag{7}$$

We considered that only $X_{residual}$ be anabolically active biomass composed of proteins and nucleic acids. Therefore, the specific PHB formation rate, q_{PHB}, based on the unit of the residual biomass is more rational than the one based on the unit of the total biomass.

$$q_{PHB} \equiv \frac{1}{X_{residual}V_{med}} \frac{d(X_{PHB}V_{med})}{dt} = \frac{1}{(X_{total} - X_{PHB})\,V_{med}} \frac{d(X_{PHB}V_{med})}{dt}\,. \tag{8}$$

2.2.2 Fed-batch Culture with Constant C/N Ratio of Feeding

With this method cellular growth and PHB production occurred simultaneously. Figure 2 shows the result of the fed-batch cultures by this method. When the solution had C/N ratio of 7.5, the growth rate was maximal and PHB scarcely accumulated. In cases of the higher values of C/N ratio of feeding than 10.0, ammonium concentration in the culture supernatant was always less than 20 ppm. The higher values of C/N ratio of feeding resulted in the smaller growth rate and the higher content of PHB. In addition, at each level of constant C/N ratio

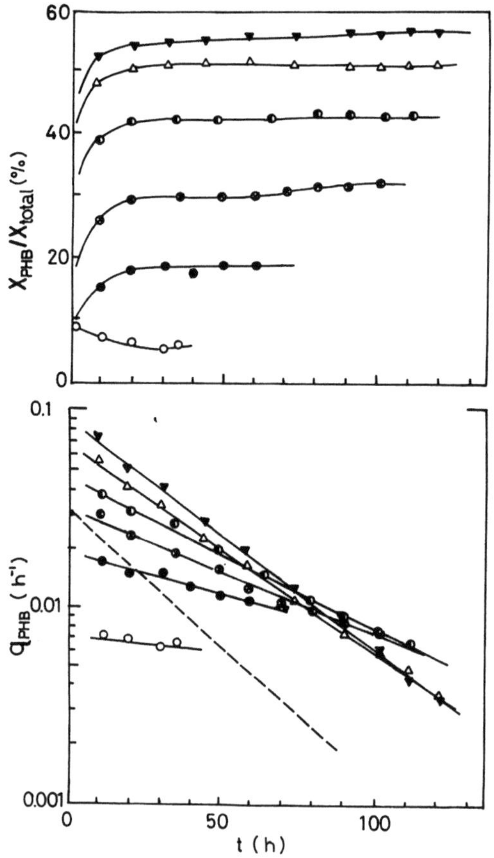

Fig. 2. Effect of ammonia feeding on PHB production by the fed-batch cultivation with constant ratio of C/N. From the initial stages of the fed-batch cultures, C/N ratios of feeding solution were preset at ○, 7.5; ●, 10.0; ⊗, 12.5; ◐, 16.3; △, 21.3; ▼, 25.0 ($mol\ mol^{-1}$). The *dotted line* shows the time course of q_{PHB} by the fed-batch culture without ammonia feeding (i.e. C/N = ∞)

of feeding PHB content, X_{PHB}/X_{total}, was kept almost constant during cultivation except during the early phase [Fig. 2]. q_{PHB} were calculated and their time courses gave Fig. 2. In every case, q_{PHB} was expressed as the first-decay model.

$$q_{PHB} = (q_{PHB})_0 \exp(-k_d t) . \tag{9}$$

Neither Heinzle and Lafferty model nor Ollis model gave straight lines like Fig. 2. The fact that q_{PHB} could be expressed by Eq. (9) verifies the soundness of our definition of q_{PHB} given by Eq. (8). Results shown in Fig. 2 suggest that the proportion of C/N in the feeding solution regulated not only the PHB content but also maintenance of PHB synthetic activity. The relationships between the C/N ratio of feeding and PHB content as well as between C/N ratio of feeding and k_d are plotted in Fig. 3, from which the following empirical equations are obtained.

$$X_{PHB}/X_{total} = -77 + 42 \ln(C/N) \tag{10}$$

$$k_d = 5.9 \times 10^{-4} (C/N)^{1.4} . \tag{11}$$

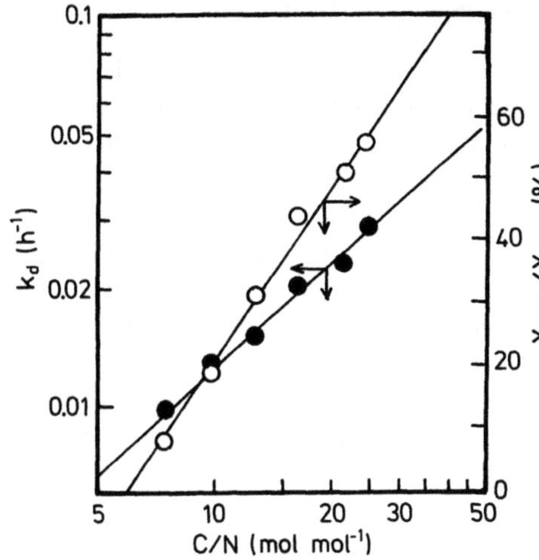

Fig. 3. Relationships between C/N ratio of feeding solution and k_d, and between C/N ratio and PHB content, X_{PHB}/X_{total}

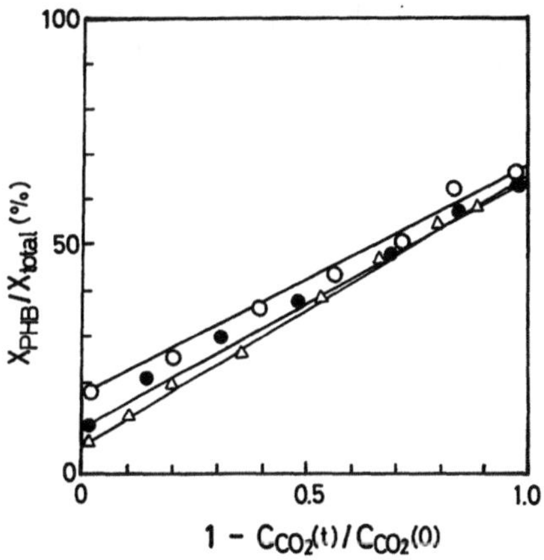

Fig. 4. Relationships between PHB content, X_{PHB}/X_{total}, and concentration of CO_2 in exhaust gas, C_{CO_2}

It was also found that there was a relation between CO_2 evolution rate and PHB content. As the flow rate of inlet gas was always kept at $1.0\,l\,min^{-1}$, the concentration of CO_2 in the exit gas, C_{CO_2}, was used instead of the CO_2 evolution rate. The results shown in Fig. 4 indicate that the increase in PHB content is linear with the decrease in C_{CO_2}.

$$X_{PHB}(t)/X_{total}(t) = \{(X_{PHB}/X_{total})_m - X_{PHB}(0)/X_{total}(0)\} \times \{1 - C_{CO_2}(t)/C_{CO_2}(0)\} + X_{PHB}(0)/X_{total}(0) \qquad (12)$$

where 0 and t represent the values at the initial time of PHB production phase and at the time t, respectively. $(X_{PHB}/X_{total})_m$ is the final maximum PHB content, which was usually about 65%.

2.2.3 Fed-batch Culture with Controlled Carbon/Nitrogen Feeding

In order to produce PHB more efficiently, k_d should be lowered with smaller value of C/N ratio of feeding. On the other hand, in order to get higher content of PHB, C/N ratio of feeding must be set at higher level. Therefore, the C/N ratio of feeding must be varied to get the highest possible X_{PHB} in the shortest possible time. Although the optimization of the C/N ratio of feeding needs more rigorous kinetics and theory, we tried to increase C/N ratio of feeding following the increase in the PHB content. The feeding policy was based upon the idea that we could retain high PHB synthetic activity by low C/N ratio at the onset of the feeding and could increase the PHB content by gradually increasing the C/N ratio of feeding, taking into account the decline in the PHB synthetic activity of the microorganism. We believed that this feeding policy would reduce the time required to reach the maximum PHB content. In order to put into effect the feeding policy described above, the PHB content at time, t, must be estimated during the cultivation. For this purpose, Eq. (12) was utilized, and C_{CO_2} was used as a control index to change the C/N ratio of feeding automatically by a microcomputer-aided process controller on the assumption that Eqs. (10) and (11) still hold even when the C/N ratio of feeding is dynamically changed. The result is shown in Fig. 5. By this strategy, intracellular PHB content was increased rapidly,

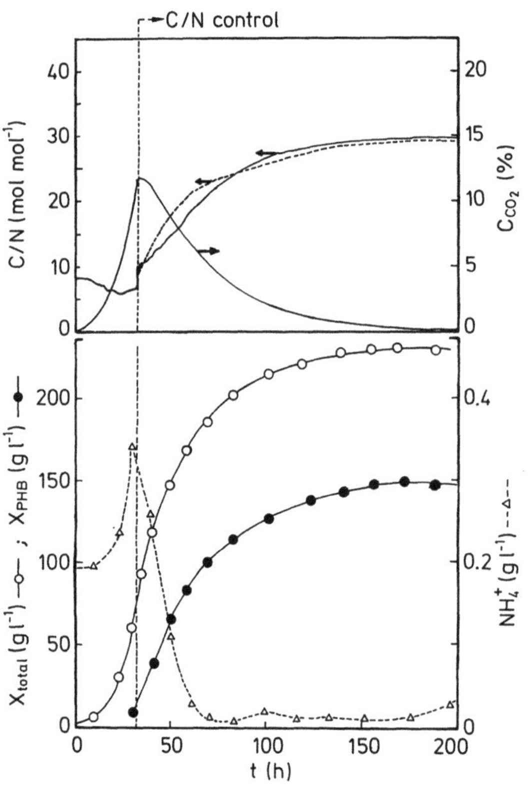

Fig. 5. Cultivation results for production of PHB by the fed-batch culture with controlled C/N feeding. Control of C/N ratio of feeding was started when the cell concentration reached 70 g l^{-1}

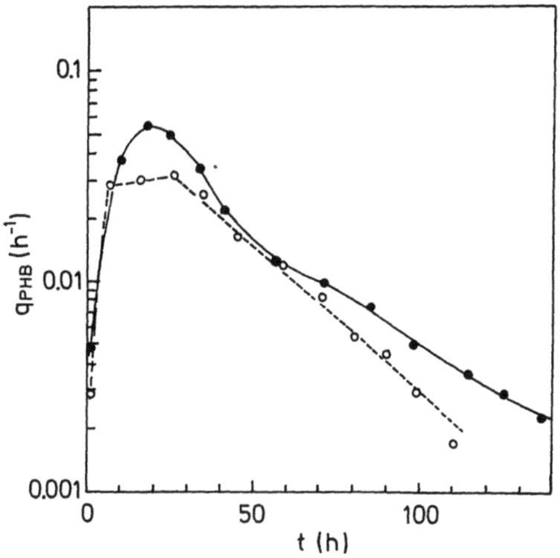

Fig. 6. Time course of q_{PHB} for PHB production by the fed-batch culture with controlled C/N feeding. The *dotted line* shows the time course of q_{PHB} in the fed-batch culture without nitrogen feeding

while microorganisms grew still vigorously even in the PHB production phase. Accordingly, X_{PHB} increased in a shorter time. At 121 h culture time, X_{PHB} reached 136 g l^{-1}. The fed-batch culture without nitrogen feeding took 175 h to attain the same level of X_{PHB} [6]. Finally at 170 h the PHB and cell mass concentrations were 149 g l^{-1} and 233 g l^{-1}, respectively. The maximum PHB content was 64% and overall PHB yield from methanol was about 20%, which was slightly higher than the yield of 18% obtained with the fed-batch culture to which no ammonia was supplied. Figure 6 shows the time course of q_{PHB} for the PHB production shown in Fig. 5. The values of q_{PHB} were maintained at higher levels than those obtained from the fed-batch culture without ammonia feeding.

3. Mass Production of Thiostrepton (TS) by Fed-batch Culture [12]

3.1 Materials and Methods

Microorganism used was *Streptomyces laurentii* ATCC 31255, which could produce and accumulate TS inside cells. Examination of effects of concentrations of medium components on both cell growth and on TS production revealed that natural nutrients such as defatted soy bean meal (SBM), corn steep liquor (CSL) and yeast extract (YE) were effective to induce TS production. The basal medium composed of glucose, these natural nutrients and mineral salts was decided from this information.

Spores were inoculated into 100 ml of seed culture and the seed culture was cultivated in a shake flask at 28 °C for 3 d. Then, 15 ml of sterilized glycerol was added into the seed culture and was aseptically distributed into six test tubes. The sealed seed cultures were stocked at −20 °C and one seed tube was used for each run of main culture. The main cultures were performed in 3 l reactor initiated with 0.8 l. The pH controller used had upper and lower limiters which could be preset manually. pH was automatically controlled around these limiters.

Dissolved oxygen (DO) in the culture liquid was controlled in the range of 3–5 ppm with a microcomputer-aided DO control system [9].

The composition of multisubstrate solution (MS solution) in the fed-batch cultures with pH-stat modal feeding of multisubstrate was determined after investigating both effects of CSL and SBM on the production of TS. The composition of MS solution thus fixed was as follows: Solution A contained 700 g of glucose, 15 g of H_3PO_4, 3 g of H_2SO_4, 8 g of $MgSO_4 \cdot 7\,H_2O$, 0.8 g of $FeSO_4 \cdot 7\,H_2O$, 0.8 g of $MnSO_4 \cdot n\,H_2O$, 0.3 g of $CoCl_2 \cdot 2\,H_2O$, 0.1 g of $ZnSO_4 \cdot 7\,H_2O$ and 0.1 g of $CuCl_2 \cdot 2\,H_2O$ in 300 g of water, while solution B contained 240 g of CSL, 130 g of SBM, 10 g of $CaSO_4 \cdot 2\,H_2O$ and 6 g of YE powder in 300 g of water. Then, the solutions A and B were separately sterilized and were mixed after cooling. Amounts of mineral ions (except for K^+ and Na^+) in the MS solution were calculated

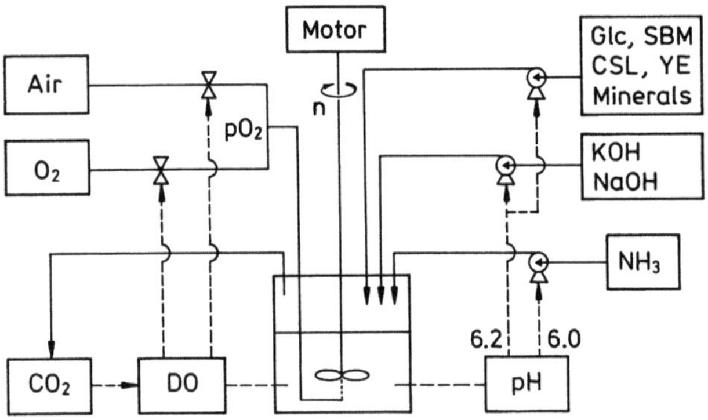

Fig. 7. Automatic nutrients supplementation system with pH-stat mode for fed-batch culture

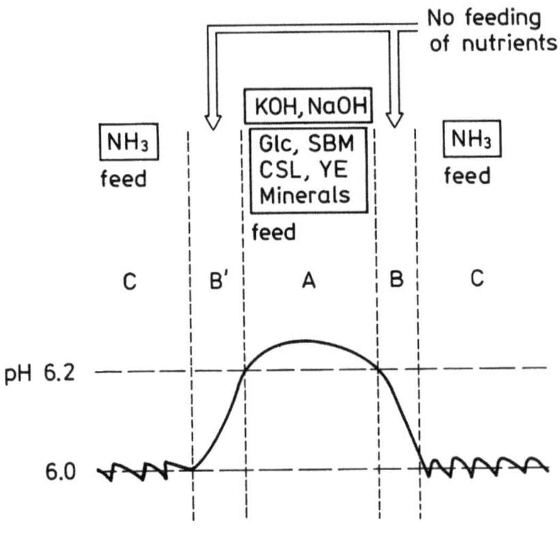

Fig. 8. Profile of pH value during the fed-batch culture phase. The following nutrient was supplied in each period: *A*, MS solution; *B* and *B'*, no nutrient feeding (dead zone); *C*, aqueous ammonia

according to the equation proposed by us [4]. Alkaline solution composed of 200 $g\,l^{-1}$ of KOH and 20 $g\,l^{-1}$ of NaOH was fed simultaneously with the feeding of MS solution. The proportion of the feed rates of these solutions (g MS solution feeding per g alkaline solution feeding) was 8.0. Feedings of both solutions were controlled by a pH controller. When pH of the culture rose above 6.2 the MS solution was supplied, and when it droped below 6.0 aqueous ammonia was supplied. Amounts of the MS solution fed during pH above 6.2 was varied by setting both ON/OFF interval times of the pH controller and the rotation speed of the tubing pumps. The nutrients feeding system of the fed-batch culture is schematically shown in Fig. 7. General profile of pH value during the fed-batch culture period is depicted in Fig. 8.

3.2 Results and Discussion

Influences of temperature and pH on growth and TS production are shown in Fig. 9. Although the maximum growth rate was obtained at 29 °C and pH 6.9, this condition was not preferable for the TS production. It was found that the yield of TS was quite pH-sensitive and the highest in very narrow range of pH 6.0–6.2.

The temperature and pH were set at 27 °C and 6.1, respectively. At this condition the growth rate was not so much smaller than the maximum value.

Effective TS production necessiated both CSL and SBM but their excessive concentrations lowered the TS yield. Maximum concentration of TS was obtained at 20 $g\,l^{-1}$ of SBM, 35 $g\,l^{-1}$ of CSL and 100 $g\,l^{-1}$ of glucose.

Figure 10 shows typical time courses of various variables in a batch culture of *Streptomyces laurentii*. After the inoculation both pH and ammonium concentration increased prominently, but meanwhile they gradually decreased. pH was not

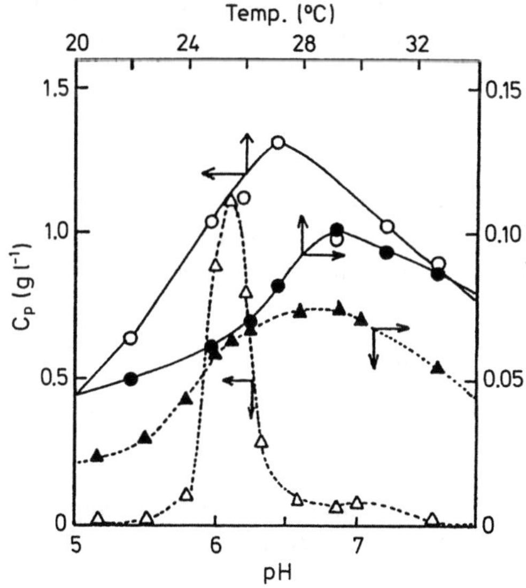

Fig. 9. Effects of temperature and pH on both cell growth and TS production. C_p and μ_m are the final cell concentration and the maximum specific growth rate, respectively

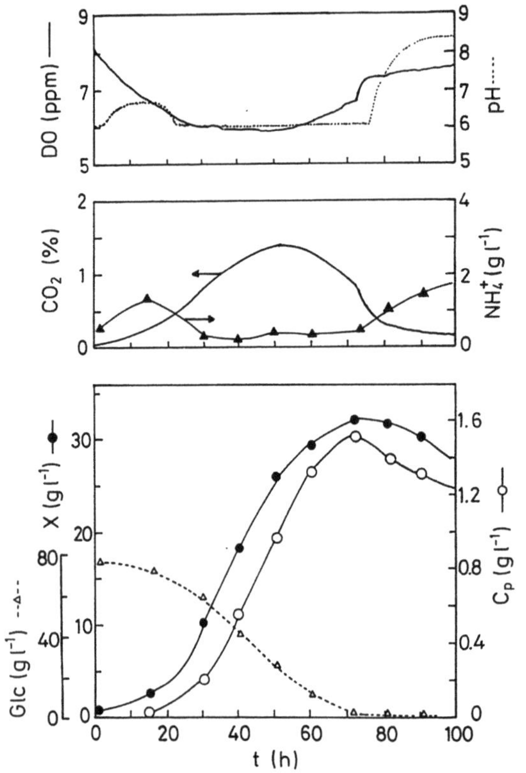

Fig. 10. Typical time courses of variables in a batch culture of *Streptomyces laurentii*. X, Glc and NH_4^+ are concentrations of dry cell mass, glucose and ammonium ion in the medium, respectively. CO_2 is the carbon dioxide concentration in the exhaust gas

controlled in this phase because excessive acid feeding should be avoided to achieve high cell mass concentration [4]. When the lower limit of pH was controlled at 6.1 with aqueous ammonia, both concentrations of biomass and TS increased until glucose was exhausted. pH rose again when glucose was exhausted. These results suggest that the nutrients should be supplemented so as not to cease TS production and that the pH rise is an effective indicator to notice glucose exhaustion.

Effects of various combinations of nutrients fed on the TS production were examined. These were 1) glucose + minerals, 2) glucose + SBM, 3) glucose + CSL and 4) glucose + SBM + CSL + minerals (MS solution). TS was most efficiently produced by 4). The result shows that natural nutrients, SBM and CSL are necessary to achieve high concentrations of both biomass and TS.

Then we tested the effect of the amounts of MS solution fed during pH above the preset upper limit (6.2) by varying 72 g, 24 g, 12 g and 2.4 g of MS solution feeding. The total amounts of MS solution fed in each cultivation was 360 g. The results are shown in Fig. 11. At greater amount of MS solution feeding, glucose was rapidly consumed but no TS accumulated, while at smaller amount of MS solution feeding, the biomass was autolized and TS was decomposed. Only the case of 12 g of the fed amount led to ceaseless prominent accumulation of TS.

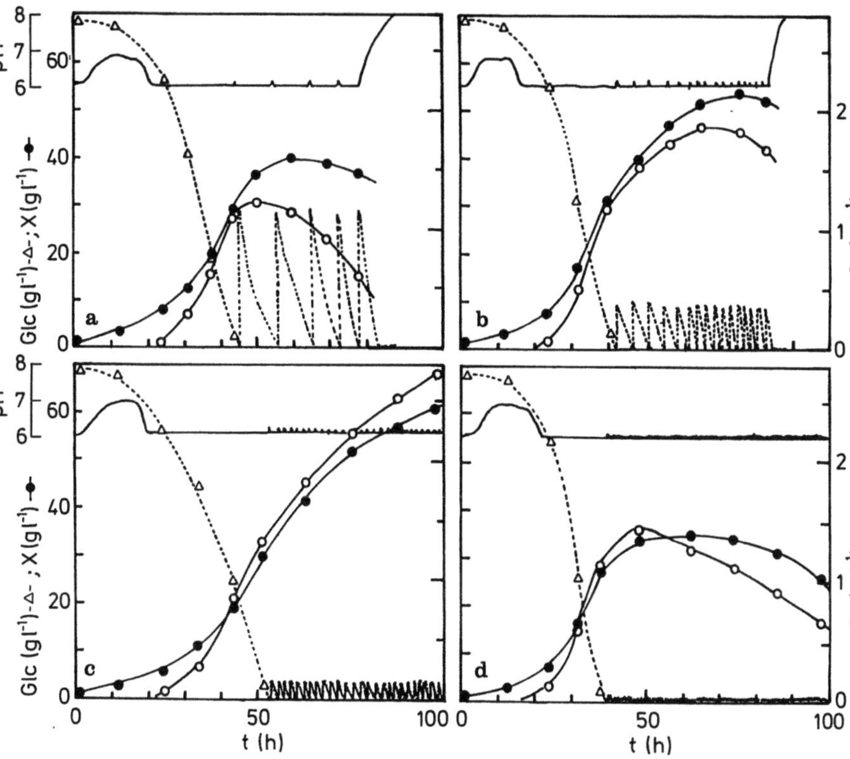

Fig. 11a–d. Effects of the amounts of MS solution fed. The amounts of MS solution (g) fed during a shoot over pH 6.2 were preset at: **a** 72; **b** 24; **c** 12; **d** 2.4

Table 1. Comparison of productivities of thiostrepton

Strain	Cultivation mode	Cultivation time (h)	Thiostrepton final conc. ($g\,l^{-1}$)	Thiostrepton productivity ($mg\,l^{-1}\,h^{-1}$)	Ref.
Streptomyces laurentii ATCC 31255	Fed-batch	240	10.5	44	12
Streptomyces laurentii ATCC 31255	Batch	65	0.015	0.23	13
Streptomyces azureus	Batch	96	0.48	5.0	14

It was finally tried to achieve high TS concentration. Figure 12 shows the time courses of the fed-batch culture. At 60 h of the batch cultivation when the glucose contained in the initial medium was consumed, MS solution was supplied so that the glucose concentration might not exceed $5\ g\,l^{-1}$ but might not become very limiting by pH-stat. At 220 h the highest concentration, $10.5\ g\,l^{-1}$, was obtained while the total cell mass concentration reached $157\ g\,l^{-1}$. During the almost entire

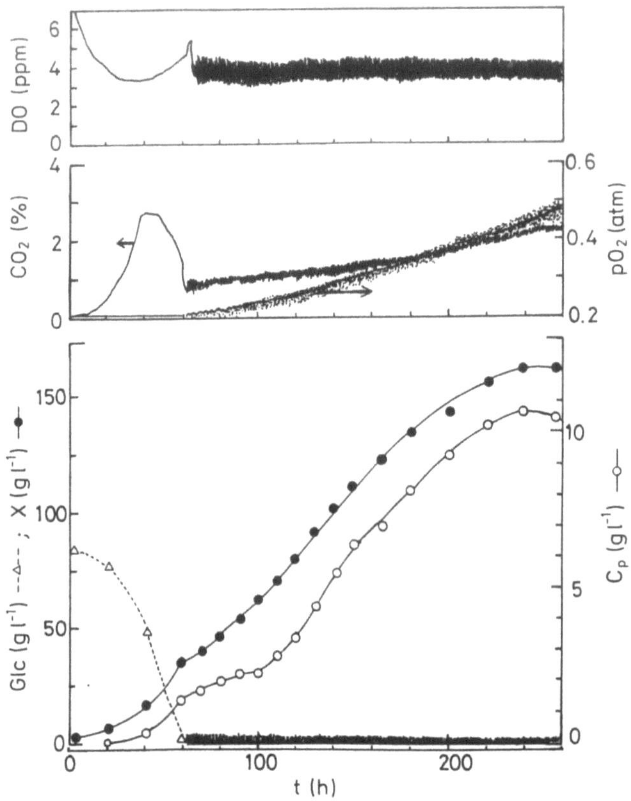

Fig. 12. Mass production of TS by a pH-stat modal fed-batch culture. pO_2 is the partial pressure of oxygen in the inlet gas

fed-batch phase the yield of TS maintained higher level than that in the batch phase.

Table 1 summarizes the comparison of TS productions among literatures. The productivity of the fed-batch culture reported here was about 190 times that in conventional batch culture of the same strain and 9 times that obtained by a batch culture of *Streptomyces azureus*.

4. Concluding Remarks

The productions of PHB from methanol and of TS from glucose, SBM and CSL reported here exemplify mass production of intracellular metabolite by automatic fed-batch culture up to high cell mass concentration.

To produce an intracellular metabolite by a fed-batch culture, maintaining of microbial synthetic activity is very important. Thus, automation of nutrients feeding is the keystone. In case of PHB production, feedings of carbon (methanol) was automated by use of a control system composed of a porous Teflon tubing sensor,

Table 2. Comparison of various culture operations

(A) Comparison of two fed-batch cultures for PHB production

Operation	Culture time (h)	Substrates used (g)	Final culture volume (l)	PHB concentration ($g\ l^{-1}$)	Yield of PHB from methanol[a] ($g\ g^{-1}$)	Productivity[b] ($g\ l^{-1}\ h^{-1}$)
No feeding of NH_3 in PHB production phase (Fig. 5 in Ref. 6)	175	1210 g of methanol 24 g of NH_3	1.6	136	0.18	0.777
Controlled C/N feeding in PHB production phase (Fig. 5, same as Fig. 5 in Ref. 8)	121	1170 g of methanol 24.5 g of NH_3	1.7	136	0.20	1.12
	175	1290 g of methanol 26.7 g of NH_3	1.7	149	0.20	0.851

[a] (PHB produced) × (final culture volume)/(methanol used)
[b] (PHB produced)/(culture time)

(B) Comparison of batch and fed-batch cultures for TS production

Operational mode	Substrate used (g)	Final culture volume (l)	TS produced (g)	Yield ($g\ g^{-1}$)	Running time (h)	Productivity ($g\ l^{-1}\ h^{-1}$)
Three runs of Batch	300 g of glucose 105 g of CSL 60 g of SBM	1.0	4.5	0.015 $g\ g^{-1}$ glucose 0.043 $g\ g^{-1}$ CSL 0.075 $g\ g^{-1}$ SBM	240	0.019
One run of fed-batch	764 g of glucose 262 g of CSL 143 g of SBM	1.8	18.9	0.025 $g\ g^{-1}$ glucose 0.072 $g\ g^{-1}$ CSL 0.132 $g\ g^{-1}$ SBM	240	0.044

a flame ionization detector and a microcomputer together with CO_2 concentration of exhaust gas. Feeding of nitrogen (ammonia) was automated also by CO_2 concentration of exhaust gas with aid of a kinetic model of PHB accumulation. Thus, automatical controls of carbon and nitrogen feedings made it possible to achieve very efficient PHB production. In case of TS production, feedings of glucose and some natural nutrients (SBM, CSL and YE) were simply automated by a pH-stat system. Automatic pH controller is the most commonly used in all kinds of microbial processes and is not a complicated automatic control system. It is a control index which can be applied to complex media containing natural nutrients.

However, to apply pH change to automatic feeding of nutrient(s), one must carefully determine the composition and feed rate of nutrients solution which should be supplied to the culture. We showed a strategy to determine both the composition of natural nutrients solution and feeding mode of the solution. The technique developed in these two studies will be hopefully applied to various microbial processes to attain high productivities of desired intracellular metabolites.

It is very clear that an intracellular metabolite production from methanol by a fed-batch culture is far more economical than a batch culture because one cannot get enough cell mass by a batch culture. Methanol is inhibitory for microbial growth at as low concentration as 0.5–1.0 $g\,l^{-1}$ [6]. The comparison of the two fed-batch cultures for PHB production, the one with no feeding of NH_3 in PHB production phase and the other with controlled C/N feeding in PHB production phase, is summarized in Table 2(A). Controlled C/N feeding does not save the raw-material cost, but can reduce the running cost by 30%.

As for TS production, a comparison of cost between the batch (Fig. 10) and the fed-batch (Fig. 12) can be made on the basis of the same length of operating period. These Figures indicate that three runs of the batch lasting 70 h each are possible in one run of the fed-batch lasting 240 h. The difference in time (30 h) is reserved for spent medium harvesting, cleaning of the bioreactor and preparation of the next run etc. among the batches. The result of the comparison is shown in Table 2 (B), from which one can say roughly that the raw material cost of the fed-batch culture is about 60% of that of the batch culture and the running cost of the fed-batch culture is about 43% of that of the batch culture. These cost down will be, however, counterbalanced a little by the cost of pure oxygen gas or oxygen-enriched air used in the fed-batch culture.

In conclusion, it can be generally stated that the production of intracellular metabolite by fed-batch culture up to high cell mass concentration is much more economical than batch culture if the metabolite synthetic activity of the microorganisms is kept high during the whole period of the fed-batch culture.

5. References

1. Kudo, T., Kato, C., Horikoshi, K.: J. Bacteriol. *156*, 949 (1983)
2. Collmer, A., Wilson, D. B.: Biotechnol. 594 (1983)
3. Nagahari, K., Kanaya, S., Munakata, K., Aoyagi, Y., Mizushima, S.: The EMBO Journal *4*, 3589 (1985)
4. Suzuki, T., Mori, H., Yamane, T., Shimizu, S.: Biotech. Bioeng. *27*, 192 (1985)
5. Yamane, T., Shimizu, S.: Adv. Biochem. Eng./Biotechnol. *30*, 147 (1984)
6. Suzuki, T., Yamane, T., Shimizu, S.: Appl. Microbiol. Biotechnol. *23*, 322 (1986)
7. *idem: ibid. 24*, 366 (1986)
8. *idem: ibid. 24*, 370 (1986)
9. *idem*: J. Ferment. Technol. *64*, 317 (1986)
10. Heinzle, E., Lafferty, R. M.: Eur. J. Appl. Microbiol. Biotechnol. *11*, 8 (1980)
11. Ollis, D. F.: Ann. N.Y. Acad. Sci. *413*, 144 (1983)
12. Suzuki, T., Yamane, T., Shimizu, S.: Appl. Microbiol. Biotechnol. *25*, 526 (1987)
13. Trejo, W. H., Pluscec, J., Brown, W. E., Meyers, E.: Japanese Patent 60-40837 (1985)
14. Ogata, S., Yamada, S., Hayashida, S.: J. Gen. Appl. Microbiol. *31*, 193 (1985)

5.5 A Database System and an Expert System for Realizing Factory Automation in the Bioindustries

Isao Endo, Hajime Asama, and Teruyuki Nagamune

Chemical Engineering Laboratory, RIKEN (The Institute of Physical and Chemical Research), Wako-shi, Saitama 351-01, Japan

Contents

1. Introduction

An industrial microorganism has been thought as an everything which is not only a very small hardware but also a software in a sense that various information is embedded in its DNA. It is the significant meaning of the new biotechnology that we are able to handle with the software part of a microorganism by means of recombinant DNA techniques. Informations in the DNA begin to walk alone parting from the DNA structure now and so that we can design genes as well as enzymes artificially.

But when we look the actual production lines in the bioindustries, we are surprised by the behindness in the technology. Namely, any information of a microorganism, in other words, even the cellular level of functions which are represented by specific rate of substrate consumption, that of oxygen uptake, that of cellular growth, that of metabolic production and so on have not been measured directly yet nor applied to the control of the bioreaction processes. This means that informational studies have not well been fortified in the bioindustries nevertheless they start to use a gene manipulated cell for their production. This situation also tells us the reason why the bioindustry is far behind the other high technology industry in factory automation.

In order to use the information which is inherent in a microorganism to bioprocessings and realize the factory automation in the bioindustries, following problems should be solved first of all:

1) Standardization of bioprocess parameters.
2) Construction of a database system.
3) Codification of expert's knowledge.

Bioproducts and Bioprocesses
Editors: Fiechter, Okada, Tanner

Concerning standardization of bioprocess parameters we explain the parameters in detail in the next chapter, and concerning the other items we discuss the problems and present the concept of our approach in the following chapters.

2. BIOACS (BIO Advanced Control System)

Recently, the authors have developed an automatic monitoring and controlling system for bioprocess (Bio Advanced Control System; BIOACS) in cooperation with Fuji Facom Co. Ltd. and Komatsugawa Chemical Engineering Co. Ltd., that was granted by Research Development Corporation of Japan [1]. During the developmental period, the authors have obtained knowledge of on-line measurement and control of the bioprocesses on the basis of microbial physiological activities.

The characteristics of BIOACS are summarized as follows:

1) On-line measurement of cell mass concentration, one for the substrate and one for the product.
2) On-line monitoring of physiological activities of cells which are represented by various specific rates.
3) Construction of database (Biodatabase), acquiring characteristic properties of bioprocess parameters such as pH-value, temperature, viscosity, etc., and physiological activities of cells.
4) Optimal control of the bioprocesses by using biodatabase.

These characteristics are supported by the following new devices and computer softwares:

1) On-line turbidisensor [2],
2) on-line sampling unit of cell free culture medium [3],
3) accurate estimation of the specific rates via Kalman filter [4,5], and
4) dynamic responses of the specific rates to fed-batch-operations [6,7].

Comparison between the automatic monitoring and controlling system developed so far and BIOACS is shown in Table 1.

The concept of BIOACS is shown in Fig. 1 [1]. The biodatabase shown in this figure is a relational database [8], which applies the mathematical concept of a relation to the system. We therefore can cooperate with other database systems via operators as and/or, large/small/equivalent. Another characteristic of the system is the possibility to add or cancel arbitrary data without minding about the program. Furthermore, we can make a necessary report by communicating with a computer. Hereafter we will discuss the expansion of the BIOACS and its biodatabase.

3. BDBS (BioDataBase System)

The term of database was already mentioned without any definition being given. We define this term as follows: A database system is a collection of stored operational data used by the application systems of some particular enterprise [9].

Table 1. Standardization of the bioprocess parameters for monitoring

	Items	Commonly used system	BIOACS
Monitoring	Temperatur	○	○
	Pressure	○	○
	pH-value	○	○
	D.O.	○	○
	Airation rate	○	○
	Agitation speed	○	○
	pO_2 in	○	○
	pO_2 out	○	○
	pCO_2 in	○	○
	pCO_2 out	○	○
	Working volume	○	○
	Torque	○	○
	Turbidity	×	◎ on-line
	Product concentration	×	◎ on-line
	Substrate concentration	×	◎ on-line
	Specific rates	×	◎ on-line
Controlling	Temperature	○	◎ optimal control
	Pressure	○	○ setpoint control
	pH-value	○	◎ optimal control
	pO_2 in	○	○ setpoint control
	Airation rate	○	○ setpoint control
	Agitation speed	○	○ setpoint control
	Feed rate	○	◎ optimal control

This means that a database system provides the enterprise with centralized control of its operational data. Examples of enterprises are:

Pharmaceutical industry,
biochemical industry,

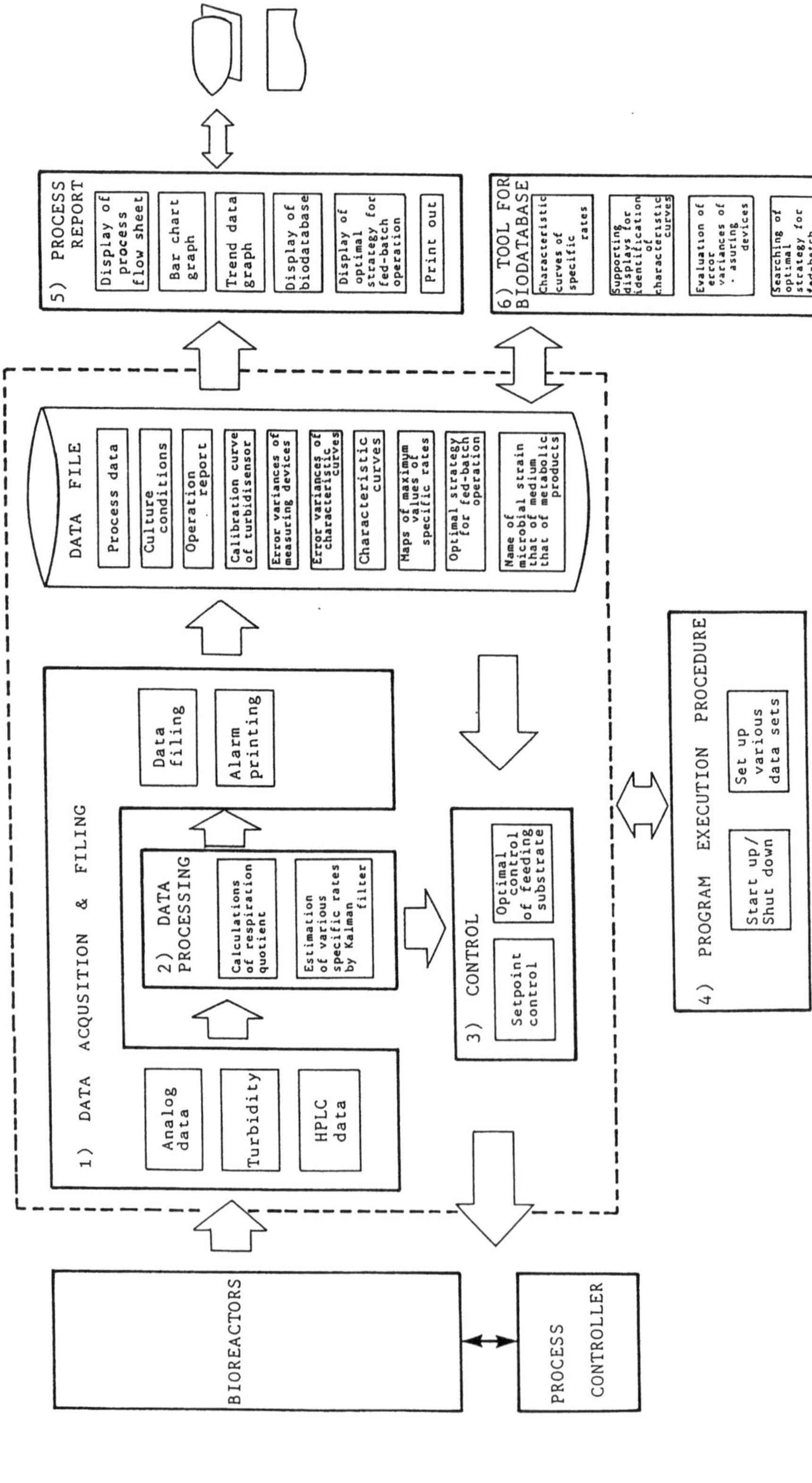

Fig. 1.
The concept of BIOACS

food industry,
universities and
governmental institutes.

The operational data for the enterprise may include

sales data,
product data,
planning data,
process data including down stream processing,
medium composition data,
culture condition data,
characteristic curves concerning various specific rates of cells,
and optimal operational strategies.

The characteristics of these data are rather dynamic than static.

The dynamic data are the fact data. From the viewpoint of a microbiologist, a screened organism is considered to be the most important technical item. But it is of no use to the bioprocess engineer, if the fact data are not accessible from the organism. A database system, which provides the enterprise with the dynamic data, namely, the fact data about bioprocesses, is called the biodatabase system (BDBS). Needless to say that the advantages of the BDBS are the same as those of usual database system. The advantages are:

Redundancy can be reduced,
inconsistency can be avoided,
the data can be shared,
standard can be enforced,
security restrictions can be applied,
integrity can be maintained, and
conflicting requirements can be balanced.

The BDBS consists of four main parts. The first part is called a BIOTRON in which hundreds of mini jar fermentors are operated under various conditions by a single robot. The second part is a supervisory computer for the BIOTRON. The third part is an analysis room for the culture media and product concentrations. The fourth part is the DB machine for acquiring and assembling the fact data.

The reason why we need the help of a robot is that the fact data are so varied and abundant that they exceed the limit of manpower. Many apparatuses, such as the refrigerator for the preservation of the microorganisms, the clean bench for inoculations and the incubator for precultivations, etc., should be improved so that they can assist the work of the robot effectively.

4. Expert Systems for Bioprocess

There are several stages in bioprocesses from upstream processing to downstream processing, and specific problems exist in the individual processes. It leads that knowledge is necessary to solve the problems, and implementation of the knowledge

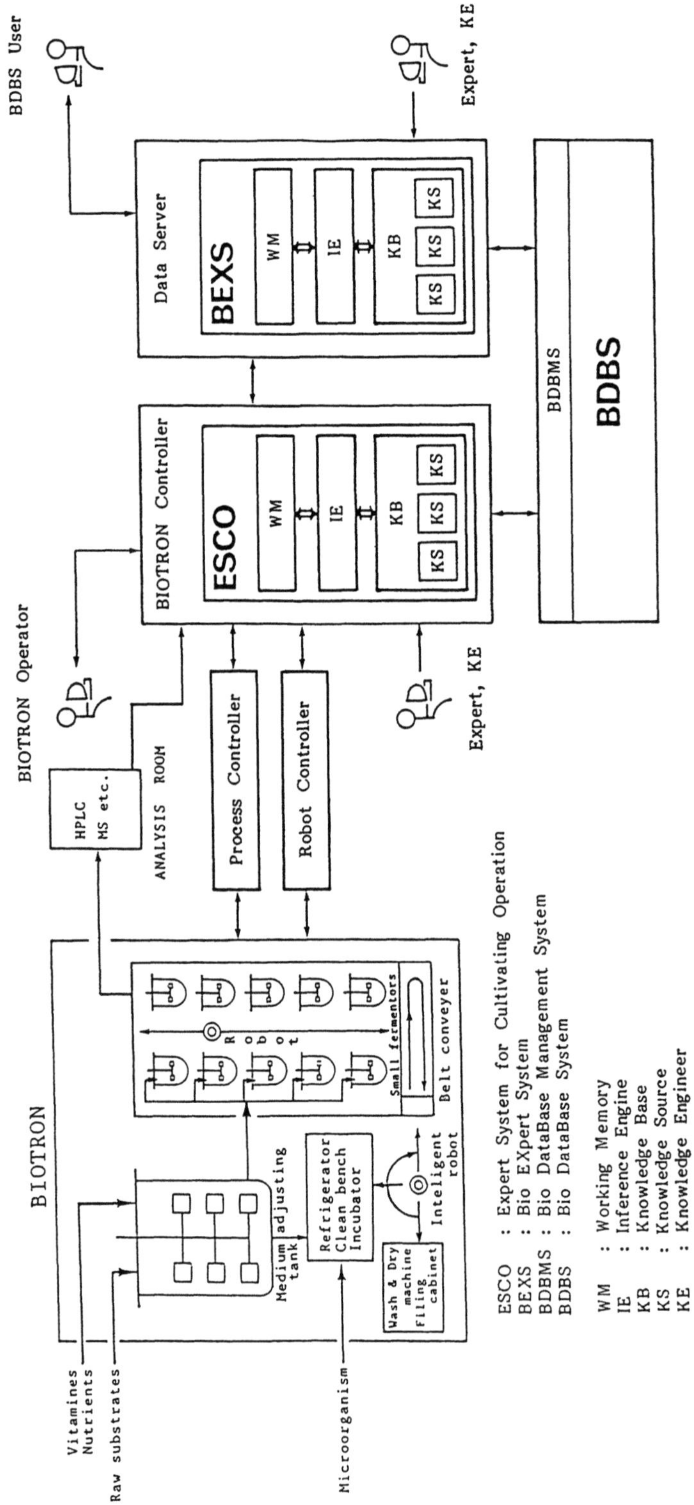

Fig. 2. The concept of biodatabase system (BDBS) and expert systems (ESCO & BEXS)

is indispensable to realize an automated system, though the knowledge is quite spread over various processes. Moreover, as factory automation can never accomplish the efficiency unless the whole processes are integrated and automated, the functions of the automated system should be extensible easily. From these points of view, we discuss the application of knowledge engineering and construction of a bio-expert system which supports the factory automation in bioengineering.

Our concept of the BDBS and two expert systems is illustrated in Fig. 2. One is an expert system for cultivating process, and the other is an expert system for providing adequate data and designing downstream processes in actual industrial operations.

4.1 ESCO (Expert System for Cultivating Operations)

We are developing ESCO aiming at automation of operations in BIOTRON, or supporting operators of BIOTRON in designing and preparing experiments, operating bioreactors in cultivation, diagnosing troubles, measuring, analyzing the obtained data, and storing the data in BDBS. This system is applicable to the factories in product development.

Figure 3 shows a flow of operations in our BIOACS as a typical example of cultivating operations of bioprocess. Some operations are procedural (but rather complicated), and can be described easily in a computer. Others, however, are non-procedural, and quite dependent on knowledge that operators of bioreactors own and apply, which should be called knowhows and heuristics. Knowledge engineering can be considered to give a solution for problems with non-procedural operations.

Experimental design is one of the typical problem which requires expert knowledge. The process of experimental design can be divided in two steps. First step of inference is optimal conditions for the best productivity, which makes cultivating parameters assumed, and second step of inference is measurability of medium components including microorganism, substrates products and by-products. The system should design all the conditions for cultivation and its measurement, e.g., medium composition, pH, temperature, agitation speed for cultivation, and concentration and flow rate of buffer proper column, measuring wave length of spectrometer of liquid chromatograph for measurement.

But when we design experiments, we usually have only such a part of information concerning cultivation as the kind of microorganism and its main products, but not detail information of by-products or composition of natural medium. Therefore experimental design system should cover the function of re-designing after acquiring sample data of past experiments to find better solutions.

Concerning the trouble diagnosing system, cultivating system consists of several devices and pipes. For example in the case of BIOACS, it consists of a reactor, heaters, coolers, pH control agent pots, pumps, air filters, various sensors (pH-sensor, thermometer, dissolved oxygen sensor, high performance liquid chromatograph, turbidisensor, form sensor), various valves, and pipes. Pipes are provided with water line, steam line, and air line. Devices are not always so reliable that we should take account of possible troubles, the method for diagnosing and recovering them.

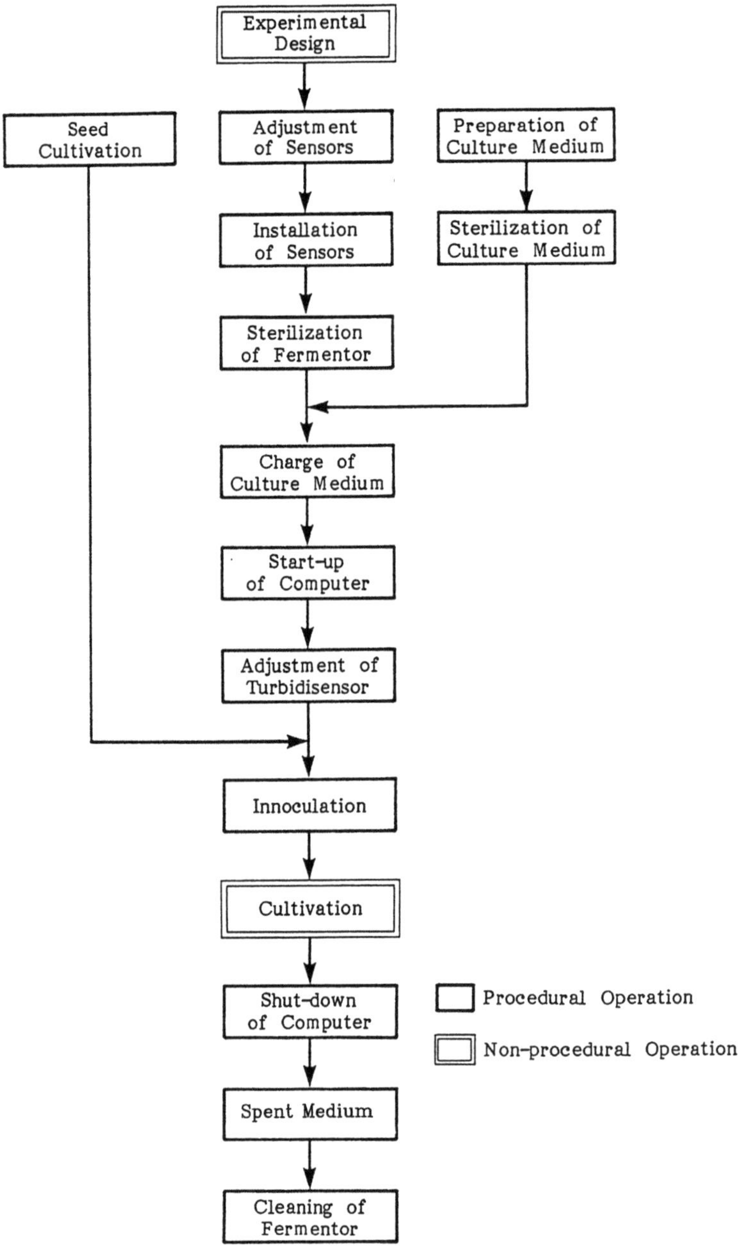

Fig. 3. A flow of cultivating operations in BIOACS

Troubles are usually recognized by monitoring measured values from sensors. In the development stage, however, it can be often the case that microorganisms behave contrary to the primary estimation. Namely we can not always know the behavior when we design the cultivating experiments as mentioned above. These cases should not be dealt with as troubles or errors, but the new data acquired in the process which introduce the new solutions with regard to cultivating conditions. Therefore expert knowledge is also required for the judgement of troubles occurrence.

4.2 BEXS (Bio EXpert System)

The fact data are of no use, if we do not know how to use them for actual industrial operations. For this reason we have planned to build an expert system by which users of the BDBS can ask for the desired data or acquire knowledge of solving emergency. This system is called BEXS.

This system searches data in BDBS and delivers adequate data to end-user, or generates commands to data generator BIOTRON to acquire insufficient data.

The data stored in BDBS should be utilized not only for the purpose of planning of cultivating conditions, but also for the purpose of industrial production. On the stage of industrial production, downstream operations with separating and purifying technologies, are desired to design. We can combine such operations to introduce a sequence for producing products, as liquid chromatography, liquid-liquid extraction, membrane separation, electrophoresis, etc. taking account of conditions of property of microorganism, its products and by-products, components of culture medium, and other parameters in cultivation.

Namely BEXS should have not only a function of searching and presenting required data in the past records, but also a function of introducing optimal procedures to synthesize the operations for production. Therefore the system should be an advanced knowledge based system with expert knowledge with regard to downstream.

5. Conclusion

Firstly, we have proposed a standardization of bioprocess parameters for monitoring and controlling and introduced our advanced control system (BIOACS), which we developed.

Secondly, a model of a database system for bioindustry (BDBS) has been presented.

Thirdly, as a tool for processing information about the effective use of the BDBS, the indispensability of expert systems (ESCO and BEXS) was discussed.

We are just at the starting point of the worldwide development of database systems and expert systems. Substantial development of these systems will realize the FMS in bioindustry.

6. References

1. Report of research development corporation of Japan. Presented by Fuji Facom Ltd. and Komatsugawa Chem. Eng. Ltd. 1985
2. Nagamune, T., et al.: Japan Patent Laid-Open Publ. No. 57-201, 954. US Patent 4, 561, 779
3. Endo, L., et al.: Japan Patent Laid-Open Publ. No. 57-68, 781. US Patent 4, 501, 161

4. Endo, I., et al.: Japan Patent Laid-Open Publ. No. 59-113113, 900
5. Endo, I.; Nagamune, T.; Inoue, I.: Ann. N.Y. Acad. Sci. *413*, 228 (1983)
6. Nagamune, T.; Endo, I.; Inoue, I.: Kagaku Kogaku Ronbunshu *9*, 455 (1983)
7. Nagamune, T.; Endo, I.; Inoue, I.: ibid, *10*, 506 (1984)
8. Date, C.J.: An. introduction to database system, fourth Ed., vol. 1, p. 96, 1985
9. Date, C. J.: ibid, vol. 1, p. 9, 1985

5.6 Recombinant Bioprocess Technology: Is It Different from Traditional Microbial Culture Process Technology

Dewey Ryu[1] and Sun Bok Lee[2]

[1] University of California, Davis, CA, USA
[2] KAIST, Seoul, Korea

Contents

1. Introduction

Recent developments and advances in the area of genetic engineering and biotechnology have shown that there are unlimited possibilities of their potential applications to various gene products and bioconversion products. Some of these biotechnology derived products are already available on the market and some others are well on the way to the market. The important front runners among these new gene products include: insulin, growth hormones, interferons, interleukin-2, tissue plasminogen activator, various vaccines, diagnostics, amino acids, and enzymes. A significant progress has been made also in the areas of bioprocess engineering or "engineering biotechnology" which includes traditional bioprocess technology, enzyme technology, bioseparation technology, bioreactor system, and biosensor technology.

In recent years, we have gained considerable experience in all of these biotechnology areas especially in the recombinant bioprocess technology area, and we are beginning to understand some of the new problems associated with the development recombinant bioprocess technology.

The recombinant DNA technology enables us to identify, clone, and transform the desired genes, and we can have the host cells express the encoded genes very efficiently. On the other hand, the bioprocess technology enables us to produce the gene products economically if and when the recombinant bioprocess is properly

Bioproducts and Bioprocesses
Editors: Fiechter, Okada, Tanner

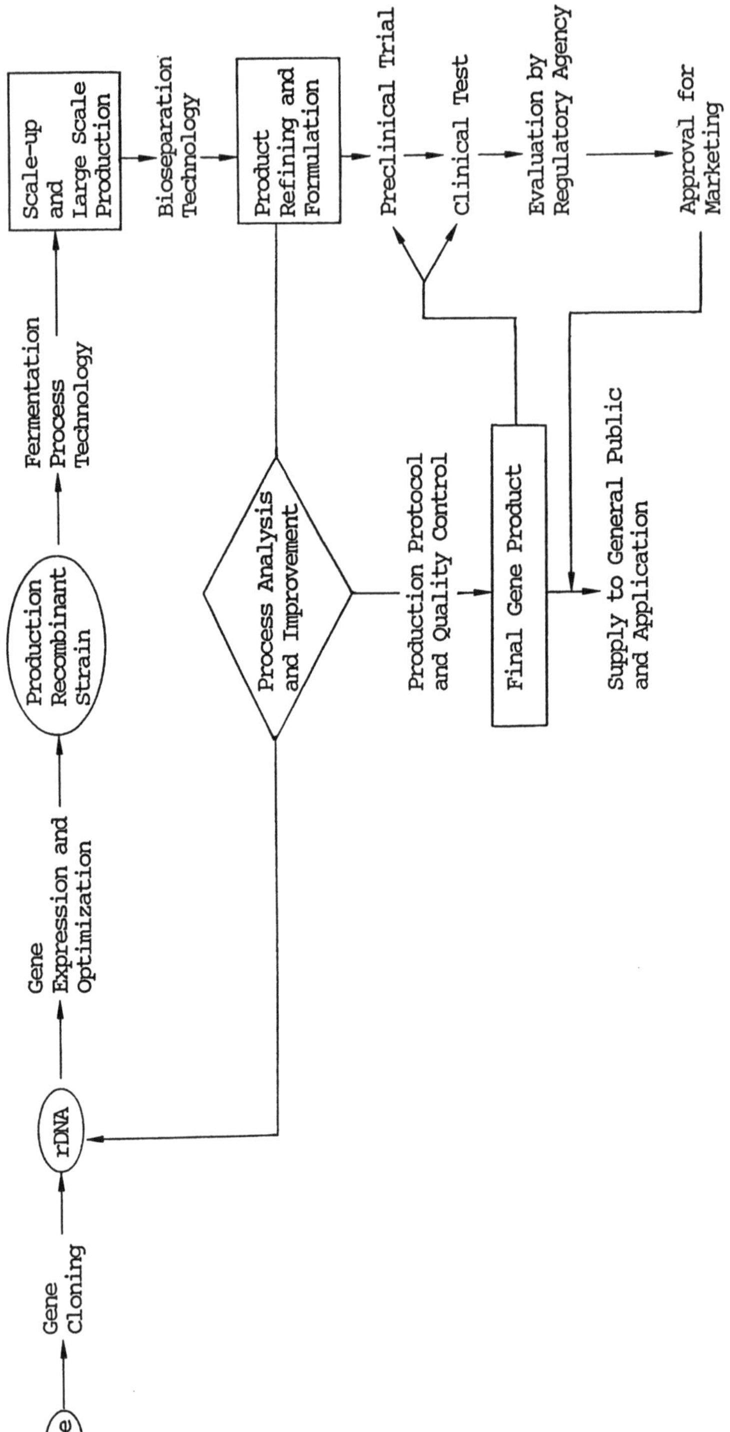

Fig. 1. Commercialization of gene products: development strategy and pathway

developed and optimized. Although many had underestimated the importance of bioprocess technology during the early period of biotechnological development, it is now well recognized and many of us are now focussing our research endeavors on these problems, namely, the scale-up and optimization problems related to the recombinant bioprocess and bioseparation of gene products. The important strategic planning task components required for the commercial development of gene products are highlighted in Fig. 1.

It is also important to consider the "system engineering" approach to the biotechnology process development and optimization. The rDNA technique used in the "front end" or "up stream" of the research endeavor (see the flow chart in Fig. 1) and the "tail end" or "down stream" of the bioprocess technology development endeavour are highly interactive, and a well coordinated and concerted research effort could bring about a very efficient and economical commercial development of the gene products. The selection of host cell and vector system, gene expression system, excretion system, and their effects on the recombinant cell culture process and the purification process of gene products are good examples of the case in point.

Recently, as more and more of the second and thirs generation gene products are being developed for commercialization, the importance of rational development of recombinant cultivation process technology is recognized and appreciated by the research and industrially community.

2. Microbiological and Bioprocess Parameters for the Genetically Engineered Recombinant Cells

Both the genetic and environmental factors or parameters affect the cell growth and product formation. The genetic factors that render unique characteristics to a given recombinant strain include DNA concentration or gene dose, DNA replication, transcription, and translation. Those environmental factors include: temperature, aeration, agitation, pressure, apparent viscosity, osmotic pressure, pH, sources of

Table 1. Microbiological process design parameters

1. Gene Dosage
   - O plasmid copy number
   - O regulation of replication
2. Transcription Efficiency
   - O promoter strength
   - O regulation of transcription
3. Translation Efficiency
   - O nucleotide sequence of ribosome binding site (Shine-Dalgarno sequence)
   - O distance between Shine-Dalgano sequence and ATG or GTG initiation codon
   - O secondary structure of mRNA
   - O codon usage
4. Stability of Recombinant DNA
   - O stability of cloned gene
   - O stability of plasmid
5. Stability of mRNA
   - O structure of mRNA
   - O nuclease
6. Stability of Protein
   - O structure of protein or gene product
   - O protease
7. Host Cell
   - O host/vector interaction
   - O metabolic activity of the host cell
8. Others
   - O protein secretion efficiency
   - O proper termination between genes

carbon and nitrogen, medium composition, growth factors, precursors, inducers, metabolites having toxic and/or inhibitory effects, etc. These parameters are also highly interactive and their relationships to the overall productivity are important. In order to optimize the recombinant process and maximize the productivity of gene product, the important genetic and environmental parameters have to be identified and the strategy for both front end and tail end bioprocess technology have to be developed. Some selected genetic and microbiological process design parameters that are important to the development of recombinant process technology are listed in Table 1.

Invariably, the recombinants are challenged to overproduce the gene products, and the problems associated with their metabolic strain or imbalance are very common phenomena. For this reason, it is most likely that we will have to find optimal set of conditions with respect to DNA replication, transcription, and translation.

2.1 Host Cell Selection

The selection of best host cell is one of the most important tasks in production of gene products. The host cells provide all the facilities and regulation required for the biosynthesis of product, including DNA replication, transcription, and translation. It is also important to understand the mechanism and regulation of gene expression in the host cell.

So far, *E. coli* has been most widely used as the host organism in recombinant DNA (rDNA) technology, and transformation are relatively easy with *E. coli* as compared with other organisms. However, *E. coli*, as a host organism, has some disadvantages. They are: (1) bacterial endotoxin has to be carefully eliminated by purification, (2) *E. coli* does not carry out post-translational glycosylation of the protein or gene products, (3) it does not eliminate methionine from the product as in the case with the human growth hormone (met-hGH), (4) it does not normally excrete gene products and accumulates intracellularly, and others.

Although there are some advantages of employing yeast and bacilli as host organisms and significant advances have been made in recent years, we know relatively little about their genetic characteristics, the gene expression mechanisms, and the plasmid instability in these host organisms. The selection of host cell influences significantly the fermentation conditions, medium composition, cell growth rate, production rate, host cell vector interaction, stability of mRNA, stability of gene product, the product recovery process, etc.

2.2 Gene Concentration

Gene concentration depends on the plasmid copy number since the desired gene is very often incorporated into the plasmid vector system when rDNA is to be constructed and cloned. The plasmid copy number ranges from a low copy number (1 ~ a few) to a high copy number (~ order of 100). The multicopy number plasmids like pBR322 and ColE1 are very widely used in practice. In some cases, "run away" plasmids are used. More recently, the method by which the copy number

Table 2. The effect of plasmid copy number on the gene product and plasmid stability. Summary of experimental data from Aiba et al. [1)]

Strain	TSase ($U\ mg^{-1}$ protein)	Copy number	Plasmid stability (%)	Tryptophan ($g\ l^{-1}$)
Tna(RP4 *trp.* 115)	36	1 ~ 3	~100	1.7
Tna(pSC101 *trp.* 115)	107	~5	~ 85	3.1
Tna(RSF1010 *trp.* 115)	215	10 ~ 50	~35	2.6
Tna(pBR322 *trp.* 115)[a]	—	60 ~ 80	—	—

[a] Stable transformants difficult to obtain

can be controlled has been developed by linking RNAII gene, which regulates the copy number of ColE1 plasmid, to P_L, *trp*, or *lac* promoter.

In principle, the productivity of gene product is expected to be proportional to the gene dose or plasmid copy number. However, the productivity does not increase beyond a certain level when the gene dose is too high since the primary metabolism of the host cell is severely strained. Thus, controlling the optimal plasmid copy number is very important in certain cases. Aiba and Imanka [1)] found that a low copy number plasmid pSC101 yielded more tryptophan as compared with other high copy number plasmid pBR322 or RSF1010 because of the instability of high copy number plasmids. Their results are shown in Table 2. They also showed that the plasmid stability decreased significantly with an increase in the plasmid copy number although the tryptophan synthetase (TSase) activity increased with the plasmid copy number. They suggested that, in view of these opposing effects, an optimal gene or plasmid concentration which corresponds to the maximum productivity of gene product should be used for production purpose.

2.3 Transcription Efficiency

The transcription efficiency depends, in large part, on the promoter strength and the regulatory mechanism in the given host organism. An ideal promoter should be a very strong one, it should be easy to control or its transcription can be easily turned on or off, and it could be induced under specific induction condition. For maximum productivity, a very strong promoter which can be easily controlled is highly desirable. The rationale for having such a promoter system is that: (1) the plasmid containing uncontrollable strong promoter is usually highly unstable, (2) some foreign protein produced in a large quantity intracellularly is often very toxic to the host organism, and (3) the degradation or instability of gene product can be minimized when the gene product is made at a higher rate with a controllable promoter as compared to a situation where the protein product is made at a lower rate for a prolonged period. The examples of good promoters which have been widely used for *E. coli* are: *lac*, *trp*, P_L, *tac* and *recA*.

One method of increasing the transcription efficiency is to increase the inducer concentration. There are evidences that the recombinant organisms respond to the optimal inducer concentrations for its maximum productivity. The well known examples are productions of hemagglutinin and immunoglobulin using *trp* promoter and indoleacetic acid (IAA) as an inducer, and production of β-interferon by using *recA* promoter and nalidixic acid as an inducer.

2.4 Translation Efficiency

The translation efficiency depends on the ribosome binding site nucleotide sequence (or Shine-Dalgano sequence), the distance between the initiation codon (ATG or GTG) for protein biosynthesis and SD sequence, the secondary structure of mRNA, and the codon usage of *cloned gene*. Recently, the ribosome binding site nucleotide sequence (TTAAAATTAAGGAGG, the lined portion of nucleotides represents the SD sequence) of which the translation efficiency was found to be very high was synthesized and applied successfully to the production of interferon, interleukin-2, and other gene products. The product yield up to 10 to 20% of intracellular protein was achieved by this approach. The optimal space between the initiation codon and SD sequence for production of alpha- and beta-interferons appears to be about 9 base pairs (bp) when *trp* promoter and *E. coli* as host cell were used [2]. Another example shows that the optimal spacing between the SD sequence and ATG initiation codon was about 10 bp for immunoglobulin production when *trp* promoter was used [3].

2.5 Stability of mRNA and Protein

The stability of mRNA and protein products depends a great deal on the growth rate of the host organism, and the rates of degradation of protein and mRNA increase with the unfavorable environmental conditions and the increasing temperature. The stability of protein depends also on the intracellular location where the product is accumulated. Insulin was found to be more stable when it was accumulated outside the periplasmic membrane as compared to the cytoplasmic site for accumulation. The half-life of the insulin accumulated in the cytoplasm was only about 2 min as compared with about 20 min outside the periplasmic membrane [4].

When multiple structural genes are put together or the protein product accumulates in the form of inclusion bodies, the rate of protein degradation was found to decrease [5].

2.6 Protein Secretion and Termination of Transcription

Although the regulatory mechanisms of protein secretion are not completely understood, some progress has been made with the signal peptide and efficient secretion of product has been achieved using other cells including bacilli, pseudomonas, and yeast. When a strong promoter is used, the replicon contained in the same vector system affects the transcription adversely and low productivity results. In this situation, a strong terminator could be used to prevent any interaction between the replicon and the vector system [6].

2.7 Instability of Recombinant Plasmids

The instability of recombinant plasmids remains to be one of the most critical problems in the practical application of many recombinants. Although some recombinants are relatively stable, others have both the structural and segregational instability which often leads to low productivity. The structural instability is often caused by deletion, insertion, recombination, and others while segregational instability is caused by uneven distribution of plasmids during cell division. Thus, it is also important to understand the nature of plasmid instability and its effect on productivity.

The degree of instability of recombinants can be estimated by the growth ratio and plasmid loss rate (or segregation rate). These two parameters determine the instability characteristics of a given recombinant plasmid. The growth ratio ($\alpha = \mu^+/\mu^-$) is defined as the ratio of specific growth rate of plasmid-harboring cells to that of plasmid free cells. The segregation rate parameter ($\theta = \Theta/\mu^+$) is defined as the ratio of the specific rate of generation of plasmid-free cells from the plasmid-harboring cells (or plasmid loss rate, Θ) to the specific growth rate of plasmid-harboring cells (μ^+). In general, the growth ratio (α) is less than 1.0. The growth rate of recombinants is adversely affected by the expression of recombinant DNA which causes stress on primary metabolism of the host cell. Some exceptions to this observation were found in some cases. The growth ratio is influenced by the medium composition, mineral ion concentration, temperature, other environmental factors, and the characteristics of the gene product.

Those instability kinetic parameters, α and θ, are affected by both the genetic and environmental factors. They include: genetic properties of plasmid vectors, interaction between the host cell and plasmids, the degree of gene expression, cultivation temperature, nutrients, dilution rate, the mode of bioreactor operation, and others. When pBR322 and pBR325 plasmids are compared, pBR325 was found to be somewhat more unstable and it was also observed that *E. coli* K12 GM31 strain was slightly more unstable as compared to *E. coli* K12 GY2354 strain [7)]. Another study with *S. lividans* harboring pIJ2 plasmid showed that the recombinants become more unstable when continuously cultivated at a higher temperature and lower dilution rate [8)].

Ryu and others showed that two-stage continuous culture system was found to be very useful in controlling and improving the plasmid stability problem [9–11)]. They used the first-stage as the growth stage under the repressed condition and the second-stage as the expression or production stage where relatively high productivity could be maintained. They also showed that a high level of the plasmid-harboring recombinant cell fraction (Φ) and a high level of gene product concentration (P_2) could be maintained in the second stage of the two-stage continuous culture system.

2.8 Strategies for Improving the Stability of Recombinants

A list of selected strategies for improving the recombinant stability is shown in Table 3.

Table 3. Strategies for improving or maintaining stability of recombinant cells

1. Addition of Antibiotics
   - ○ ampicillin, tetracycline, etc.
   - ○ novobiocin
2. Use of an Auxotrophic Mutant as Host Cell
   - ○ amino acid auxotroph
3. Use of Genes Modulating Stable Maintenance of Plasmid
   - ○ *par* loci
   - ○ *cer* gene
4. Use of Genes Involving Natural Fitness
   - ○ λ *cos* site
5. Use of Internal Selection Pressure
   - ○ λ lysogens
   - ○ streptomycin dependency
6. Use of Recombinant-deficient Host-Strains
   - ○ *recA*, etc.
7. Regulation of Gene Expression
   - ○ proper induction/repression
8. Design of Bioreactor Operating Mode
   - ○ two-stage continuous culture

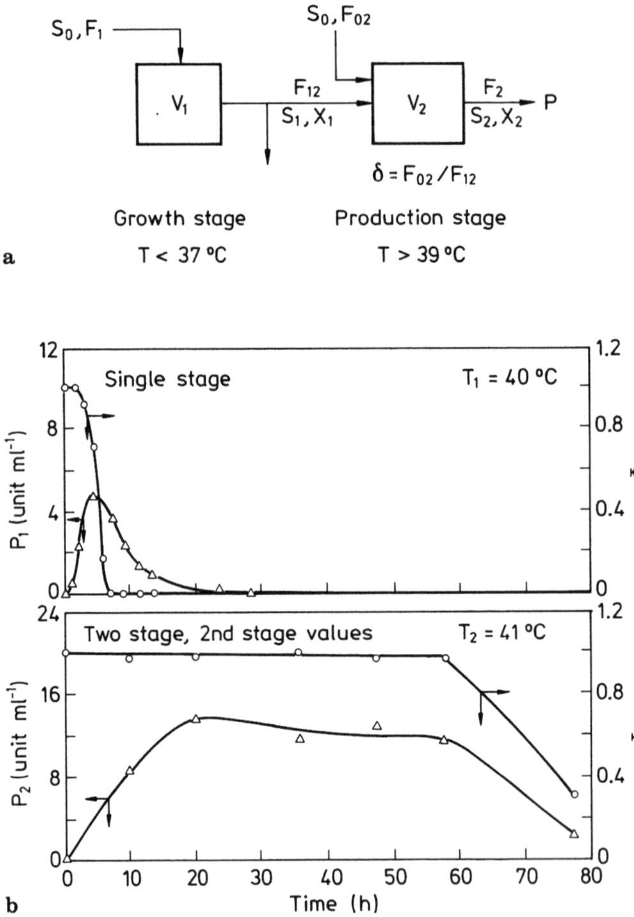

Fig. 2. **a** Two-stage continuous culture system. **b** Comparison of plasmid stability (Φ) and product formation (P)

Antibiotics are widely used to prevent growth of plasmid-free cells because it is convenient, although it could be a problem in a large scale operation.

Auxotrophic mutant as the host organism in combination with the gene coded for the required amino acid can be used to improve the stability of recombinant cell population. This approach also presents some problems.

In nature, very stable plasmids exist and certain mechanism by which the stability of plasmids are maintained have been studied. The *par* (partitioning) function of pSC101 plasmid and *cer* (ColE1 resolution) function of ColE1 plasmid have been identified. The *par* function is for equal partitioning of plasmids and *cer* makes monomer plasmid from multimer by multimer resolution at the time of cell division. Certain plasmid containing *trp* operon was stabilized by *par* function, but only with a limited success [12-15].

Control of gene expression by gene switching system could be employed. The expression and repression of a promoter can be controlled by separating the growth-stage and production stage. Ryu and others [9,10,16] illustrated a possibility of applying this principle to a two-stage continuous culture system (Figs. 2a and 2b).

Other strategies include insertion of λcos site into plasmid, use of λ lysogen which makes suicidal cell when the host cells lose plasmids, streptomycin dependency, use of recombination-deficient host strains *recA*, etc.

3. Kinetics of Recombinant Bioprocess

When the recombinant cultivation process is to be optimized and/or improved, many genetic parameters of recombinant, as discussed in this paper, have to be dealt with. Thus, in-depth understanding of recombinant process and quantitative analyses of bioprocess kinetics will be required for the optimization and improvement of recombinant process.

Recent advances in molecular biology enhanced our understanding of regulatory mechanisms of microbial metabolism and enabled us to analyze these more accurately than ever before. The analysis of molecular model provides us with some insights into these metabolic regulations taking place at the molecular level. One may develop a strategy which gives the best gene expression efficiency. The *trp* or P_L promoter and the autorepressor control system was found to give a high gene expression efficiency. Another strategy would be to clone the repressor gene with the target gene on the plasmid and put the repressor formation under the autorepressor control system.

The recombinant culture system contains both the plasmid-harboring and plasmid-free cells and a heterogeneous cell population must be evaluated for the design and analysis of recombinant process technology. Only the plasmid-harboring cell fraction is productive and a good understanding of the dynamics of heterogeneous cell population is also important to the process optimization.

3.1 Bioprocess Kinetics

Based on the transcription and translation kinetic model, a mathematical model for gene product formation, by which the efficiency of gene expression can be evaluated, have been recently developed and reported [9,10,16]. The specific production rate of gene product (q_P) was derived as a function of gene concentration (G_P), efficiency of gene expression (ε), the specific growth rate of plasmid harboring cells (μ^+), overall rate constant for gene product formation (k_0), and other lumped specific rate parameter which is implicit in contribution of nongrowth associated parameter toward the product biosynthesis (b).

$$q_p/G_p = k_0\varepsilon(\mu^+ + b)$$

The value of $k_0\varepsilon$ was then estimated from the experimental results and it was found to be about 5,700 protein molecules (as gene product) per molecule of DNA. This result shows that the gene expression efficiency of the λP_L promoter system which the authors [9,10,16] used in combination with pPLc23*trp*A1 could be estimated. This example illustrates how we can evaluate and estimate the gene expression efficiency and the performance of recombinant processes based on molecular and cellular biology and apply this to practical recombinant bioprocess design and development.

3.2 Recombinant Bioprocess Variables

Basically the same process variables or parameters that are important to the traditional microbial cultivation processes should be considered for the recombinants. These parameters are related to the bioreactor system design and operation, media design and optimization, and the state and control variables involved in the recombinant processes.

For the commercial scale microbial processes, one has to consider the scale-up of the bioreactors, the mode of operation, and the strategy for optimal process control. When recombinant processes are to be scaled up, one of the critical problems recognized has been the instability of recombinant organisms harboring unstable plasmids. For such unstable recombinants, the inoculum development protocol and the medium design have to be carefully examined.

The recombinants showed more sensitivity to the conditions of cellular environment, and more accurate control strategy and technique will be required for the optimal process control especially when the gene expression is controlled by a temperature shift or inducer addition.

The recombinants appear to have higher specific uptake rate, in large part, due to the faster or higher specific production rate. The faster metabolic rate or recombinants requires that the response time of biosensors employed for the control of recombinant processes will have to be shorter and the measurement needs to be more accurate.

In some cases when hazardous recombinants are used, post-treatment such as pasteurization or sterilization of recombinants and broths will be required in accordance with the biosafety guidelines.

Temperature optimization is still an important area of investigation for the recombinant process development. While growth optimal temperature for the host organism, *E. coli*, is 37 °C, the optimal production temperature of both the interferon and insulin was found to be 30 °C [17–19]. In the case of insulin production, the productivity was increased by 3-fold when the cultivation temperature was controlled at 30 °C as compared with the productivity for cultivation controlled at 37 °C. This increase was attributed to a significant reduction in the product degradation at a lower cultivation temperature. Thus, the stability of gene product is significantly affected by the operating temperature.

Recent studies showed that an early induction of gene expression during the exponential growth phase resulted in about 5-fold increase in the productivity of EcoRI restriction enzyme as compared to the productivity resulted from a late induction during the stationary phase [20]. In this case, a significant increase in oxygen uptake rate coincided with the induction of gene expression and production of the enzyme. These results suggest that the selection of the best time of induction for gene expression during the growth cycle and a sufficient oxygen supply to the recombinant cells especially during the production phase immediately following the induction or gene expression are very important to the optimization of the recombinant processes.

The productivity can also be improved by using immobilized cell culture, high density cell culture, or continuous culture. Usually the cellular content of chromosomal DNA tends to increase slightly with an increase in the specific growth rate. However, the cellular content of certain plasmid DNA tends to decrease with an increase in the specific growth rate. Such examples include pPLc23*trp*A1, ColE1, and pBR322 plasmids [9,21,22]. On the other hand, pLP11 plasmid contained in *B. stearothermophilus* showed an increasing tendency with an increasing dilution

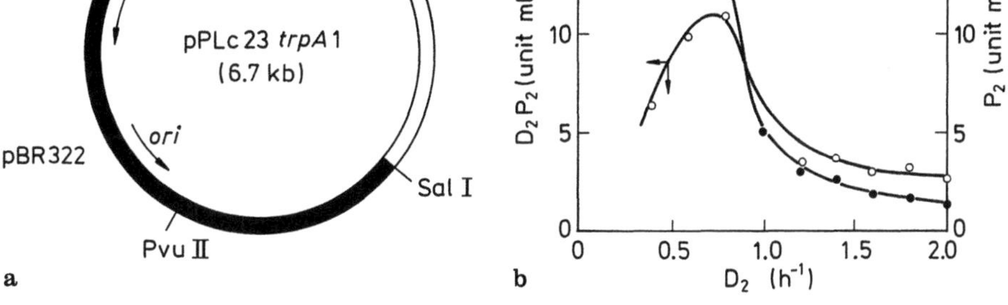

Fig. 3a, b. Production of tryptophan synthetase using pPLc23*trp*A1 in a two-stage continuous culture system. **a** the plasmid pPLc23*trp*A1. **b** productivity of the gene product as a function of dilution rate in the second-stage at a given first-stage dilution rate

rate [23]. In the case of pPLc23*trp*A1, there appears to be an optimal dilution rate which corresponds to the maximum productivity [10,11]. In their study, they have used a two-stage continuous culture system and the gene expression was switched on by shifting the temperature in the second-stage to 41 °C while maintaining the growth stage under the repressed condition at 37 °C (Figs. 3a and 3b).

4. Acknowledgement

Parts of this lecture material are presented in our contribution to the book, "Horizons of Biochemical Engineering" (ed. S. Aiba), which was published by the University of Tokyo Press in 1987.

5. References

1. Aiba, S., Tsunekawa, H., and Imanaka, T.: Appl. Environ. Microb. *43*, 289 (1982)
2. Shepard, H. M., Yelverton, E., and Goeddel, D. V.: DNA *1*, 125 (1982)
3. Kurokawa, T. et al.: Nucleic Acids Res. *11*, 3077 (1983)
4. Talmadge, K., and Gilbert, W.: Proc. Natl. Acad. Sci. *79*, 1980 (1982)
5. Shen, S.-H.: Proc. Natl. Acad. Sci. *81*, 4627 (1984)
6. Gentz, R., Langer, A., Chang, A. C. Y., Cohen, S. N., Bujard, H.: Proc. Natl. Acad. Sci. *78*, 4936 (1981)
7. Noack, D. et al.: Mol. Gen. Genet. *181*, 121 (1981)
8. Roth, M., Noack, D., and Geuther, R.: J. Basic Microbiol. *25*, 265 (1985)
9. Seigel, R., and Ryu, D. D. Y.: Biotech. Bioeng. *27*, 28 (1985)
10. Lee, S. B., Siegel, R., and Ryu, D. D. Y.: Biotech. Bioeng. (in press)
11. Ryu, D. D. Y., Siegel, R., and Lee, S. B.: Paper presented at the Annual Meeting of Am. Inst. Chem. Eng. (1985)
12. Meacock, P. A., and Cohen, S. N.: Cell *20* 529 (1980)
13. Summers, D. K., and Sherratt, D. J.: Cell, *36*, 1097 (1984)
14. Hakkaart, M. J. J., Veltkamp, E., and Nijkamp, H. J. J.: Mol. Gen. Genet. *188*, 338 (1982)
15. Skogman, G., Nilsson, J., and Gustafsson, P.: Gene *23*, 105 (1983)
16. Ryu, D. D. Y.: Proc. BioExpo 85, p. 351 (1985)
17. Bachmann, K. and Ptashine, M.: Cell *13*, 54 (1978)
18. Meyer, H.-P. et al.: Proc. 3rd Eur. Cong. Biotechnol., Vol. I, p. 499 (1984)
19. Emerik, A. W. et al.: Bio/Technol. *2*, 165 (1984)
20. Botterman, J. H. et al.: Biotech. Bioeng. *27*, 1320 (1985)
21. Engberg, B., and Nordstrom, K.: J Bacteriol. *123*, 179 (1975)
22. Stueber, and Bujard, H.: EMBO J. *1*, 1399 (1982)
23. Koizumi, J.-I., Monden, Y., and Aiba, S.: Biotech. Bioeng. *27*, 721 (1985)

5.7 Effect of Shear Stress on Xanthan Gum Production

Hitoshi Funahashi[1], Toshiomi Yoshida[2], and Hisaharu Taguchi[2]

[1] Department of Fermentation Technology, Faculty of Engineering, Osaka University, Yamada-oka, Suita-shi, Osaka 565, Japan
[2] International Center of Cooperative Research in Biotechnology, Faculty of Engineering, Osaka University, Yamada-oka, Suita-shi, Osaka 565, Japan

Contents

The effects of agitation by a flat-bladed turbine impeller on the microbial production of xanthan gum were studied. The results showed that the intensity of shear flow of an impeller was particularly important compared with other physical factors relevant to agitation. It was also observed in an equipment of defined shear rate that the specific production rate of xanthan gum and the specific oxygen uptake rate (OUR) were influenced by the shear stress. Both of these rates increased with increasing shear stress in the range of 0–40 Pa, beyond which they were constant. Increases in the glucose concentration, as the driving force of glucose transfer from the medium into the cells, resulted in increases of the specific production rate of xanthan gum. These results suggested that the shear stress was particularly important for xanthan gum production and that microbial production of xanthan gum might be limited by the mass transfer of the materials from the medium into the cells at low shear stress.

1. Introduction

Xanthan gum produced by *Xanthomonas campestris* is an extracellular biopolymer. It is a hetero-polysaccharide and its aqueous solution has some useful properties in that it is stable against pH, temperature and existence of salts in the solution and

Bioproducts and Bioprocesses
Editors: Fiechter, Okada, Tanner

is highly viscous even at low concentration [1]. Use of this biopolymer as a thickening, suspending, stabilizing and gelling agents in both food and industrial applications have major importance [2,3]. Moreover, xanthan gum has been expected to be used for the secondary oil recovery [3,4].

Moraine et al. [5,6] suggested that the agitation was one of important operational factors for the microbial production of xanthan gum. It was speculated that the shear flow caused by the impeller rotation removed the xanthan gum layer around the cells. But the effects of agitation have not been studied quantitatively. Hence an investigation was carried out in detail about the effects of the agitation by a flat-bladed turbine impeller and the physical factors relevant to the agitation on the xanthan gum production. In addition, the effects of the shear stress on the specific production rate of xanthan gum and the specific oxygen uptake rate, and the effects of glucose concentration, as a driving force of the glucose transfer, on the specific production rate of xanthan gum (SPR), were studied.

2. Materials and Methods

2.1 Organism

Stock culture of *Xanthomonas campestris* ATCC 13951 was maintained at −80 °C in a medium. 1% yeast extract, 2% peptone and 2% glucose at pH 7.0. Preculture was done by inoculating 2 ml of the stock culture into a shake-flask which contained 100 ml of the above medium and incubating for 2 days at 30 °C on a reciprocating shaker.

2.2 Medium for Xanthan Gum Production

A modified medium of Lilly et al. [7] was used as the medium for the xanthan gum production. 1 l of the medium contained 50 g of glucose, 7 g of sodium fumarate, 2 g of peptone, 2 g of KH_2PO_4, 0.5 g of $MgSO_4 \cdot 7\,H_2O$ and 2.0 ml of salt solution, which contains 0.723 g l^{-1} of $Fe(NO_3)_3 \cdot 9\,H_2O$, 0.440 g l^{-1} of $ZnSO_4 \cdot 7\,H_2O$ and 0.203 g l^{-1} of $MnSO_4 \cdot 4\,H_2O$, and 1 ml l^{-1} of antifoam (Adekanol Lg-109, Asahi Denka Ind. Co. Ltd.).

2.3 Bioprocess

A reactor of 10 l capacity (B. E. Marubishi Co., Ltd., MODEL MD500) was used for the cultivation. The working volume was 6.0 l. Agitation was performed by a 6-flat-bladed turbine impeller shown in Fig. 1, placed at the half point of the liquid depth. The impeller diameter and the blade width were varied within the range of 0.087–0.14 m and 0.01–0.08 m, respectively. The blade length and the disk diameter were 0.025 m and 0.066 m, respectively. The vessel diameter was 0.20 m. Agitation speed was varied in the range of 3.33–16.67 rps, air was supplied at 1.0×10^{-4} m^3 s^{-1} into the reactor, temperature was kept at 30 °C and pH was not controlled during

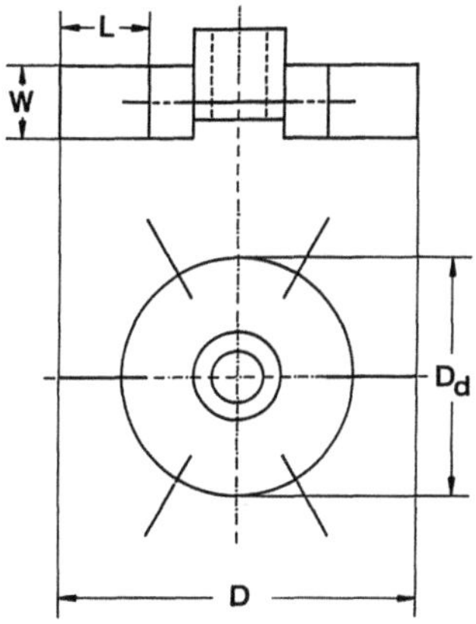

Fig. 1. Dimensions of a 6-flat-bladed turbine impeller

the cultivation. The dissolved oxygen tension (DO) was controlled by changing the partial pressure of oxygen (pO_2) in supplied air. This was done to investigate the effects of DO in the specific production rate of xanthan gum.

2.4 System for Investigating Effects of Shear

A schematic diagram of the equipment used to investigate the effects of the shear rate and the shear stress is shown in Fig. 2. The medium for SPR measurement was prepared as follows. Cells collected by centrifugation were suspended at 2 g kg^{-1} cells per volume in 50 mM n-tris(hydroxymethyl)methyl-2-aminoethane sulfonic acid (TES) buffer (pH 8.0) containing 4% glucose and was incubated at 30 °C until the concentration of xanthan gum reached the desired value. SPR was measured after introducing this reaction mixture into the reactor. Pure oxygen gas was supplied into the open space inside the reactor, keeping the inner pressure at 3×10^5 Pa to avoid oxygen deficiency during the measurement of SPR. Oxygen supply was done only by backmixing from the open surface and no air was introduced by a sparger. The buffer was suitable to maintain the cell concentration without loss of production activity throughout the experiment and the value of SPR was of the same order as in the ordinary production medium. In the case where effects of shear stress on oxygen uptake were investigated, oxygen was sparged into the medium, which was quickly transferred into the reactor and changes of dissolved oxygen concentration were measured under unaerated condition at a low rotating speed to avoid introducing bubbles. The oxygen transfer rate (OTR) from the interfacial area was very small compared with the oxygen uptake rate (OUR) and was negligible in this study. For example, the OTR was less than 5.7×10^{-6} kg m^{-3} s^{-1} even using water at 7 rps rotating speed and the values of the OUR in the experiments were from 0.5×10^{-4} to 1.0×10^{-4} kg m^{-3} s^{-1}.

Shear rate and shear stress at a distance of R from the center can be calculated by the following equations [8,9],

$$\dot{\gamma} = (R_{in}/R)^{2/n} \{2\omega/(1 - 1/k^2)\}$$

$$\tau = cK\dot{\gamma}^n$$

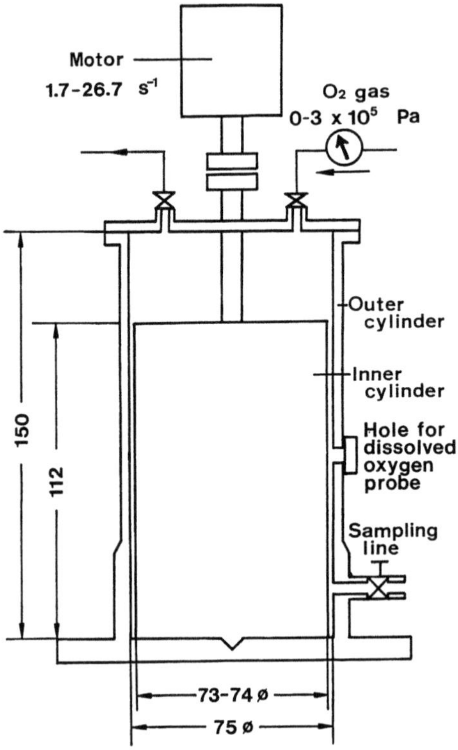

Fig. 2. Dimensions of a rotating drum reactor. Unit of the dimension is $\times 10^{-3}$ m

where c is a correction factor according to the effects of Taylor vortex. Value of c is 2.7 in the supercritical region, where Taylor number (N/υ) is from 100 to 1000 [10].

2.5 Analysis

The cell concentration was estimated from medium turbidity at 660 nm. Xanthan gum was assayed by the method of Moraine et al. [5] in the case of high xanthan gum concentration. When the xanthan gum concentration was low, it was assayed by means of the procedure shown in Fig. 3.

The dissolved oxygen tension (pO_2) was measured by a galvanic type probe (Oriental Electric Co., Ltd.) and its unit was expressed in percent of the concentration at saturation. The volumetric k_La was measured by the use of a polarographic type probe (Fieldlab Oxygen Analyzer, Beckman-Toshiba, Ltd.). The probe was placed near the tip of the impeller. k_La was measured by the dynamic method [11]. Solution of actual xanthan gum (from SIGMA Chemical Co., St. Louis, MO USA) was used to measure k_La, because it has the same properties as the reaction solution.

The consistency coefficient and the flow behavior index were estimated by use of a Brookfield type viscometer (Tokyo Keiki Seizousho) at 30 °C.

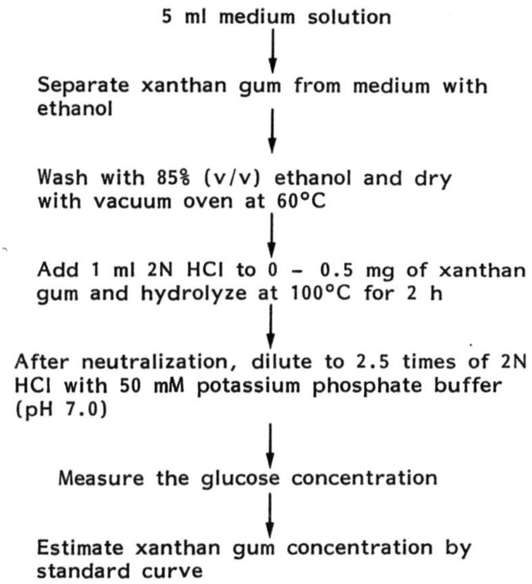

Fig. 3. Procedure for assay of xanthan gum at low concentration

3. Results and Discussion

3.1 Effects of Agitation Speed

Batch culture for the production of xanthan gum were performed at various agitation speeds with a flat-bladed turbine impeller; the impeller diameter and the blade width were 0.10 m and 2.0×10^{-2} m, respectively. The effects of agitation speed on SPR are shown in Fig. 4. Increases in agitation speed increased SPR, which became 7.5×10^{-5} kg kg^{-1} s^{-1} (xanthan per dry cell matter per time) at the agitation speed of 16.67 rps. This result shows that agitation speed is one of the important operational factors for SPR.

3.2 Effects of Physical Factor Relevant to Agitation

In order to discuss the effects of the agitation on the microbial production of xanthan, the effects of pO_2, k_La, discharging performance (ND^2W) and intensity of shear flow (ND) were investigated as the physical factors relevant to the agitation on SPR.

3.2.1 Effects of Dissolved Oxygen Tension (pO_2)

The effect of pO_2 on SPR was investigated by varying the partial pressure of the oxygen in the inlet gas at fixed agitation speed. The results are shown in Table 1. The experiments were conducted at low xanthan gum concentration, less than 1.0%, by avoiding dead space in the reactor and to have uniform pO_2. The values of SPR shown in Table 1 were larger than those in Fig. 4, because of the good mixing at low

Table 1. Effect of dissolved oxygen tension on the specific production rate of xanthan gum

Agitation speed (s^{-1})	Dissolved oxygen tension (% to saturation)	Specific production rate of xanthan gum × 10^5 (kg kg^{-1} s^{-1})
10.00	100 — 80[a]	10.3
10.00	50	10.0
10.00	6[a]	9.7
6.67	0[b]	5.0
3.33	0[b]	0.6

[a] The dissolved oxygen tension was controlled by varying the oxygen partial pressure of aerated gas (1.0×10^{-4} m^3 s^{-1}).
[b] The DO meter indicated 0 under these agitation condition with 1.0×10^{-4} m^3 s^{-1} of air supply.

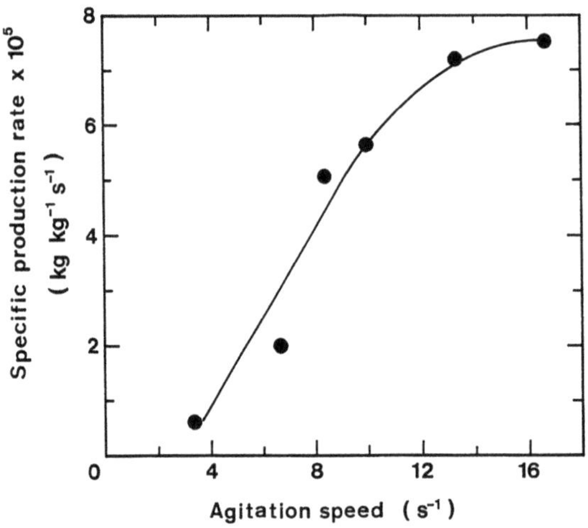

Fig. 4. Effects of agitation speed on specific production rate of xanthan gum. The cultivation was done by the use of 10 l reactor. The specific production rate of xanthan gum was estimated by the data during the period of the constant cell concentration, and the ranges of xanthan gum concentration was from 1 to 2.8%

concentration of xanthan gum. It is concluded from the results in Table 1 that SPR was independent of pO_2 as long as the latter was more than 6%. On the other hand, when the DO meter indicated 0, i.e. the oxygen supply into the solution was not enough, SPR became low according to oxygen supply. Of course no xanthan gum was produced and the organisms died when nitrogen gas was sparged instead of the air (data is not shown).

3.2.2 Effects of Volumetric Oxygen Transfer Coefficient (k_La)

SPR was found to be independent of pO_2. The next question is how the OTR affects SPR, because SPR might depend on the rate of oxygen supply for energy requirement to synthesize the biopolymer. The oxygen transfer from the gas bubbles to the medium depends on the transfer coefficient (k_L) and the interfacial area per unit volume (a)[12]. The k_La depends on the agitation speed. Narayanan et al.[13] suggested that the small bubbles would coalesce with large free rising bubbles and reduce k_La in highly viscous liquids. The effects of k_La on SPR are shown in Fig. 5.

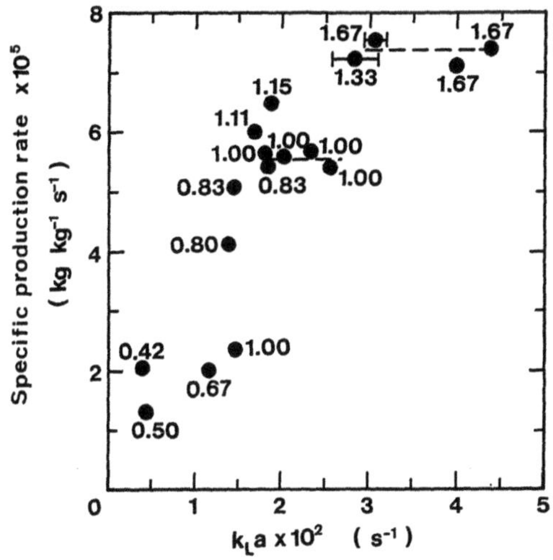

Fig. 5. Effects of volumetric oxygen transfer coefficient on specific production rate of xanthan gum. The figures besides symbols stand for the values of the shear flow intensity (m s^{-1})

SPR decreased at $k_La < 1.67 \times 10^{-2}$ s^{-1}. Also DO meter indicated 0 in these cases. However, SPR was independent of k_La and was constant, at a constant value of the shear flow intensity.

3.2.3 Effects of Discharging Performance

Other roles of agitation are mixing and homogenizing the reaction solution.

There are two types of the mechanism of mixing e.g. mixing by shear or by secondary circulation flow.

The discharging performance of an impeller is an important factor for the xanthan gum production as well. Its effects on SPR are shown in Fig. 6. The open and closed circles in this figure stand for values of shear intensity 1.67 m s^{-1} and 1.0 m s^{-1}, respectively. At low values of the discharging performance, DO meter indicated 0.

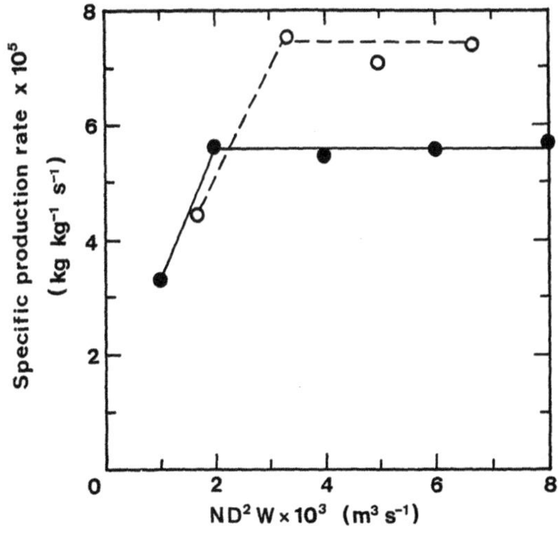

Fig. 6. Effect of discharging performance on specific production rate of xanthan gum. ○, ND = 1.67 m s^{-1}; ●, ND = 1.0 m s^{-1}

SPR was affected by a low discharging performance. It became independent and constant however, if the latter was high — with DO > 0 under the agitation conditions applied.

These results suggest that the intensity of the shear flow is an important factor for xanthan gum production under well mixed conditions.

3.2.4 Effects of Shear Flow Intensity

The shear flow in a stirred tank is important for transferring oxygen from the bubbles to the medium, as described above. Also, it is important for substrate transfer onto and removal of product from the cells. The effects of shear flow intensity on SPR are shown in Fig. 7. Under agitating condition of the shear flow intensity <0.667 m s^{-1}, DO meter indicated 0 and SPR decreased with decreasing the shear flow intensity because of the low oxygen supply.

Under agitating condition of the shear flow intensity >0.667 m s^{-1}, SPR also decreased with decreasing the shear flow even if $pO_2 \neq 0$.

These results suggested that the shear flow caused by the impeller movement was an essential factor not only for OTR from the bubbles to the medium and other actions.

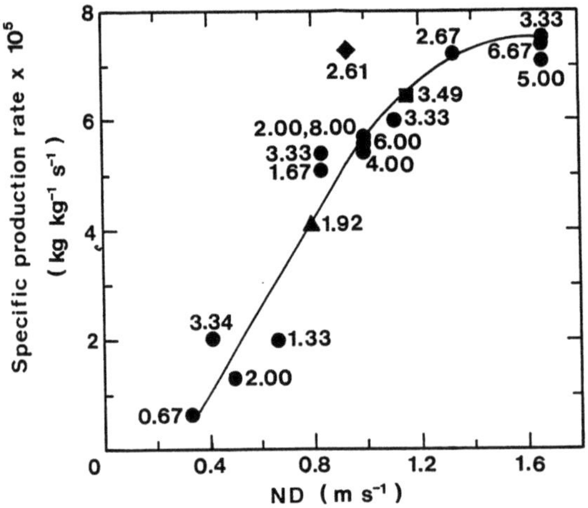

Fig. 7. Effects of shear flow intensity on specific production rate of xanthan gum. The figures besides symbols stand for the values of ND^2W (10^{-3} m^3 s^{-1}). The differences of symbols stand for the sizes of impeller diameter, ▲, D = 0.087 m; ●, D = 0.100 m; ■, D = 0.120 m; ◆, D = 0.140 m

3.3 Shear and Xanthan Gum Production

Metzner et al. [14] reported that in highly viscous solution, the shear flow is restricted near the impeller and there is little shear flow in the region far from an impeller. They also reported that this phenomenon became significant in non-Newtonian fluids. In other words, *ND* stand only for the representation of the shear flow intensity in the vessel. There is a nonisotropic distribution of the shear flow in the vessel, and from the effect of the shear flow intensity on SPR alone it is impossible to know whether this effect is caused by the shear rate or the shear stress. Therefore, the effects of the shear rate and the shear stress on SPR and the specific oxygen uptake rate were investigated by the use of a rotating drum reactor (Fig. 2).

3.3.1 Effects of Shear Stress and Shear Rate on Specific Production Rate and Specific Oxygen Uptake Rate

Figures 8 and 9 show the effects of the shear stress on SPR and the specific oxygen uptake rate. Both SPR and OUR increased with increasing shear stress when the shear stress was less than 40 Pa. However, both were constant when it was more than 40 Pa. As shown in Figs. 8 and 9, SPR and OUR did not change at constant shear stress in spite of the difference in the shear rate. This means that SPR and the specific oxygen uptake rate were independent of the shear rate. These results suggested that transfer of oxygen and glucose from the medium into the cells may limit xanthan gum production when the shear stress is less than 40 Pa.

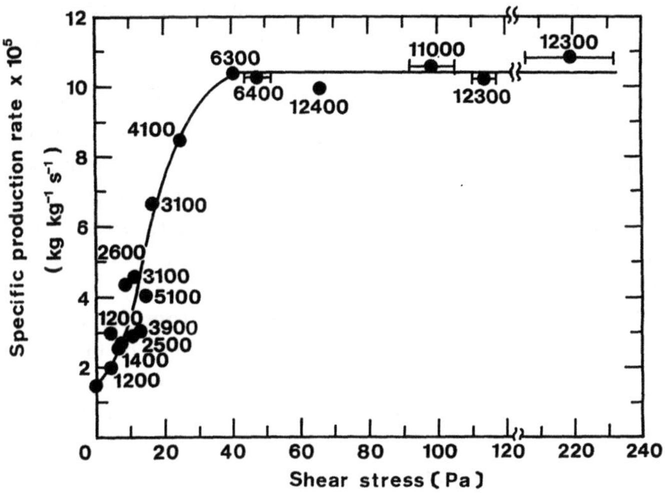

Fig. 8. Effects of shear stress on specific production rate of xanthan gum. The figure besides symbols stand for the values of shear rate (s^{-1})

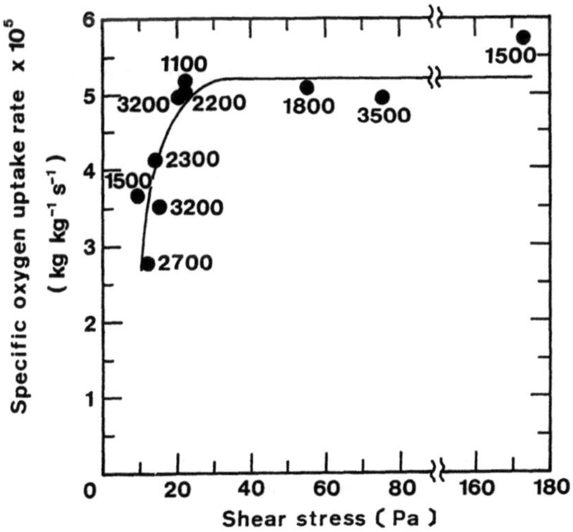

Fig. 9. Effects of shear stress on specific oxygen uptake rate. The figures besides symbols stand for the values of shear rate (s^{-1})

3.3.2 Effects of Glucose Concentration as Driving Force of Glucose Transfer on Specific Production Rate of Xanthan Gum

The mass transfer rate from the medium into the cells changes with the concentration of the materials in the medium. If the mass transfer is the limiting step of the production and not the reaction, the concentration of the material, e.g. oxygen or glucose, is important as the driving force of the mass transfer. The effects of DO on OUR could not be investigated, because it was difficult to obtain a constant pO_2 over a wide range. To investigate the effects of the glucose concentration on SPR at a constant shear stress (10–12 Pa), a rotating drum reactor was used. The results are shown in Table 2. SPR was 1.11×10^{-4} kg kg^{-1} s^{-1} (xanthan per dry cell matter per time) at 7.43×10^{-2} kg kg^{-1} of the highest glucose concentration, and its value decreased with decreasing glucose concentration. In these cases, the intracellular glucose concentration was negligible and the concentration itself was the driving force of the glucose transfer.

This result suggests that the shear stress influences the transfer of glucose for the xanthan gum production and cell maintenance. Also, this influence may be occurring in the transfer of oxygen from the medium into the cells. Finn [15] has shown that the liquid film around the cells offers no appreciable resistance to the diffusion of oxygen. But in our case the mass transfer coefficient was smaller, because the diffusion coefficient in the xanthan gum layer around the cells may be very small compared with that in water.

These results suggested that the mass transfer of the materials from the medium into the cells might limit xanthan gum production and that SPR might be increased by reducing the resistance of the mass transfer around the cells in a high shear stress. However, it is difficult to interpret functionally the effects of the shear stress on mass transfer due to the lack of methods to obtain the concentration and the thickness of the xanthan gum layer around the cells during the cultivation. This information is necessary for estimation of the diffusion coefficient in the xanthan gum layer around the cells during the cultivation so as to interpret the mass transfer around the cells.

Table 2. Effects of glucose concentration as a driving force of glucose transfer from medium into cells on specific production rate of xanthan gum

Glucose concentration per weight of medium $\times 10^2$ (kg kg^{-1})	Shear stress[a] [Pa]	Specific production rate of xanthan gum $\times 10^5$ (kg kg^{-1} s^{-1})
7.43	10.1	11.1
5.59	12.2	7.2
3.91	11.6	4.4
3.28	11.6	3.1
1.27	12.0	2.8
0.68	11.6	2.2

[a] Shear stress was controlled at about 12 Pa by xanthan gum concentration and the rotating speed of a rotating drum reactor.

4. Conclusion

It was found that the agitation was one of the important factors for the xanthan gum production. SPR increased with increasing agitation speed, and the shear stress was one of the most important physical factors caused by the agitation in the microbial production of xanthan gum. Moreover, it was suggested that the xanthan gum layer around the cells might limit the mass transfer of the materials, e.g. oxygen or glucose, from the medium into the cells and that SPR might be increased by reducing the resistance of the mass transfer around the cells by a high shear stress.

In future, the distribution of the shear stress in the stirred tank should be investigated at the various xanthan gum concentration to simulate the xanthan gum synthesis by a model considering the volume of high shear stress and the distribution of circulation time to come back to the region.

5. Nomenclature

c = correction factor of shear stress in the rotating drum reactor [—]
D = impeller diameter [m]
D_d = disk diameter [m]
K = consistency coefficient in Power law equation [Pa s^n]
k = ratio between R_{out} and R_{in} [—]
k_La = volumetric oxygen transfer coefficient [s^{-1}]
L = blade length [m]
N = agitation speed [s^{-1}]
n = flow behavior index in Power law equation [—]
R = distance from the center of cylinder [m]
R_{in} = radius of the inner cylinder [m]
R_{out} = radius of the outer cylinder [m]
W = blade width [m]
$\dot{\alpha}$ = shear rate [s^{-1}]
υ = kinematic viscosity [$m^2\ s^{-1}$]
τ = shear stress [Pa]
ω = angular velocity [rad s^{-1}]

6. References

1. Sandford, P. A., Pittsley, J. E., Knutson, C. A., Watson, P. R., Cadmus, M. C. and Jeans, A.: Extracellular Microbial Polysaccharides (P. A. Sandford and A. Laskin, Eds.) (ACS Washington, D. C.), Series No. *45*, 192 (1977)
2. Andrew, T. R.: ibid., Series No. *45*, 231 (1977)
3. Wells, J.: ibid., Series No. 45, 299 (1977)
4. Sandvic, E. I. and Maerker, J. W.: ibid., Series No. *45*, 242 (1977)
5. Moraine, R. A. and Rogovin, P.: Biotech. Bioeng. *13*, 381 (1971)

6. Moraine, R. A. and Rogovin, P.: Biotech. Bioeng. *15*, 225 (1973)
7. Lilly, V. G., Wilson, H. A. and Leach, J. G.: Appl. Microbiol. *6*, 105 (1958)
8. Bird, R. B., Stewart, W. E. and Lightfoot, E. N.: "Transport Phenomena" 71, John Wiley & Sons, Inc. 1960
9. Harris, J.: "Rheology and non-Newtonian Flow" 24, Longman Inc., New York 1977
10. Stuart, J. T.: J. Fluid Mech. *4*, 1 (1958)
11. Taguchi, H. Humphrey, A. E.: J. Ferment. Technol. *44*, 881 (1966)
12. Moo-Young, M. and Blanch, H. W.: Advances in Biochemical Engineering *19*, 1 (1981)
13. Narayanan, S., Goossens, L. H. and Kossen, N. W. F.: Chem. Eng. Sci. *29*, 2071 (1974)
14. Metzner, A. B. and Taylor, J. S.: AIChE J. *6*, 109 (1960)
15. Finn, R. K.: Bact. Rev. *18*, 254 (1954)

5.8 The Biological Component of Expert Systems

H. R. Bungay and S. P. Fraleigh

Rensselaer Polytechnic Institute, Troy, NY 12180, USA

Contents

Expert systems for biochemical engineering do not yet incorporate biochemical and biological information of sufficient sophistication for dealing with some key situations encountered in practical processes. Of particular concern is relating specific growth rate coefficient to concentrations of limiting nutrients. Some common equations proposed for modeling limitations by several nutrients are invalid, so another approach is proposed. We also demonstrate an expert system that analyzes interactions in microbial ecosystems.

1. Introduction

A computer that carries out functions of an experienced human is an "expert system". An expert system employs an algorithm that makes IF—THEN decisions that may have quite complicated and profound interdependencies. A fully mature system should be aided by an information library and should be able to learn and progress to new situations. There is no reason for constructing an expert system that only codifies and duplicates the decision process of a human expert. The computer program can avoid human frailties and can incorporate tools that human experts are not trained to use as well as those that are too difficult or time consuming for a human. For example, an ecologist today is certainly not ignorant of mathematics, but few are likely to employ modern chemical engineering methods. The goal is not to mimic a human but to construct a computer sytem that equals or outperforms a human.

Algorithms for biotechnological expert systems suffer from inadequate biological relationships. Classical biology may add little to an expert system — what is needed is biology based on kinetics, dynamics, and control. Unfortunately, very little biological research is based on systems engineering. The computer models need functions for growth rates, product formation rates, competing reactions, and the

Bioproducts and Bioprocesses
Editors: Fiechter, Okada, Tanner

like, but the required fundamental understanding is not readily available to those persons who are constructing expert systems. Systems engineering applied to microbiological processes can provide time constants needed by engineers, but much more sophisticated use of the information should result from unravelling the details of cellular control. We would like to be able to dissect the time history of a run into its unit processes and quantitative interconnections.

An immediate practical application of expert systems is operation and control of bioreactors. Built in should be relationships that interpret diauxie, mass balances, pathways, and energetics. More subtle effects such as mutation and selection may be appreciated qualitatively, but mathematical description may not be practical yet. A very important advantage claimed for expert systems is the ability to learn, but this will be hampered to a very serious degree if the biological foundations are shaky.

One area of great concern is a reliable equation for growth rate coefficient. The Monod equation has served well in the past, but its lack of time dependency and the shift in coefficients during a process are nearly fatal flaws. While cell mass is important, the state of this cell mass is crucial. A bioprocess accomplishes little if the cells are dead or lacking in the required enzymes or cellular machinery. Another complication for growth rate is a shift from one limiting nutrient to another. There is some controversy about the best method to handle this mathematically, but there are scant experimental data on which to judge the merits of the models.

Systems engineering should play an important role in biology because of the complexity and the interactions and networks that are analogous to the physical systems already handled by engineers. Analyzing responses that highlight the dominating slowest steps provide time constants. There are ways to handle nonlinear systems, but a very strong beginning to systems biology can be based on much simpler techniques for analyzing linear systems. Although many biological steps are not linear, it is usually possible to restrict the magnitude of forcing so that the response is not seriously distorted by nonlinearities. If we can ascertain the order of the system and the coefficients in the differential equations governing the responses, complications due to nonlinearities can be included later. These coefficients and relationships help to make an expert system more deterministic.

1.1 Nutrient Limitation

Monod laid the basis for several decades of advances in the understanding of microbial processes by recognizing a growth-limiting nutrient in lowest stoichiometric proportion in the medium and by formulating a growth rate equation analogous to the Michaelis-Menten equation. The Monod equation is:

$$\mu = \frac{\hat{\mu} S}{Ks + S}$$

where μ is the specific growth rate coefficient, $\hat{\mu}$ is the maximum specific growth coefficient, Ks is the half-saturation constant, and S is the concentration of limiting nutrient.

The Monod equation lacks time-dependency and should be applied with caution when microbial systems are changing rapidly because growth rates may not adjust instantaously. Furthermore, it does not account for changes in coefficients during bioprocess as the population shifts from young to old cells. Consumption, fresh ditions, or changes in the culture may cause the limitation to shift from one trient to another.

For several years, modelers have combined Monod expressions to relate specific ɔwth rate coeffient to concentrations of two nutrients, A and B, either of which can limiting. For example:

$$\mu = \frac{\hat{\mu}A}{Ka + A} \frac{B}{Kb + B}$$

ıere μ is the specific growth rate coefficient, $\hat{\mu}$ is the maximum specific growth te, and Ka and Kb are half-saturation coefficients.

though Bader [1] has criticized growth models, a number of recent papers are opagating the practice of multiplying Monod equations. Perhaps a reiteration of the gic of nutrient limitation will discourage use of such a poor model. The concept a single limiting nutrient comes from stoichiometry. When various elements com- ne to form microbial mass, the one provided in lowest proportion should ɪminate.

Microbial systems in nature seldom operate at high substrate concentrations, so e error from multiplying Monod relationships can be great. This is illustrated by nsidering a balanced medium — a rare instance where all ingredients are in rfect stoichiometric ratios. Although there may be more, let us assume that the niting ingredients are C, N, O, P, Mg, Mn, S, K, biotin, and niacin for a total ten, each with its own Monod equation. If this balanced medium is diluted so at each Monod equation shows 70% of the initial growth rate, multiplying 0.7 ten nes would indicate a specific growth rate coefficient of 0.02689 times the original ɪefficient. The actual coefficient should be fairly close to 0.7, and the very low rate ridiculous. This practice could be excused as expedient if no better appraoch ɪuld be conceived. However, computers make it very easy to implement a selection which component is limiting [2]. One approach is to evaluate separate expressions the actual concentrations, and the lowest value of growth rate is taken. In other ords, even though each is supplied at a concentration below that corresponding to .turation, the specific growth rate will still be dominated by the component ɾailable in lowest proportions. We are attempting to generate growth rate data sufficient accuracy for two or more limiting nutrients so that equations can be ɜrived for the small effects of non-limiting nutrients on the specific growth rate ɔefficient obtained from the Monod equation for the limiting nutrient. Our model ith weighting coefficients that correct for the ratio of concentration to Ks seems to ork well [3].

2 Time Delay

ur group has long been incorporating time delay in growth rate adjustment [4], ıd others are proposing delay functions [5]. A rapid, immediate small increase in rowth rate followed by a gradual rise has been observed for a step change to microbial culture [6]

1.3 Ecosystem Analysis

Natural populations of microorganisms are highly dynamic, and an organism that predominates for a period of time may become undetectable later as other organisms take over. While relationships in defined laboratory mixed cultures have some value, the most urgent need is understanding of natural mixed cultures. There are some valuable tools such as following energy flux, but no really good approach is available for codifying and following the extremely complicated interactions in multicomponent microbial systems.

2. Methods

We are using FORTH, a flexible language especially well-suited for process control, on a routine basis for our research with yeast. Our procedure-oriented control and analysis routines are now augmented with a template program for expert systems

Table 1. Rules for the ecological expert

Rule 1. IF the concentration of a population component shows no peaks or valleys, THEN no decision is possible.
Explanation: Analysis depends on times when the derivative of concentration equals zero.

Rule 2. IF the maxima and minima of a component concentration correlate with a certain concentration of a different component,
THEN the former preys on only the latter.
Explanation: Two models are considered:
A. $dP/dt = KVP - DP$ (Lotka-Volterra)

B. $dP/dt = \mu P - DP$ (Monod)

μ is a monotonic function of V
where P = predator concentration, V = victim concentration, t = time, D = dilution rate, μ = specific growth rate coefficient for predators, K = coefficient relating collisions to growth

In either case, when $dP/dt = 0$, V, the other component cancels. The model equations can be rearranged and used to determine constants once the predator relationship is identified.

Rule 3. IF maxima and minima of one component correlate with the weighted sum of two other components, THEN the first preys on both of the others.
Explanation: This holds for the model:
$dP/dt = K1\,V1\,P + K2\,V2\,P - DP$

When the derivative = 0, $K1\,V1 + K2\,V2 = D$

Rule 4. IF the maxima and minima for an organism correlate with a certain sugar concentration, THEN this organism does not associate with other organisms except for competing for sugar.
Explanation: When the derivative goes to zero,

$$dC/dt = \mu C - DC = 0$$

$$D = \mu = \frac{\hat{\mu} S}{Ks + S} \quad \text{(Monod eqn.)}$$

and the only variable is S.

(EXPERT-2, Mountain View Press, Mountain View, CA). This provides a production or rule-based expert system shell with a collection of FORTH words that constitute an inference engine.

Simulations of ecosystems in continuous culture were conducted with SIMBAS, a BASIC program that handles ordinary differential equations and is well suited to personal computers [7]. Files of numerical data for population behavior and sugar concentration versus time were supplied to the expert system in FORTH. The data generated by the computer were coarsly spaced to resemble real data that can be collected only once or twice per day. Analytical subroutines in the hybrid FORTH system handled pattern matching calculations to answer the antecedents (IF-clauses) in the knowledge base. Typical inference rules are shown in Table 1.

3. Results and Discussion

3.1 Nutrient Limitation

We are exploring two different situations — nutrient limitation by substrates such as glucose or ammonium compounds that figure in the cell stoichiometry and by potassium ion that involves membrane energy and ionic balance. The details of this research will be reported elsewhere, but preliminary data shown in Table 2 discredit the model which multiplies Monod equations for systems with multiple nutrient limitations. Table 1 has data for *Saccharomyces cerevisiae* growth rate versus concentration of glucose and potassium ion. When both are below saturation levels, the specific growth rate is in good accord with the rate predicted from either Monod equation and is greater than the rate calculated by multiplying one Monod rate by the other. The agreement with a model that selects one or the other growth coefficient depending on which nutrient is limiting is good but not fully satisfactory.

Table 2. Growth rate under different limitations

Concentration	Specific growth rate
Excess glucose and K^+	$0.29\ h^{-1}$
Glucose at reduced level	0.14
K^+ at reduced level	0.20
Both at reduced level	0.13
Rate calculated by multiplying	0.09

3.2 Analysis of Ecosystems

An example hypothetical ecosystem is shown as Fig. 1. The session of an expert system that analyzes data files from these simulations is shown as Table 3. Of course, we were able to tune our programs so that the analysis system was able to give correct results for the simulated ecosystem. More importantly, real ecosystem data can be tested, and success with real systems will validate the models. In other words, ecologists can begin to use expert systems to unravel complicated ecosystems if their

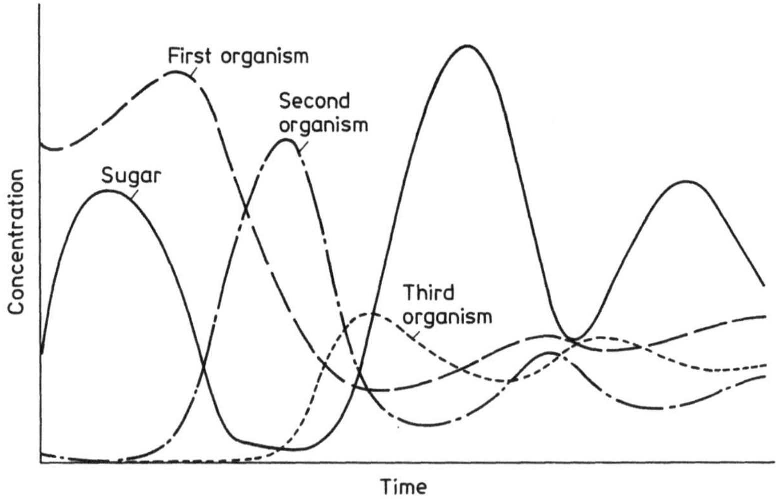

Fig. 1. Concentration changes in hypothetical ecosystems

models for the associations are correct [8,9]. Doing this without an expert system would be prohibitive in terms of effort and cost, and the usual languages such as FORTRAN, PASCAL, and BASIC have no special features to sort out complicated decision trees. There is now the opportunity to test different models of biological interactions by developing rules for expert systems that handle real data.

Table 3. Analysis by ecological expert system

Postulated		Decision by computer
Predator	Prey	
1	2	Not in accord with rules, postulate unlikely
1	3	Not in accord with rules, postulate unlikely
2	1	Agrees with rules, probably true.
2	3	Not in accord with rules, postulate unlikely
3	1	Not in accord with rules, postulate unlikely
3	2	Not in accord with rules, postulate unlikely
Postulated competitor		
	1	Poor correlation with sugar, postulate unlikely
	2	Poor correlation with sugar, postulate unlikely
	3	Agrees with rules, probably independent except for competition for sugar.

4. Conclusions

Multiplying Monod expressions is demonstrably false when extended to many limiting substrates. Selecting a single substrate as the critical limit is a better practice, but even better would be some means to account for the effects of the other substrates.

Our expert system looks at population behavior and diagnoses possible interactions. There are many ways in which better rules can be added. For example, size relationships could tell if it is feasible that one ingests the other. In the future, a library of literature information may tell if one produces a substance that is likely to inhibit the other. Hypothetical situations are accurately analyzed by our program to find the relationships that we have built into a separate program for simulation.

5. Acknowledgement

Research by Trupti Mankad was the basis for Table 1.

6. References

1. Bader, F. G.: Biotech. Bioeng. *20*, 183 (1978)
2. Fraleigh, S. P., and Bungay, H. R.: J. Gen. Microbiol. *132*, 2057 (1986)
3. Mankad, T.: M. S. Thesis, Rennselaer Polytechnic Institute, 1987
4. Bungay, H. R., and Paynter, M. J. B.: AIChE Symposium Ser. *124*, 68, 220 (1972)
5. Wang, N. S., and Stephanopoulis, G.: Biotech. Bioeng. Symp Series *14*, 635 (1984)
6. Mateles, R. I., Ryu, D. Y., and Yasuda, T.: Nature, *308*, 263 (1965)
7. Bungay, H. R.: "Computer Games and Simulation for Biochemical Engineering", Wiley, NY 1985
8. Pave, A., and Rechenmann, F.: in "AI Applied to Simulation" (E. J. H. Kerckhoffs, G. C. Vansteenkiste, and B. P. Zeigler, eds.) The Society for Computer Simulation, San Diego, pp. 52–66, 1986
9. Muetzelfeldt, R., Bundy, A., Uschold, M., and Robertson, D.: ibid, pp. 67–70.

MIX
Papier aus verantwortungsvollen Quellen
Paper from responsible sources
FSC® C105338

If you have any concerns about our products,
you can contact us on
ProductSafety@springernature.com

In case Publisher is established outside the EU,
the EU authorized representative is:
Springer Nature Customer Service Center GmbH
Europaplatz 3, 69115 Heidelberg, Germany

Printed by Libri Plureos GmbH
in Hamburg, Germany